D1064490

TOPICS IN MATHEMATICAL SYSTEM THEORY

INTERNATIONAL SERIES
IN PURE AND APPLIED MATHEMATICS

William Ted Martin and E. H. Spanier, *Consulting Editors*

QA
102
K32

TOPICS IN
MATHEMATICAL SYSTEM THEORY

R. E. Kalman
Stanford University

P. L. Falb
Brown University

M. A. Arbib
Stanford University

McGraw-Hill Book Company

New York   San Francisco   St. Louis   Toronto   London   Sydney

174579

TOPICS IN MATHEMATICAL SYSTEM THEORY

Copyright © 1969 by McGraw-Hill, Inc. All rights reserved.
No part of this publication may be reproduced, stored in a retrieval
system, or transmitted, in any form or by any means, electronic,
mechanical, photocopying, recording, or otherwise, without the
prior written permission of the publisher.

Printed in the United States of America.

Library of Congress catalog card number: 68-31662

1234567890   MAMM   754321069

33255

74.75

//8/01

To
CONSTANTINA, FORTUNA, and PRUDENCE

Preface

The child of three fathers and borne of circumstances, this book does not pretend to be a systematic treatise. Rather, it aims to present Mathematical System Theory as it is today—a lively, challenging, exciting, difficult, confused, rewarding, and largely unexplored field, one which is already very important and yet holds the promise of still bigger discoveries.

Limitations of space, time, and human affairs have made it utterly impossible to give a systematic account of the revolution which occurred in this field in the late 1950s and early 1960s. Nor is such an account to be expected in the near future, from us or any one else. We have tried, therefore, to emphasize, and perhaps overemphasize, just four of the main ideas: *states*, *control*, *optimization*, and *realization*. We have also tried to convey the fundamental notion that system theory is not simply a branch of applied analysis, but provides a source of problems and intuition for *a rich interplay between algebra and analysis*. On the other hand, many important topics such as stability, qualitative theory of differential equations, algebraic linguistics, practical control theory, switching circuits, automata theory from the standpoint of logic, etc., etc. are barely mentioned here or are ignored altogether.

Each of the four sections of the book is the independent work of one author, as indicated in the table of contents. No attempt has been made to eliminate duplications (there are many, and we hope that they prove helpful to the reader); there is no standardized notation (impossible if the reader is not to be misled about the diversity of the literature!); the marks of individuality of each author have not been edited out of the book.

The book is intended to be useful for self-study, especially for those who have no personal contact with the "in-groups" in the field. The material is organized so that the book may be used for a variety of seminars and one-semester or one-quarter courses such as optimization, automata theory, or algebraic theory of linear systems. The research worker or graduate student will find here many fresh points of departure for future research. The choice of subjects has been governed only by the authors' interests (ca. 1965) and the demands of exposition. Major portions of the text have been "classroom" tested or presented in special lectures.

Lectures by all three of us during the ASEE-NASA Summer Institute of 1965 at Stanford University formed the nucleus of the book. Various portions of the text were written at various times. Most of PLF's Part Two was completed in the spring of 1966 and is not, therefore, entirely representative of his current views. His Section 3.8 and Appendix 4.A are much more recent. MAA's exposition of automata theory was available in 1965 and has been but lightly revised after that. REK's Part One was written in 1966 and revised in 1967; of course, its contents were known in research circles in the early 1960s. The core of the material for Part Four represents research done in 1965, but the exposition was finalized only during the latter part of 1967.

A complex project such as this cannot reach completion without the help of others, and in this case there were many. In the first place, we wish to express our sincere thanks for continued and generous research support of the United States government, particularly the Office of Scientific Research of the Air Force (through Contract 49(638)-1440 with REK, through Grants 814-66 and 693-66 with PLF, and through Grant AF-1198-67 with MAA), as well as to the National Aeronautics and Space Administration (through Grant NgR-05-020-073 with REK). It was Professor Max Anliker who arranged the summer course of 1965. He has encouraged us steadfastly to complete the manuscript. REK is indebted to the University of Minnesota (1965), the University of New Mexico (1966), and Imperial College, London (1967) for invitations to deliver special lecture series, and of course also to Stanford University where some of the material has been presented in graduate courses (1965–1968). PLF wishes to thank Harold Kushner, Elmer Gilbert, and David Kleinman for fruitful discussions. MAA is deeply grateful to Rowena Swanson for her support of his research, and to Professor John Westcott of Imperial College, London, and Professor John Blatt of the University of New South Wales for providing visiting lectureships in 1964–1965 when much

of Part Three was taking shape. Each of us have had important con-
versations with Professor H. P. Zeiger whose influence is felt in many
passages. Dr. P. Faurre read a large part of the manuscript and suggested major improvements. The expert and untiring assistance of
Mrs. M. Newcomb, John Parsons, Kate Nolan, and Mrs. K. Omura is
deeply appreciated. We are also very much indebted to the publishers
for their cheerful and understanding cooperation.

Finally, praise and blame for the idea of a project of this type rests
on the publishers, who persuaded PLF and MAA to persuade REK,
while complaints about misnumbered references, contradictory notations,
wrong proofs and similar failings should be unfailingly addressed to REK,
who acted as the coordinating editor.

Woodside Heights   **R. E. Kalman**
Cambridge   **P. L. Falb**
Los Altos Hills   **M. A. Arbib**

Washington's Birthday, 1968

Table of contents

Part four   **ADVANCED THEORY**
                 OF LINEAR SYSTEMS
                 R. E. Kalman

General introduction

R. E. Kalman, P. L. Falb, and M. A. Arbib

1 Reader's guide

To provide some unity in the exciting but chaotic new field of system theory, we present in this chapter a general survey of the various interrelated topics of this book. The essential preamble to everything that follows is Section 1.1, which contains a rather lengthy formal definition of a system (more accurately, dynamical system). Except for reliance on the common terminology set up in this section, the four parts of the book are self-contained. Part One (Chapter 2) is very easy. Its purpose is to prepare the reader for one of the three specialized topics: optimization (Part Two, Chapters 3 to 5), system structure and description (Part Three, Chapters 6 to 9), and advanced theory of linear systems (Part Four, Chapter 10).

1.1 Systems and states

In this book we shall treat system theory as the study of dynamical relationships. The terms "systems," "systems concepts," "systems approach," and "systems science" are used so widely and so broadly today that they tend to connote fuzzy thinking. For us, however, a

system, or more correctly, a *dynamical system,* is a precise mathematical
object; the study of system theory is then largely, although not entirely,
a branch of mathematics.

Before proceeding to a technical definition of a dynamical system,
let us review briefly the intuitive ideas involved.   We regard a system as
a structure into which something (matter, energy, or information) may
be put at certain times and which itself puts out something at certain
times.   For instance, we may visualize a system as an electrical circuit
whose input is a voltage signal and whose output is a current reading.
Or we may think of a system as a network of switching elements whose
input is an on/off setting of a number of input switches and whose output is the on/off pattern of an array of lights.   In the first case we may
think of events in the system as taking place continuously in time (since
electric signals usually vary continuously in time); in the second case it is
more natural to think of the time as discrete (the input switches are set
every five seconds, say).

In either case, our concept of the system Σ includes an associated
time set T.   At each moment of time $t \in T$ our system Σ receives some
input $u(t)$ and emits some output $y(t)$.   We assume that the values of
the input are taken from some fixed set U; that is, at any time t the
symbol $u(t)$ may be chosen from U.   In general the input segments of a
system are *not* allowed to be arbitrary functions $\omega: (t_1, t_2] \to U$, but must
belong to a restricted class Ω.   The choice of Ω may be inferred from
physical considerations, but more often it is dictated by mathematical
expediency.   Each value of the output $y(t)$ belongs to a fixed set Y, and
only mild restrictions are imposed on the "allowable" output segments
$\gamma: (t_2, t_3] \to Y$.

Now, we may not be able to say what $y(t)$ is without knowing more
than just what the present input $u(t)$ is.   The past history of inputs to
the system may have altered Σ (for example, by energy storage in our
first example or by setting internal switches in the second example) in
such a way as to modify the output.   In other words, the output of Σ in
general depends on both the present input of Σ and the past history of Σ.
It is best not to treat the present input differently from the past inputs.
We say, therefore, that the present output depends on the *state* of Σ, and
we define the (present) state of Σ intuitively as that part of the present
and past history of Σ which is relevant to the determination of present
and future outputs.   In other words, we think of the state of Σ as being
some (internal) attribute of Σ at the present moment which determines
the present output and affects the future outputs.   Intuitively, the state
may be regarded as a kind of information storage or memory or an accu-
mulation of past causes.   We must, of course, demand that the set of
internal states of Σ be sufficiently rich to carry all information about the

history of Σ needed to predict the effect of the past upon the future. We do not insist, however, that the state be the *least* such information, although this is often a convenient simplifying assumption.

To merit the label "dynamical" Σ must have one more property: knowledge both of the state $x(t_1)$ and of the input segment $\omega = \omega_{(t_1, t_2]}$ is necessary and sufficient to determine the state $x(t_2) = \varphi(t_2; t_1, x(t_1), \omega)$ whenever $t_1 < t_2$.† Notice that this requires that the time set T be an ordered set, that is, there must be preferred direction of time. We customarily assume that the ordering is such that the past precedes the future. Notice also that with this convention "dynamical"‡ has roughly the same meaning as "causal": the past influences the future, but the converse is not true. In short, the mathematical notion of a dynamical system serves to depict the flow of causation from past into future.

We are now ready to formalize our discussion as follows:

(1.1) **Definition.** *A* **dynamical system** Σ *is a composite mathematical concept defined by the following axioms:*

(a) *There is a given* **time set** *T, a* **state set** *X, a set of* **input values** *U, a set of acceptable* **input functions** *$\Omega = \{\omega: T \to U\}$, a set of* **output values** *Y, and a set of* **output functions**

$$\Gamma = \{\gamma: T \to Y\}.$$

(b) *(Direction of time). T is an ordered subset of the reals.*

(c) *The input space Ω satisfies the following conditions:*

    (1) *(Nontriviality). Ω is nonempty.*

    (2) *(Concatenation of inputs). An* **input segment** *$\omega_{(t_1, t_2]}$ is $\omega \in \Omega$ restricted to $(t_1, t_2] \cap T$. If ω, $\omega' \in \Omega$ and $t_1 < t_2 < t_3$, there is an $\omega'' \in \Omega$ such that $\omega''_{(t_1, t_2]} = \omega_{(t_1, t_2]}$ and $\omega''_{(t_2, t_3]} = \omega'_{(t_2, t_3]}$.*

(d) *There is given a* **state-transition function**

$$\varphi: T \times T \times X \times \Omega \to X$$

whose value is the state $x(t) = \varphi(t; \tau, x, \omega) \in X$ resulting at time $t \in T$ from the **initial state** *$x = x(\tau) \in X$ at* **initial time** *$\tau \in T$ under the* **action of the input** *$\omega \in \Omega$. φ has the following properties:*

    (1) *(Direction of time). φ is defined for all $t \geq \tau$, but not necessarily for all $t < \tau$.§*

---

† We do not wish to consider here probabilistic dynamical systems (conditional Markov processes), where the present state $x(t_1)$ and input ω yield not $x(t_2)$, but merely its probability distribution $Pr(t_2; t_1, x(t_1), \omega)$.

‡ Some authors prefer the neologism "nonanticipatory," which we interpret as a synonym for "dynamical."

§ Slight abuse of the notation $\varphi: T \times T \times X \times \Omega \to X$.

(2) (*Consistency*). $\varphi(t; t, x, \omega) = x$ for all $t \in T$, all $x \in X$, and all $\omega \in \Omega$.

(3) (*Composition property*).   For any $t_1 < t_2 < t_3$ we have

$$\varphi(t_3; t_1, x, \omega) = \varphi(t_3; t_2, \varphi(t_2; t_1, x, \omega), \omega)$$

for all $x \in X$ and all $\omega \in \Omega$.

(4) (*Causality*).   If ω, $\omega' \in \Omega$ and $\omega_{(\tau,t]} = \omega'_{(\tau,t]}$, then

$$\varphi(t; \tau, x, \omega) = \varphi(t; \tau, x, \omega').$$

(e) *There is given a* **readout map** $\eta: T \times X \to Y$ *which defines the output* $y(t) = \eta(t, x(t))$. *The map* $(\tau, t] \to Y$ *given by* $\sigma \mapsto \eta(\sigma, \varphi(\sigma; \tau, x, \omega))$, $\sigma \in (\tau, t]$, *is an* **output segment,** *that is, the restriction* $\gamma_{(\tau,t]}$ *of some* $\gamma \in \Gamma$ *to* $(\tau, t]$.†

It is customary and convenient to use a picturesque language in referring to some of the objects introduced under the formal definition (1.1). The most common locutions are the following: the pair (τ, x), with $\tau \in T$ and $x \in X$, is an *event* (or *phase*) in Σ. The set $T \times X$ is the *event space* (or *phase space*) of Σ. (The state space is sometimes also called a phase space, especially in physics, but this is archaic and ambiguous.)   The state transition function φ (or its graph in the event space) has many more-or-less equivalent names: *trajectory, motion, orbit, flow, solution* (of an ordinary differential equation), *solution curve*, etc. We say also that the input or *control* ω *moves, takes, transfers, carries, transforms* the state x (or the event (τ, x)) to the state $\varphi(t; \tau, x, \omega)$ (or to the event $(t, \varphi(t; \tau, x, \omega))$). By the *motion of the system* we are referring (vaguely) to φ.

We say that a system Σ is *free* if Ω has but one element; this case corresponds to the situation where Σ is isolated from inputs and where the "environment" is the only input.   Example: the solar system, as represented by the equations of celestial mechanics, is a free system since the only forces acting are those determined by gravitation and they are a function only of the state of the system (more precisely, of the positions of the various planets).   When Σ is linear (see below), it is customary to take Ω as a zero-dimensional vector space, that is, the set $\{0\}$. Σ is said to be *reversible* if the transition function is defined for all values of t and τ, not merely for $t \geq \tau$.

Sometimes we shall refer to a dynamical system as an octuple $\Sigma = (T, X, U, \Omega, Y, \Gamma, \varphi, \eta)$.   We shall insist on the adjective "dynamical" only as a matter of special emphasis.

The notion of a dynamical system as just defined is far too general.   Such a definition is needed to set up terminology, to analyze

† The notation $a \mapsto b$ means that b corresponds to a under the map $A \to B$.

and refine concepts, and to perceive unity in a diversity of applications, but it is not sufficiently coherent to bear a large array of deep mathematical theorems or useful practical deductions. To get good theorems and interesting applications, we must particularize and impose additional structure. As we study in this book the standard problems of system theory—stability, control, state reconstruction, optimization, equivalence, structure, decomposition, and synthesis—we shall find it convenient or necessary to restrict attention to various special classes of systems.

Let us review now the most important special properties of systems which we shall often use later.

A system is *constant*, or *time invariant*, if its response to a given input segment, when in a given state, is independent of the time interval in which the trial takes place. In other words, the defining relations (structure) of the system do not change with time. In precise language, we have

(1.2) **Definition.** *A* dynamical system Σ *is* **constant** (*or* **time invariant**) *iff*†
   (a) *T is an additive group* (*usual addition of the reals*).
   (b) Ω *is closed under the* **shift operator** z^τ: $\omega \mapsto \omega'$ *defined by*

$$\omega'(t) = \omega(t + \tau)$$

   for all τ, $t \in T$.
   (c) $\varphi(t; \tau, x, \omega) = \varphi(t + s; \tau + s, x, z^s\omega)$ *for all $s \in T$.*
   (d) *The map $\eta(t, \cdot): X \to Y$ is independent of t.*‡

Most of the literature of system theory is concerned with constant systems, since they are much more amenable to study than nonconstant systems. We shall follow this pattern and state our definitions and results mainly for the constant systems, leaving the reader to judge for himself which topics are amenable to generalization. It is well to bear in mind, however, that in many cases, especially in fundamental questions such as the definition of controllability, the formulation of the optimal control problem, the equivalence of systems, etc., the constancy assumption does not result in an essential simplification. The technique of proof often becomes much more subtle, but the key ideas remain the same. Thus in Part Two we generally do not assume constancy. In Part One constancy is not necessary, but it is a considerable simplification. In Parts Three and Four constancy is essential, since our emphasis will be primarily algebraic.

† We use "iff" as shorthand for "if and only if."
‡ The notation $f(\cdot, y)$ means a *function* (of the suppressed argument).

Another dichotomy of systems is given by

(1.3) **Definition.** *A dynamical system* Σ *is* **continuous time** *iff* $T = reals;$ Σ *is* **discrete time** *iff* $T = integers.$

In many applications of system theory the distinction between continuous-time and discrete-time systems is not critical; the choice is likely to be governed by mathematical convenience. Continuous-time systems correspond to the models of classical physics, while discrete-time systems arise in a natural way whenever digital computers are part of the system. In Part One we study continuous-time systems (differential equations) because of a long tradition in physical applications. In Part Two we study the same systems because they are the proper objects for applying the calculus of variations and related methods. In Parts Three and Four, where the emphasis is on algebra, discrete-time systems are the natural object of investigation.

The most important measure of complexity of a system is the structure of its state space. Hence it is sensible to adopt the following

(1.4) **Definition.** *A dynamical system* Σ *is* **finite dimensional** *iff* X *is a finite-dimensional linear space;* Σ *is* **finite state** *iff* X *is a finite set;* Σ *is* **finite** *iff* X, U, *and* Y *are finite sets and, in addition,* Σ *is constant and discrete time. Thus* $\dim \Sigma \triangleq \dim X_\Sigma$.

A finite system Σ is traditionally known as a (finite) automaton; it is the simplest general class of systems which has been studied. The theory of automata requires only the finitary methods of logic and algebra, as will be seen in Part Three.

The assumption of finite dimensionality is essential if we want concrete numerical results. Even in dealing with distributed systems, which are, of course, infinite dimensional, the final step in the numerical analysis involves finitary approximations; roughly speaking, this amounts to an implicit assumption of finite dimensionality. From the theoretical point of view, however, it may be more natural to work with the tools of functional analysis and take X as, say, a Banach space.† This is the point of view taken in Part Two. In Part One and to some extent in Part Four finite dimensionality coupled with linearity is the chief assumption. Part Three explores the implications of finiteness without linearity.

Throughout this book linearity will play a very big role. First, because we want to tap the vast amount of information available in

---

† Of course, if the state space is infinite dimensional, it is essential to agree on the kind of topologies to be used. The specification of these must be added to the basic definition of Σ; systems which have the same sets but carry different topologies must be regarded as different. We shall avoid problems of this sort because their theory is underdeveloped.

"linear mathematics." Second, because we need linear system theory to study the local behavior of continuous nonlinear systems. From the point of view of engineering, the latter is the main practical reason for learning linear theory.

So tempting is the assumption of linearity that elementary texts in system theory often begin by explaining the "principle" (rather than assumption) of superposition,† without due attention to causation or states, and without even a clear definition of what a system is. Whatever pedagogical virtue this approach might have at the beginning, the student will run into serious confusion later when he learns about other parts of system theory, such as nonlinear mechanics, stability, automata theory, etc. Because of our emphasis on the unity of system theory, we prefer to define a linear system in the fundamental way.

(1.5)   **Definition.**   *A dynamical system* Σ *is* **linear** *iff*

    (a) *X, U,* Ω*, Y, and* Γ *are vector spaces* (*over a given arbitrary field K*).
    (b) *The map* $\varphi(t; \tau, \cdot, \cdot)\colon X \times \Omega \to X$ *is K-linear for all t and* τ.
    (c) *The map* $\eta(t, \cdot)\colon X \to Y$ *is K-linear for all t.*

If, in addition to linearity, we demand that X, U, and Y be finite dimensional and also that Σ be constant, then system theory begins to look much like classical linear algebra. This is the point of view taken in Part One. A closer look at the results of Part One shows, however, that classical linear algebra is not adequate to express system-theoretic facts in a really nice way. We are then led to the study of finitely generated modules over a polynomial ring (Part Four). This evolution parallels the shift of emphasis from vector spaces to modules (see [BOURBAKI, 1962]) in the treatment of linear algebra in pure mathematics. However, the reasons are entirely different.

The definitions of system properties given so far have involved only very primitive set-theoretic or algebraic notions. If we wish to avail ourselves of sophisticated mathematical tools such as calculus and analysis, then the definition of Σ must include some kind of continuity. Therefore we must assume that the various sets (T, X, U, Ω, Y, Γ) are topological spaces and that the maps φ and η are continuous with respect to the appropriate (product) topologies. Moreover, it is important to know how smoothly the state transitions occur with time. We do not wish to go into detail here, since these matters are largely mathematical technicalities. In Chapter 3, Definition (2.1), we give a precise definition of a smooth dynamical system which is especially suitable for

---

† The term "superposition principle" seems to be borrowed from quantum mechanics. This is archaic and misleading. In quantum mechanics "superposition" means that the states are a linear space; in system theory "superposition" means that the input/output function is linear.

the treatment of optimization problems.   A preliminary definition of
"smoothness" is the following:

(1.6)   **Definition.**   *A dynamical system* Σ *is* **smooth** *iff*
    (*a*) $T = \mathbf{R}$, *the real numbers* (*with the usual topology*).
    (*b*) X *and* Ω *are topological spaces.*
    (*c*) *The transition map* φ *has the property that* $(\tau, x, \omega) \mapsto \varphi(\cdot\,; \tau, x, \omega)$
    defines a continuous map $T \times X \times \Omega \to C^1(T \to X).$†

In Section 2.1 we present a theorem which shows that under suitable
specific assumptions smooth dynamical systems are governed by differential equations.   All the classical dynamical systems are of this type.
It is, of course, largely a matter of taste whether (1) we say that smooth
dynamical systems are those governed by differential equations (as in
Chapter 3, Definition (2.1)), or (2) we define smooth dynamical systems
as in (1.6) and then prove the theorem: *If* Σ *is smooth, its transition function* φ_Σ *satisfies a differential equation* (see Chapter 2, Theorem (1.1)).

Let us emphasize that "inputs" and "outputs" are just as essential
in our definition of a system as are the "states."   We are interested in
the interaction between system and environment (described by ω and γ)
just as much as we are in the internal behavior of the system (described
by φ).

Especially in Parts Three and Four, we shall encounter many examples of systems whose primary definition is given in terms of their *input/
output behavior*.   By this term we mean precisely the following: given
any initial event (τ, x), an input segment $\omega_{(\tau,t_1]}$ acting on the system Σ
produces an output segment $\gamma_{(\tau,t_1]}$; that is, we have a map

$$f_{\tau,x} \colon \omega_{(\tau,t_1]} \to \gamma_{(\tau,t_1]}.$$

If we assume as known the system structure in the sense of Definition
(1.1), then the output at $t \in (\tau, t_1]$ is given by

(1.7) $f_{\tau,x}(\omega_{(\tau,t_1]})(t) \,=\, \eta(t, \varphi(t; \tau, x, \omega)).$

Conversely, any family of functions which have the same properties as
the functions defined by (1.7) (that is, they are compositions of functions
φ and η which satisfy (1.1*d*) and (1.1*e*)) can be viewed as defining a
dynamical system in the input/output sense.   The formal definition is
the following:

(1.8)   **Definition.**   *A* **dynamical system** Σ **(input/output sense)**
    is a composite mathematical concept defined as follows:
    (*a*) *There are given sets* T, U, Ω, Y, *and* Γ *satisfying all the properties
    required by Definition* (1.1).

† $C^1(T \to X)$ denotes the family of C^1 functions $T \to X$.

(b) *There is given a set A indexing a family of functions*

$$\mathfrak{F} = \{f_\alpha \colon T \times \Omega \to Y,\ \alpha \in A\};$$

each member of $\mathfrak{F}$ is written explicitly as $f_\alpha(t, \omega) = y(t)$ which is the output resulting at time t from the input ω under the **experiment** α. *Each f_α is called an* **input/output function** *and has the following properties:*

(1) *(Direction of time). There is a map $\iota\colon A \to T$ such that $f_\alpha(t, \omega)$ is defined for all $t \geq \iota(\alpha)$.*

(2) *(Causality). Let τ, $t \in T$ and $\tau < t$. If ω, $\omega' \in \Omega$ and*

$$\omega_{(\tau, t]} = \omega'_{(\tau, t]},$$

then $f_\alpha(t, \omega) = f_\alpha(t, \omega')$ for all α such that $\tau = \iota(\alpha)$.

According to this definition, a dynamical system may be considered as an abstract summary of experimental data. The "experiments," labeled by the abstract parameter α, consist of applying an input and observing the resulting output (stimulus → response). Of course, we must approach the description of these experiments in the scientific spirit and not impose any conditions which may implicitly constrain the results of an experiment before it is performed. Condition (b1) merely tells us when the experiment started; condition (b2) requires us to label two experiments differently if they yield different results under seemingly identical circumstances.

The *problem of realization* is that of constructing a dynamical system in the sense of Definition (1.1) from data provided by Definition (1.8). This is simply an abstract way of looking at the problem of scientific model building. And on an abstract level the solution is surprisingly easy; we shall indicate here the main ideas and leave the details as a nontrivial exercise. Later we shall give a complete theory of the effective solution of the realization problem in certain important special cases (Chapter 10).

The set A clearly corresponds to a subset of $T \times X$. However, A may not be large enough; some experiments could be conceivably performed that are not yet listed in A. So we must enlarge $\mathfrak{F}$ to include all functions obtained as follows: If $\tau < t_1 < t_2$, ω, $\omega' \in \Omega$, and

$$\omega_{(t_1, t_2]} = \omega'_{(t_1, t_2]},$$

then the function g defined by

$$g(t, \omega') = f_\alpha(t, \omega), \qquad t \in (t_1, t_2], \qquad \iota(\alpha) = \tau$$

must be in $\mathfrak{F}$. We should call $g = f_\beta$, where $\beta \notin A$; then we set $\iota(\beta) = t_1$ and replace A by the bigger set $A \cup \{\beta\}$. Then we define the state space at time τ as

$$X_\tau = \{\alpha \in A : \iota(\alpha) = \tau\}$$

and set

$$X = \bigcup_{\tau \in T} X_\tau.$$

Now the definitions of the functions φ and η satisfying (1.1d) and (1.1e) can be readily induced from the f_α. For instance, the composition property (1.1d3) will follow from the completion of $\mathfrak{F}$ explained above. There is one difficulty: φ may not be defined everywhere on $T \times X$, but only for (τ, x) with $x \in X_\tau$. This is an inherent limitation of any experimental setup; it can be avoided (standard practice in the natural sciences) by assuming that the system is constant. The construction $\{A, F\} \to \{X, \varphi, \eta\}$† is especially important in automata theory (Chapter 7).

There has been much confusion in the literature about the realization problem. For instance, [ZADEH and DESOER, 1963] attempt to use a basic system definition which is much closer to Definition (1.8) than to (1.1), but they still require certain "self-consistency" conditions (reminiscent of (1.1d3)) which the results of the experiments must satisfy in order to qualify as a "dynamical" system. These "self-consistency" rules are unnecessary if we adopt a definition like (1.8) (see also [WINDEKNECHT, 1967]). The real issue here is applied-mathematical (effective construction of the state space X), not philosophical (is there a state space at all?).

This concludes our description of the framework which the reader should have before him in interpreting the specialized studies in various parts of the book. He will encounter many different techniques: differential equations, calculus of variations, functional analysis (in problems of control theory), as well as combinatorics, semigroups, groups, modules (in problems of automata theory). He will observe the interplay between analytic and algebraic techniques, similar to the interplay in group theory between the algebraic theory of finite groups and the analytic theory of Lie groups. And we hope that he will finish the book with the ability to apply to *his* system problems whatever technique is most appropriate, irrespective of tradition or habit.

(1.9) **Historical note.** Our definition of a dynamical system is a straightforward generalization of the classical mathematical one (see [NEMYTSKII and STEPANOV, 1960]). The latter is an idealization of the properties of ordinary differential equations, whose role in physics originated with Newton's model of the solar system. The solutions of differential equations enjoy two special properties: they define dynamical systems which are *smooth* and *reversible* (φ is defined for all t, not just for

† A fat arrow → means "induce."

$t > \tau$). So, in our language, the *classical dynamical systems* are continuous, time, constant, smooth, reversible, and (usually) finite dimensional, are free (Ω has but one element), and have a trivial readout map $\eta(t, x) \equiv x$.

The "dynamical polysystems" of [BUSHAW, 1963] and [HALKIN, 1964] generalize the classical concept of a dynamical system by allowing inputs. A similar generalization was made by [ROXIN, 1965], who cites many earlier investigations. Our definition was formalized in [KALMAN, 1963c] and [WEISS and KALMAN, 1965]. Quite independently of classical dynamics, in the young field of automata theory the definition of a system (or "machine") always included inputs and outputs. That dynamical systems and automata belong to the same subject has been vigorously argued by [ARBIB, 1965] (see also Chapter 6). That this viewpoint is gaining popularity is attested by recent textbooks in system engineering [ZADEH and DESOER, 1963; WYMORE, 1967; WINDEKNECHT, 1967], which tend to be more and more abstract and general.

There is no universal agreement at present on what the primary definition of a system should be. See [ARBIB, 1965; KALMAN, 1965b] for a comparison of alternate possibilities. Obviously, this is a topic in which no one can yet claim to have the final word.

1.2   Elementary control theory

The raison d'être of Chapter 2 is to provide a quick and painless introduction to a part of system theory where mathematical ideas predominate. Our topic is the theory of regulators for a linear plant. Our tools are linear algebra and the elementary theory of ordinary differential equations. The mathematical level is modest, but we expect some conceptual sophistication from the reader. The results are actually quite subtle and have a long history. Many of the topics of later chapters can be viewed as natural generalizations of the elementary problems presented in this chapter.

Chapter 2 gives rather leisurely explanations of the conceptual and practical background of the mathematical formalism of system theory. From time to time the reader may return to this chapter for motivation and orientation. Most of Chapter 10 may be regarded as a deep generalization of the controllability criterion (Theorem (4.3) of Chapter 2), while Section 10.13 is concerned with the question of controllability and constructibility of nonconstant plants, which is the principal condition for the existence of a regulator (see Section 2.7).

The main ideas discussed in Chapter 2 are the following. We consider the problem of controlling a fixed physical object (airplane, factory, chemical process, etc.), called the *plant*. For our purposes, "plant" is synonymous with "dynamical system." We write down the obvious abstract properties of the plant which are necessary and sufficient for the control problem to make sense and to have a solution (controllability and constructibility). When the plant is assumed to be *linear*, these properties can be expressed equivalently as properties of certain linear operators. When we assume in addition that the plant is *constant*, the criteria for controllability and constructibility can be stated very elegantly in closed form involving only certain algebraic properties of the plant. Coupling this with a technique related to classical canonical forms of matrices, we then proceed to give a completely explicit solution of the regulator problem.

The whole story is a beautiful illustration of the power of mathematics. We show how to go from completely abstract considerations to completely practical results, which are, in the end, quite simple. This inherent simplicity of the solution was *not known* in control theory before the late 1950s, that is, prior to the "proper" abstract mathematical formulation of the problem.

1.3 Optimal control theory

Optimal control theory is the basic subject matter of Chapters 3, 4, and 5. We shall be interested in developing and describing the scientific foundations of control in these chapters. Because of the general orientation provided in Chapter 2, we can proceed immediately to an axiomatic formulation. Our approach will be mathematically precise. We shall generally consider smooth, continuous, dynamical systems with state spaces of unrestricted dimension.

We begin by observing that control theory is a body of mathematical results and techniques relevant to control problems. The basic intuitive elements of any control problem are

1. A system to be controlled.
2. A desired output or objective.
3. A set of admissible controls (or inputs).
4. A measure of the cost or effectiveness of control actions.

These intuitive elements are given a precise meaning, and we then develop various theorems which are at the heart of optimal control theory. These theorems are aimed at providing answers to the following fundamental questions:

1. Do optimal controls exist?
2. If there are optimal controls, then how can these optimal controls be determined?

Various approaches, such as Hamilton-Jacobi theory, the Pontryagin principle (necessary conditions), functional-analysis methods, and numerical techniques have been used. We shall devote some time to each of them.

After fixing the context within which we work, we derive sufficiency conditions (the so-called Hamilton-Jacobi theory) using a lemma of Caratheodory in a manner suggested by [KALMAN, 1963a]. The notion of regularity for the Hamiltonian functional and the Hamilton-Jacobi partial differential equation are the key elements here. The main result is Theorem (4.14).

A very important and useful area of application of Hamilton-Jacobi theory is the situation in which our dynamical system is linear and our cost functional is quadratic. This case is of signal interest because of its relation to conventional control methods (classical compensation techniques) and because of its application in correcting small deviations from a nominal path (so-called second-variation methods). A crucial point is the introduction of the (operator) Riccati equation (Lemma (5.11)). We derive various properties of the solution of the Riccati equation and, from these properties, deduce that the linear-system quadratic cost problem has a solution (Theorem (5.42)). Since the optimal system has a feedback structure, the results are of major practical significance.

We conclude Chapter 3 with a discussion of the Kalman-Bucy filter. In other words, we examine the question of determining the "best" linear filter, in an expected-squared-error sense, for a "signal" generated by a linear stochastic differential equation on a Hilbert space. We derive a Wiener-Hopf type equation for the optimal filter (Theorem (6.41)). Then we deduce from this equation the stochastic differential equation satisfied by the optimal estimate. The two crucial points in our treatment are (1) definition of the covariance as a bounded linear transformation and (2) use of a Fubini type theorem involving the interchange of stochastic and Lebesgue integration. We note that the optimal filter may be viewed as generating an optimal regulator for a linear dynamical system which is "dual" to the original stochastic system. Thus our results on the linear-system quadratic cost problem can be used to derive properties of the optimal filter. We rely heavily on [FALB, 1967] and [KALMAN and BUCY, 1961]; this section of the book requires a reasonable degree of mathematical sophistication.

In Chapter 4 we turn to necessary conditions for optimality. We begin with a discussion of first-order necessary conditions which are analogous to the familiar Euler equations of the calculus of variations. We use a perturbation approach. The hamiltonian functional plays a crucial role. An important aspect of the development is the introduction of the linear-perturbation equation (1.5), which represents the behavior of our system "near" a nominal trajectory. The main result (Theorem (1.12)) is that in order for a control to be optimal it is necessary that the hamiltonian have an extremum.

Next we study the maximum principle of Pontryagin and his collaborators, which represents a considerable improvement of the necessary conditions in the case of a finite-dimensional dynamical system. The Pontryagin principle is quite useful in control-system design and has been frequently applied in practical problems. A careful statement of the maximum principle is followed by a heuristic proof based on the notion of a complete set of perturbations and involving the separation of a suitable convex cone and the ray of decreasing cost by a hyperplane. The conditions of the maximum principle involve the canonical system, minimization of the hamiltonian functional, and various boundary conditions.

Since use of necessary conditions depends upon the existence of optimal controls, a simple theorem on the existence of optimal controls is presented. This theorem depends on the notion of attainability and on the well-known fact that a lower-semicontinuous real-valued function on a compact set has a minimum. In general the crucial question of the existence of optimal controls is quite deep, and we shall only scratch the surface here.

Control problems can be viewed as special cases of the problem of minimizing a functional defined on a normed linear space subject to constraints. We conclude Chapter 4 with a brief indication of various beautiful and powerful generalizations of the maximum principle due to [HALKIN, 1967; NEUSTADT, 1966-67; CANON, CULLUM, and POLAK, 1968].

Some of the ways in which the theoretical considerations of the previous chapters are applied to control-system design problems are discussed in Chapter 5. We begin by examining the simple problem of time-optimal control for a double-integrator plant, as this is a problem which can be solved analytically. The approach used is based on the Pontryagin principle, since we can readily deduce the existence of an optimal control from the results of Chapter 4. We then examine the various stages of the necessary-condition approach to deriving optimal controls in a general control problem and indicate many of the difficulties which may be encountered.

Since the actual practical solution of most control problems requires the use of numerical techniques, the remainder of Chapter 5 is devoted to these methods. We start with some general remarks on iterative methods; these remarks are aimed at indicating the general ideas underlying the various techniques used in practice. Then we turn our attention to indirect methods in control problems, with particular attention to ways in which the two-point boundary-value problem arising out of the necessary conditions for optimality has been attacked. We examine the gradient method, the Newton-Raphson method in a "neighboring-optimum" approach to solving the two-point boundary-value problem, and the successive-approximation method in a quasi-linearization approach. Again our emphasis is on the underlying ideas rather than on explicit details.

We conclude our discussion of numerical methods with an examination of direct methods. Here the key idea is the notion of a minimizing family. Among the results presented is an approximation theorem [COURANT and MOSER, 1962] which provides a theoretical basis for the use of penalty functions. We shall also consider the well-known Ritz method, which, in our opinion, has not been adequately exploited in control-system design.

1.4 Automata

The commonality of different approaches to system theory has been repeatedly stressed. In Chapter 6 we explore in detail the way in which new light is shed on control theory by examining such concepts as controllability, linearity, and optimality in the context of finite-state systems. In particular, we shall see that notions such as observability which are well understood for linear systems may exhibit many different properties in the realm of nonlinear systems.

The remainder of Part Three is devoted wholeheartedly to a discussion of algebraic methods in the study of finite-state systems (automata), with special emphasis on techniques for replacing a system by a combination of simpler systems.

Let us recall† that an *automaton* (or *machine*) is described abstractly as a quintuple $M = (X, Y, Q, \lambda, \delta)$, where X is a finite set of *inputs*, Y is a finite set of *outputs*, Q is a (not necessarily finite) set of *states*, $\lambda: Q \times X \to Q$ is the *next-state* function, and $\delta: Q \times X \to Y$ is the *next-output function*. In the language of Section 1.1, an automaton is a discrete-time, constant dynamical system. Usually it is also finite state.

We introduce X^*, the set of all finite sequences of input symbols, and include in it Λ, the "empty string" of symbols. X^* is a semigroup under concatenation (that is, concatenation is associative, $(x_1 x_2)x_3 = x_1(x_2 x_3)$, and has identity Λ). We then extend the applicability of λ and δ so that λ maps $Q \times X^*$ into Q, while δ maps it into Y.

We may associate with each state q of a machine M the way it pro-

---

† We choose here a notation more consonant with those of automata theorists than that of other sections of our introduction. The reader is warned that there are almost as many systems of notation as there are authors in this young field.

duces an output for each input string.   This is expressed by the function
$M_q: X^* \to Y$, where $M_q(x) = \delta(q, x)$.   Clearly, M behaves the same way
if started in two states q and q', with the same input/output function,
that is, if $M_q(x) = M_{q'}(x)$ for all input strings x.   So if our interest in M
is in its external behavior, we may replace it by its *reduced form*, which
has one state for each distinct function M_q.

   With such a machine we associate a semigroup, the collection of
transformations of its state set induced by the input strings.   This semigroup is finite if and only if the set of states is finite.

   Given a semigroup S and a function $i: S \to Y$, we may then define
a new "semigroup machine" with "state output" i as

$$M(S, i) = (S, S, Y, \lambda, \delta),$$

where $\lambda(s, s') = s \cdot s'$ (with $\cdot$ denoting semigroup multiplication) and
$\delta(s, s') = i(s \cdot s')$.

   We say that the machine M *simulates* the machine M' if, provided
we encode and decode the input and output appropriately, M can process strings just as M' does.   We require the encoder and decoder to be
memoryless (that is, to operate symbol by symbol) in order to make M'
do all the computational work involving memory.   If M simulates M'
we write $M'|M$ and say that "M' *divides* M." If both $M|M'$ and $M'|M$
we say that M and M' are *weakly equivalent*.   The utility of semigroups
in finite automata theory is emphasized by the result that if M has semigroup S, we may define a function i so that $M(S, i)$ is *weakly equivalent*
to M.

   It turns out that semigroups, as well as machines, have a natural
concept of divisibility.   The important tieup between the machine and
semigroup concepts of divisibility is that, modulo one extra condition,
*the machine M_1 divides the machine M_2 if and only if the pair (S_1, i_1) corre-
sponding to M_1 divides the pair (S_2, i_2) corresponding to M_2.*

   It is well known that any finite automaton may be simulated by a
network of modules (that is, *one-state* finite automata), provided we allow
loops in the network; such constructions form the central theme of any
text on switching theory.   In fact, we may use copies of just one module
(the Sheffer-stroke module) to build such a network.   In other words,
a very simple set of components suffices for the construction of arbitrary
finite automata *if loops are allowed*.   We now describe some theorems on
the restrictions following from the outlawing of loops between machines
(of course, there may be loops in the internal structure of the machines
we are combining).

   Given machines M' and M and an appropriate map Z, we define
$M' \times_Z M$, the *cascade of M' and M with connecting map Z*, by the diagram shown in Figure 1.1.   To get series composition (albeit preserving

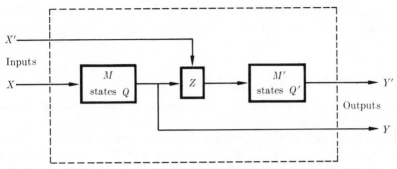

Figure 1.1

the output of M) we make Z independent of $\tilde{X}$; to get *parallel* composition we take $Z(\tilde{x}, y)$ to be just $\tilde{x}$.

Our precise notion of loop-free composition is that of repeated formation of cascades as well as memoryless codings to obtain machines simulable by such combinations.

We call M an *identity-reset machine* if an input either acts as a reset, returning the machine to a state determined by that input ($\lambda(q, x) = q_x$, $q \in Q$) or else acts as an identity, leaving the state unchanged ($\lambda(q, x) = q$, $q \in Q$). Of special importance is the two-state identity-reset machine, which we call the *flip-flop F:* It has states $\{q_0, q_1\}$ and inputs $\{e, x_0, x_1\}$ with $\lambda(q, e) = q$ (e is the identity), and $\lambda(q, x_i) = q_i$ (x_i resets to state q_i); and with state and output equal. We note that any identity-reset machine may be obtained by loop-free synthesis from copies of F.

We owe to [KROHN and RHODES, 1965] the important result: *Given a machine M, we can realize it by loop-free synthesis from flip-flops and from the machines of (not necessarily all) the simple groups which divide the semigroup of M.*

The crucial point of this discussion is that we can get really significant insight into the structure of our systems only by applying powerful algebraic methods.

Now, finite automata are relatively unstructured objects, and we can get only partial information about machine decomposition from their semigroups. If we have a restricted class of systems, we should be able to impose more structure on the corresponding algebraic system, and then apply algebra to this more structured algebraic system to yield more detailed information about the original system.

The plausibility of this general thesis is underscored by Part Four. If we restrict ourselves to linear systems, we may add to the semigroup structure the (new) operations of convolution and addition. The resultant algebraic structure is called a module—and we do in fact see that the algebraic theory of modules is a powerful tool in the analysis of linear systems.

1.5 Algebraic theory of linear systems

Textbook writers in elementary linear system theory have always claimed that Laplace transforms are used to "algebraize" the process of solving linear differential equations with constant coefficients. This is certainly substantiated by the developments of the last twenty years. At present, the definition of the Laplace transform by means of an integral tends to be relegated to a footnote, while algebra (devices such as partial-fraction expansion and canonical forms of matrices) is becoming more and more important. It has also been known for a long time that these methods are not restricted to continuous-time systems; the so-called "z-transform" theory of difference equations and sampled-data systems is now well established and much used [FREEMAN, 1965].

Chapter 10 develops *ab initio* the inherent algebraic structure of linear system theory by merging three different streams of ideas:

1. Elementary constructions of automata theory, especially the definition of the state as an equivalence class of inputs (see Chapter 7).

2. The modern format of linear algebra, which emphasizes modules as a generalization of vector spaces. (See [VAN DER WAERDEN, 1931] for an early use of modules to obtain canonical forms of matrices.)

3. The general "know-how gained" in the development of sampled-data systems during the decade of the 1950s [FREEMAN, 1965]. (We mention this only as a matter of history; no special background is required of the reader.)

Most of this theory is new [KALMAN, 1967]. It puts states, inputs, and outputs on an equal footing; they are represented as polynomials or formal power series over an arbitrary field K. By taking K as a finite field, we *could* obtain most of the results of the theory of finite-state linear sequential circuits. The states form a module, in which the inputs play the role of the operator ring. The description of the action of inputs on states in a linear system by means of a module is similar to the use of semigroups in the Krohn-Rhodes theory, but neither is a special case of the other.

Within the new setup, the transfer function of a linear system can be defined without any reference to transforms or generating functions. Transfer functions emerge with even greater importance than before. They are especially useful in computing the structure of the module associated with a linear system.

Now let us outline the principal mathematical ideas involved.

We define the input/output function f of a constant linear system Σ in an unusual but very convenient way. Let Ω be the space of all input functions which have compact support and which, in addition, vanish identically for $t > 0$. Similarly, let Γ be the space of all output functions, which are defined only for $t > 0$. Then the map $f: \Omega \to \Gamma$ is interpreted as follows: apply an input ω until time 0 and then observe the output resulting after time 0. It is easily seen that in the constant case this definition of the input/output function is no restriction of generality.

Of course, if Σ is linear its input/output function f_Σ should be a K-linear function (K = arbitrary field). So Ω and Γ must be endowed with the structure of a K-vector space. As it is linear, f has a canonical factorization $f = f_3 f_2 f_1$, where f_1 = onto, f_2 = isomorphism, and f_3 = one-to-one. Writing X = image f_1 and Ξ = image f_2, we get the commutative diagram

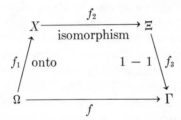

Exactly the same diagram is obtained if we start from a linear differential (or difference) equation with state space X and output space Γ. This suggests that the factorization of the input/output map f corresponds to the realization of f by a dynamical system. We can make this idea precise, as follows:

1. We note that Ω is isomorphic to a free $K[z]$-module, and that every other space in the diagram can be endowed with the structure of a $K[z]$-module without destroying commutativity. (We write $K[z]$ = ring of polynomials in the indeterminate z with coefficients in the field K.)

2. The action in the $K[z]$-module X (same notation as when X is merely a K-vector space) is *not* the dynamical action of the inputs on states. It turns out, however, that the set X is precisely the same as the set of Nerode equivalence classes used in automata theory (see Section 7.2). Hence X qualifies as a state space, at least in the intuitive sense. It is then easy to construct the dynamical system from the module X. For example, the module map $z: x \mapsto z{\cdot}x$ gives the free motions of the dynamical system.

There is an essential difference between this procedure and the construction of semigroups in the Krohn-Rhodes theory: here the multiplication on the input space is convolution; there it is concatenation. Convolution is a much more familiar kind of multiplication, and it has long been part of the intuition of everyone in linear system theory. So the "new" algebraic theory is entirely compatible with the "old" Laplace-transform treatment. In addition, the new theory provides sharper results in several areas, especially with regard to the structure of complex systems. The word "onto" in the commutative diagram is equivalent with "complete reachability." By duality, "one-to-one" is equivalent with "complete observability." Bearing these analogies in mind, and *using only the commutative diagram*, we deduce the important result that every minimal realization of f is completely reachable and completely controllable. Previously, this was obtained (see Section 10.B) by lengthy calculations which depended critically on linearity. Thus "canonical factorization" is equivalent to "minimal realization." Many other interesting results can be obtained directly from this diagram; for instance, the existence of a transfer function is demonstrated by examining the image in Γ of the generators of the module Ω and exploiting the assumption that X is a torsion module.

The two kinds of multiplication used in the linear theory and the Krohn-Rhodes theory are in effect two different "functors" each of which replaces a dynamical system by an algebraic structure. The results are in fact vaguely similar although not at all identical. The two theories overlap if and only if Σ is linear over a *finite* field (then Σ is finite state); but the "module primes" and the "Krohn-Rhodes primes" turn out to be quite different.

PART ONE

Elementary control theory from the modern point of view

R. E. Kalman

2 Theory of regulators for linear plants

The first purpose of this chapter is to obtain the basic results of the regulator problem in classical control theory. We shall strive for mathematical clarity and shall not stress the physical or engineering side of the problem. This is in keeping with our second objective, which is to introduce some of the other topics of the book: system description and structure, optimization, the role of linearity. Each of these topics will arise naturally as we develop the solution of the regulator problem. In a sense this chapter presents a "case history" which serves as the starting point for various specialized investigations in system theory.

We shall study a rather narrow topic: the analytical design of regulators for constant (time-invariant) linear plants. In the interests of fairly rapid exposition we shall generally gloss over the deeper problems, such as optimality, noise, etc., and concentrate on exhibiting a class of control schemes which satisfy the basic requirement of stability.

The central feature of this part of control theory is the fact that the existence as well as the actual construction of regulators follows from two fundamental properties of the plant: *complete controllability* and *complete constructibility*. The first property means that every state can be moved to zero by the application of a suitable input. The second means that the (internal) state of the system can always be determined from the knowledge of past (external) outputs and inputs. These properties are obvious necessary conditions for the existence of a regulator. It is a surprising and relatively recent (1959) discovery that controllability and constructibility are also sufficient for the task at hand. From the modern point of view, the success of classical control theory is due to the fact that the models used by the classical theory are intrinsically completely controllable and constructible.

The class of regulators considered here is by no means special or arbitrary. In fact, their *structure* is exactly the same as that prescribed by stochastic optimal control theory. Only the actual values of the feedback coefficients are different. To choose these numbers in a rational way, we must be given a criterion of optimality relative to the environment in which the system is to operate. The actual calculation of the optimal values of the feedback coefficients requires the full power of optimal control theory (see especially Sections 3.5, 5.4, and 5.5).

In this chapter, the demands placed on the reader's background are quite modest: only the elementary theory of linear differential equations and linear algebra is needed. Special knowledge of control theory is not assumed. Some collateral reading in modern control theory (such as [PESCHON, 1963]) will he helpful in appreciating the breadth of applications of the mathematics.

2.1   Formulation of the control problem

The purpose of control is to alter the dynamical behavior of a physical system in accordance with man's wishes. The problem splits quite naturally into two very distinct parts:

1. We must get a mathematical description of the dynamical properties of the physical object (the *plant*) to be controlled.

2. We must find a "scheme" for accomplishing the desired controlled behavior.

The first problem is essentially one of model building: we must be able to calculate the dynamical behavior of the plant from our mathematical model with an accuracy which is at least within the tolerances allowable under control. This model of the plant is a dynamical system in the sense of Section 1.1. The model is obtained from physical measurements or known physical laws; its actual determination is generally in the domain of the physical scientist and outside the domain of control theory or even system theory.

Once this model is given, we proceed to the second problem, which is essentially mathematical. The "schemes" by which control is accomplished are based on a highly developed technology, usually centered on a computer, in which a mathematical specification is often (but, of course, not always), the same as an engineering blueprint. In other words, the second half of the problem is one which requires a mathematical result (a theorem) for its solution; conversely, any control scheme is in effect a mathematical result.

To put it more bluntly, *control theory does not deal with the real world, but only with mathematical models of certain aspects of the real world; therefore the tools as well as results of control theory are mathematical.*

There is a close analogy between this situation and the evolution of probability theory into a strictly mathematical discipline.

Let us take a concrete example of some current interest.

Suppose we wish to send a spacecraft to the moon, where it must land on a small prescribed spot. The control theorist would proceed to formulate the problem as follows:

1. To land exactly on a given spot the space craft must be injected into orbit with very high precision and follow a precalculated path without any deviation until the landing. This is impossible to accomplish without control because of the difficulty of injecting in orbit, because of random forces acting on the spacecraft between the earth and the moon, etc.

2. A mathematical model for the motion of the spacecraft is provided by the equations of celestial mechanics, which are known to be highly accurate. The physical characteristics of the spacecraft can be assumed to be known, since *we* are building it. The inputs to our system, with which control must be accomplished, will be rockets and jets of various types. The outputs will be measurements of position, velocity, etc., by optical, inertial, or other means.

3. Now we have a precise definition of the plant as a dynamical system. From the properties of the equations of celestial mechanics, it follows that the system is finite dimensional, continuous time, smooth, and (unfortunately) nonconstant and nonlinear.

4. As a working hypothesis, we consider only small deviations from a given flight path; this allows us to approximate the plant (in a precise mathematical sense, to be sure) by means of a nonconstant linear dynamical system.

5. We apply (or extend) the machinery of linear control theory to our problem, assuming, if necessary, even constancy in order to get explicit results.

6. We analyze the solution process and attempt to generalize it, in order to remove the restrictions of constancy or linearity.

Let us now look more closely at the desired behavior of the spacecraft under control. Ideally, we should like the spacecraft to execute arbitrary motions in response to commands by its occupants. This is too difficult (at the current state of technology), so we settle for a simpler objective: the spacecraft is to follow a preassigned path until it lands on the moon; the deviations from this given path must be as small as possible during the entire flight and especially at landing. This is the *regulator problem;* it amounts to forcing the plant to behave according to a pre-assigned pattern.

The solution of the regulator problem consists in giving a *control law*, which prescribes the values of the controlling inputs as a function of the measured deviation from the preassigned path. The physical embodiment of a control law, called the *regulator*, is usually some type of computer or electronic device. We shall be content to obtain the explicit equations describing the regulator, and we shall regard this process as synonymous with "constructing" the regulator.

Now, the all-important point is this: in control theory it is immaterial what physical object is represented by the dynamical system under consideration; what matters is the mathematical structure of the plant (this is the reason that control theory is *not* subdivided according to whether the plant is a spacecraft, aircraft, or seacraft). So the problems are strictly mathematical.

In modern treatments of control theory it is customary to begin with the statement:

Assume that the plant is governed by the ordinary differential equation

$$\frac{dx}{dt} = f(t, x, u(t)),$$

where $t \in \mathbf{R}^1$, $x \in \mathbf{R}^n$, $u(t) \in \mathbf{R}^m$, and $f: \mathbf{R}^1 \times \mathbf{R}^n \times \mathbf{R}^m \to \mathbf{R}^n$ is a continuous function which is at least lipschitzian in x, etc.†

It is more pleasing, however, to launch control theory by deducing this equation from abstract assumptions on the plant regarded as a dynamical system in the precise sense of Definition (1.1). The following theorem will be adequate for our purposes. The key condition is (c), which implies that Σ is "smooth."

(1.1)   **Theorem.**   *Let Σ be a smooth dynamical system in the sense of Definitions (1.1) and (1.6) of Chapter 1. More precisely, assume that*
(a) *$T = \mathbf{R}$, X, and U are normed spaces,*
(b) *Ω is the normed space of continuous functions $T \to U$, with*
$$\|\omega\| = \sup_{t \in T} \|u(t)\|,‡$$

---

† Throughout the book **R** denotes the real numbers.

‡ We shall write the values of ω as $u(t)$; sometimes we shall use $u(\cdot)$ instead of ω.

(c) $\varphi(\cdot; \tau, x, \omega) \in C^1(T \to X)$ *for each* τ, x, *and* ω, *and the map* $T \times X \times \Omega \to X$ *given by* $(\tau, x, \omega) \mapsto \dot{\varphi}(t; \tau, x, \omega)$† *is continuous for each* t, *with respect to the product topology.*

Then the transition function φ *of* Σ *is a solution of the differential equation*

(1.2)
$$\frac{dx}{dt} = f(t, x, \pi_t \omega),$$

where the operator π_t *is a map* $\Omega \to U$ *given by* $\omega \mapsto u(t) = \omega(t)$

The proof is an easy exercise in the definition of continuity; we omit the details, as they are not germane to our present purposes.

(1.3) **Remark.** Although this seems to be a very general theorem, it *does* rule out certain kinds of systems. For instance, [ZADEH and DESOER, 1963] argue (nonrigorously) that the definition of π_t "should be"

$$\pi_t^* : \omega \mapsto (u(t), \dot{u}(t), \ldots, u^{(q)}(t)).$$

To get this result, we must modify condition (b) and take the norm of Ω as

$$\|\omega\| = \sup_{t \in T} [\|u(t)\| + \|\dot{u}(t)\| + \cdots + \|u^{(q)}(t)\|].$$

We regard this as an unnatural assumption.

We see that the setup in Theorem (1.1) is precisely what we need to assure that the right-hand side of the differential equation (1.2) depends on ω "pointwise." Any other dependence on ω would mean that Σ is not smooth enough to be characterizable by a differential equation. For example, suppose Σ is given by

(1.4)
$$y(t) = u(t), \qquad t \in T, \qquad U = Y = \mathbf{R}^1$$

(that is, Σ is given in the input/output sense; see Definition (1.8) of Chapter 1). This is a well-defined system also in the sense of Definition (1.1), Chapter 1: we must set $\eta =$ identity and write

$$\varphi(t; \tau, x, \omega) = \pi_t \omega = u(t).$$

But then

(1.5)
$$\dot{\varphi}(t) = \frac{d}{dt} (\pi_t \omega) = \pi_t \dot{\omega};$$

this means that (1.4) *can be regarded as a smooth system only if the original input space* Ω *is replaced by* $\dot{\Omega}$. The intuitive reason for these observations is very clear. System (1.4) has no "memory," in that the input at any instant of time provides all the information necessary to emit

† The dot is shorthand for differentiation with respect to time.

the output; on the other hand, system (1.5) "remembers" $u(t)$, and the information provided by the input is the minimal amount (namely, $\dot{u}(t)$) required to update the memory.

Let us note also that the requirement that Σ be smooth prevents the input from having an instantaneous effect on the output (as in (1.4)). This is a very desirable assumption, because we want the flow of causation (input implies output) to be unambiguous. (Note that in (1.4) we could interchange the role of input and output, something that is almost never possible in real systems.)

Now that we have agreed that the plant should be represented by a differential equation, we can give a more mathematical definition of the

(1.6) **Regulator problem.** *Let the plant be governed by the differential equation*

$$\frac{dx}{dt} = f(t, x, u(t)),$$

let $\varphi^*(t) = \varphi(t; \tau^*, x^*, \omega^*)$ be some fixed motion (solution) of the plant, and let (s, x), where $x \neq \varphi^*(s)$, be some fixed event. We regard this event as a perturbation at time $t = s$ of the fixed motion of the plant and seek a control (function) ω_x such that $\varphi(t; s, x, \omega_x) \to \varphi^*(t)$ as $t \to \infty$. Moreover, we require (precise definition to be given in each case) that

(a) *Either $\varphi(t_1; s, x, \omega_x) = \varphi^*(t_1)$ for some $t_1 > s$ (convergence in finite time).*

(b) *Or the controlled motion is stable (insensitive to further disturbances) and converges to φ^* as fast as possible.*

(1.7) **Remark.** This formulation involves only a small part of the control problem, which is the only part that we shall study in depth in this chapter. We have here a *deus ex machina* picture of a perturbation: "somehow" the state is moved from $\varphi^*(s)$ to x at time s; it would really be more appropriate to describe this process by some statistical model. We shall not even explain how $\varphi^*(t)$ is specified in a practical case; in fact, we shall usually assume (for simplicity) that $\varphi^*(t) = 0$.

2.2 Smooth linear systems

Let us narrow our attention to dynamical systems Σ which are (1) finite dimensional, (2) linear, and (3) smooth. Because of the desired linearity, the state space must be a vector space; so we take $X = \mathbf{R}^n$. Since the transition function must be linear on $X \times \Omega$, we may write it as

$$(2.1) \qquad \begin{aligned} \varphi(t; \tau, x, \omega) &= \varphi(t; \tau, x, 0) + \varphi(t; \tau, 0, \omega), \\ &= \Phi(t, \tau)x + \Theta(t, \tau)\omega. \end{aligned}$$

The linear map $x \mapsto \Phi(t, \tau)x$ defined by the first term is the *transition map* of Σ; it determines the *free motions* of Σ (that is, the motions under the influence of the trivial input $\omega = 0$). To analyze the second term in (2.1) we need the fact that Σ is smooth; so we take $T = \mathbf{R}$. By Theorem (1.1), φ satisfies the differential equation

$$(2.2) \qquad \frac{d}{dt} \varphi(t; \tau, x, \omega) = f(t, \varphi(t; \tau, x, \omega), u(t)).$$

Since φ is linear in (x, ω), it is clear that the right-hand side is linear in the second and third arguments. So we may introduce the functions F: $T \to \{n \times n \text{ matrices}\}$ and G: $T \to \{n \times m \text{ matrices}\}$ and write

$$(2.3) \qquad f(t, x, u(t)) = F(t)x + G(t)u(t).$$

With these notations, (2.2) means that φ satisfies the differential equation

$$(2.4) \qquad \frac{dx}{dt} = F(t)x + G(t)u(t);$$

in particular, the *transition map* (or *transition matrix*) Φ satisfies the differential equation

$$(2.5) \qquad \frac{d}{dt} \Phi(t, \tau) = F(t)\Phi(t, \tau)$$

as well as the additional relation

$$(2.6) \qquad \Phi(\tau, \tau) = I = \text{unit matrix}.$$

Note that (2.5) and (2.6) together determine Φ uniquely; the second identity corresponds to axiom (1.1*d*2) in the general definition of a dynamical system (see Section 1.1). Finally, since η must be linear on X, we have

$$(2.7) \qquad y(t) = \eta(t, x(t)) = H(t)x(t),$$

where H: $T \to \{p \times n \text{ matrices}\}$.

The fact that Σ is smooth implies that F, G, and H are continuous functions. This is technically important, but we shall not dwell on it in the sequel, since we want to keep the discussion as simple as possible.

We summarize these observations as follows:

(2.8) Theorem. *Every continuous-time, finite-dimensional, linear, smooth dynamical system Σ obeys the relations*

$$(2.9) \qquad \begin{aligned} \frac{dx}{dt} &= F(t)x + G(t)u(t), \\ y(t) &= H(t)x(t). \end{aligned}$$

Conversely, we have

(2.10)   **Theorem.**   *Given the relations* (2.9), *there is a unique dynamical system* Σ *which has all the properties mentioned in Theorem* (2.8) *and which is, in addition, even reversible.*

Proof.   If Φ is determined by (2.5) and (2.6), then it is well known that the general solution of the differential equation (2.9) may be expressed as

$$(2.11)\qquad \varphi(t;\tau, x, \omega) = \Phi(t,\tau)x + \int_{\tau}^{t}\Phi(t,\sigma)G(\sigma)u(\sigma)\,d\sigma;$$

it is easily verified that φ has all the properties required of a transition function.

Now, it is important to remember that (2.11) holds for *all* t and τ, and not merely for $t \geq \tau$. So (2.11) defines a reversible dynamical system. In other words, linearity plus smoothness implies reversibility. $\square$

(2.12)   **Remark.**   In view of these two theorems, all calculations, proofs, etc., may be based directly on relations (2.9) whenever we deal with a finite-dimensional, linear, smooth system. This permits us to use powerful applied-mathematical methods based on the calculus and the theory of differential equations (see especially Chapters 3 to 5). We must not forget, however, that methods are, after all, of secondary importance; the essential point is that our results are a consequence of the assumptions "finite-dimensional, linear, smooth." The applied-mathematical methods employed here may be supplanted eventually by more efficient ones, but it is extremely unlikely that we shall soon find a class of systems which constitute simpler or more useful models for physical plants.

For the remainder of this section we shall omit the adjectives 'dynamic," "finite-dimensional," and "smooth" and subsume these properties under the term "linear."

(2.13)   **Definition.**   *An event* (τ, x) *in a linear system* Σ *is said to be* **reachable (from the zero state)** *iff there is some time* $s \leq \tau$ *and some input* ω *which carries* $(s, 0)$ *into* (τ, x).

Note that s and ω may both depend on both τ and x. By causality, it is sufficient to give the function ω only on the set $[s, \tau]$.

The strict counterpart of "reachability" under reversal of time is given by the following

(2.14)   **Definition.**   *An event* (τ, x) *in a linear system* Σ *is* **controllable (to the zero state)** *iff there is some time* $t \geq \tau$ *and some input* ω *which carries* (τ, x) *into* $(t, 0)$.

We speak of *complete reachability* (or *complete controllability*) *at time* τ iff every event (τ, x), where τ is fixed and $x \in X$, is reachable (or controllable). If the time τ is not mentioned, these properties must hold for all τ.

(2.15) **Remark.** Caution! Even if a linear system is completely reachable and completely controllable, it does not follow that any event (τ, x) may be transferred to any other event (τ_1, x_1), $\tau_1 \geq \tau$, by suitable choice of ω. A sufficient condition is: there is a $\tau_2 \in [\tau, \tau_1]$ such that simultaneously $\tau_2 = t_{\tau, x} = s_{\tau_1, x_1}$.

Our definition of controllability is rigged in such a way that it is an obvious necessary and sufficient condition enabling us to design a regulator. Recall Problem (1.6): the task of a regulator is to move an arbitrary initial state x of Σ to a fixed desired state x, which is usually zero. Once we have found an explicit criterion for complete controllability, in a sense we have solved the mathematical (existential) part of the problem of building a regulator.

A convenient general criterion for controllability [KALMAN, 1960a] is the following

(2.16) **Theorem.** *An event (τ, x) in a linear system Σ is controllable if and only if x belongs for some t to the range of the linear transformation*

$$W(\tau, t) = \int_\tau^t \Phi(\tau, \sigma) G(\sigma) G'(\sigma) \Phi'(\tau, \sigma) \, d\sigma.$$

Here the prime designates the adjoint operator or matrix transposition.

(2.17) **Remark.** If $W(\tau, t)$ has maximal rank for some τ for every t, then Σ is completely controllable.

(2.18) **Remark.** The rank of $W(\tau, t)$ is nondecreasing as the range of integration is increased. To see this, recall from linear algebra that a matrix A is nonnegative definite if and only if there is a matrix B such that $A = BB'$. Since the integrand is nonnegative definite, so is the integral $W(\tau, t)$. So *rank* $W(\tau, t)$ is nondecreasing with $t \to \infty$ or $\tau \to -\infty$, and *rank* $W(\tau, t) \leq n$. It follows that, given τ, there is some $t_1 = t_1(\tau)$ such that *rank* $W(\tau, t) =$ constant for all $t > t_1(\tau)$. This means that in a *finite-dimensional* linear system either some event (τ, x) is not controllable or else every event (τ, x) can be transformed into $(t_1(\tau), 0)$.

(2.19) **Remark.** In the general linear case the integral cannot be evaluated in closed form, and from the practical point of view it is to be regarded as a prescription for numerical determination of controllability. From a theoretical point of view this result is very useful in proving more explicit criteria for controllability under special assumptions.

Proof of Theorem (2.16). The form of W is suggested by the calculus of variations. In fact, there is an explicit formula based on W for the function ω which carries (τ, x), $x \in range\ W(\tau, t)$, to $(t, 0)$ in an *optimal* fashion. See [KALMAN, Ho, and NARENDRA, 1963] for the details.

A simple direct proof of Theorem (2.16) is as follows:

Sufficiency. Suppose that $x \in range\ W(\tau, t)$ for some t; that is, $x = W(\tau, t)z_x$, with $z_x \in X$. Define $\omega_x \colon \sigma \longmapsto u_x(\sigma)$ by

$$(2.20) \qquad u_x(\sigma) = -G'(\sigma)\Phi'(\tau, \sigma)z_x, \qquad \sigma \in [\tau, t].$$

Substitute into the definition of φ (see (2.11)) and verify that

$$\varphi(t; \tau, x, \omega_x) = 0.$$

In this verification we must use the *composition property* of transition maps

$$(2.21) \qquad \Phi(t, \sigma) = \Phi(t, \tau)\Phi(\tau, \sigma),$$

which holds for all real σ, τ, and t as a consequence of the uniqueness of solutions of the differential equation (2.5) defining Φ. Note that (2.21) is the linear version of axiom (1.1d3) in Section 1.1. Note also that Φ is never singular; in fact, by (2.21),

$$(2.22) \qquad \Phi^{-1}(t, \tau) = \Phi(\tau, t).$$

Necessity. We recall from linear algebra that X can be written as the direct sum

$$X = \text{range } W \oplus \text{kernel } W$$

(this is true for any symmetric operator sending any finite-dimensional vector space into itself). If (τ, x) is not controllable, then $x = (x_1, x_2)$ must have a component $x_2 \neq 0$ in *kernel W*. If (τ, x_2) is controllable, by linearity every event of the type $(\tau, x_1 + x_2)$, $x_1 \in range\ W$, is also controllable. Hence it suffices to prove that a nonzero vector in *kernel W* is not controllable.

Suppose, then, that $x_2 \neq 0$ belongs to *kernel $W(\tau, t)$* for some t and that $\omega_2 \in \Omega$ carries (τ, x_2) to $(t, 0)$. Then

$$x_2'W(\tau, t)x_2 = \int_\tau^t \|G'(\sigma)\Phi'(\tau, \sigma)x_2\|^2\, d\sigma = 0.$$

Since the integrand is nonnegative and continuous, it follows that

$$(2.23) \qquad G'(\sigma)\Phi'(\tau, \sigma)x_2 \equiv 0 \text{ on } [\tau, t].$$

On the other hand,

$$\Phi(t, \tau)x_2 + \int_\tau^t \Phi(t, \sigma)G(\sigma)u_2(\sigma)\, d\sigma = 0,$$

or, using (2.20),

$$\int_\tau^t \Phi(\tau, \sigma)G(\sigma)u_2(\sigma)\, d\sigma \;=\; -x_2.$$

Multiplying by x_2', we obtain

$$x_2' \int_\tau^t \Phi(\tau, \sigma)G(\sigma)u_2(\sigma)\, d\sigma \;=\; -x_2'x_2 \;=\; -\|x_2\|^2.$$

By (2.23), the left-hand side is zero; so $\|x_2\|^2 = 0$, which contradicts the assumption that $x_2 \neq 0$. $\square$

The reader is invited to prove the analogous criterion for the reachability of (τ, x):

(2.24) Theorem. *An event (τ, x) in a linear system Σ is reachable if and only if x belongs for some s to* range $\hat{W}(s, \tau)$, *where*

$$\hat{W}(s, \tau) \;=\; \int_s^\tau \Phi(\tau, \sigma)G(\sigma)G'(\sigma)\Phi'(\tau, \sigma)\, d\sigma.$$

(2.25) Example. Consider a periodic linear system (that is, $F(t)$ and $G(t)$ are periodic functions of t). By Floquet's theorem,† the transition map has the representation

$$\Phi(t, \tau) \;=\; P(t)e^{(t-\tau)Q}P^{-1}(\tau),$$

where Q is a constant $n \times n$ matrix and $P: T \to \{n \times n \text{ matrices}\}$ is a nonsingular continuous periodic function of t. Moreover, P is nonsingular for each t, because Φ is never singular.

W is not necessarily periodic. It is easy to prove, however, that its rank attains its maximum within one period. In other words, we have

(2.26) Proposition. *If a periodic linear system is completely controllable, then every state can be transferred to zero in an interval of time no greater than the period of the system.*

If Σ is merely almost-periodic or quasi-periodic, Floquet's theorem is false, and the argument concerning the rank of W fails. Hence even in this "slightly" different case the explicit determination of $t_1(\tau)$ (see Remark (2.18)) is a nontrivial problem.

2.3 Constant linear systems

In "classical" control theory we almost always deal with plants which are not only linear, but constant. Intuitively speaking, a *constant* (or *time-invariant*) dynamical system is one whose defining properties (tran-

---

† See [CODDINGTON and LEVINSON, 1955, chap. 3, sec. 5]. The reader should translate the results given there into the present notation.

sition function, readout map) are independent of the initial time. From the discussion in Chapter 1 (Definition (1.2)) and Theorem (2.8) of the preceding section, we obtain immediately the obvious explicit criterion for constancy:

(3.1) **Proposition.** *A linear system Σ is constant iff the matrices F, G, and H are all constants.*

It is clear that a constant linear system Σ is equivalent to the system of equations

$$\frac{dx}{dt} = Fx + Gu(t),$$
(3.2)
$$y(t) = Hx(t),$$

where F, G, and H are constant matrices.

Our term "constant" stems from the older nomenclature "constant coefficients." For the next two sections "linear" will carry the implication not only of "dynamic," "finite-dimensional," and "smooth," but also of "constant." This flagrant abuse of language is entrenched, at least in the literature of control engineering.

(3.3) **Remark.** Often we *must* assume that the plant is constant (at least over reasonably short intervals of time), since otherwise there may be no way of determining a model for it; in an arbitrary nonconstant system past measurements may not reveal anything about future behavior.

Recall that the transition matrix of a (constant) linear system is given by

$$\Phi(t, \tau) = \exp\ (t - \tau)F = e^{(t-\tau)F} = \sum_{k=0}^{\infty} \frac{(t - \tau)^k F^k}{k!}.$$

The exponential function $e^{(t-\tau)F}$ is well defined (the power series converges absolutely) for all t, τ, and F (see [CODDINGTON and LEVINSON, 1955, chap. 3, sec. 1]). From the notation $e^{(t-\tau)F}$, properties (2.21) and (2.22) are immediately obvious—at least intuitively.

For (constant) linear systems there is a very simple criterion for complete controllability:

(3.4) **Theorem.** *A linear system Σ of dimension n is completely controllable if and only if*

$$\text{rank } C = \text{rank } [G, FG, \ldots, F^{n-1}G] = n.$$

Here C is an $n \times mn$ matrix made up of the columns of the matrices $G, FG, \ldots, F^{n-1}G$.

(3.5) **Corollary.** *If a linear system Σ is completely controllable, then for any τ, any x, and any $\epsilon > 0$ there is an input ω which carries (τ, x) into $(\tau + \epsilon, 0)$.*

(3.6) **Corollary.** *If rank $G = r$, then the complete-controllability criterion simplifies to*

$$\text{rank } C = \text{rank } [G, FG, \ldots, F^{n-r}G] = n.$$

These results are essentially algebraic. They are all interrelated. A further interesting consequence is obtained as follows:

(3.7) **Definition.** *The* **characteristic polynomial** χ_F *of a square matrix F is the polynomial* $\det(zI - F)$. *The* **minimal polynomial** ψ_F *of F is the monic polynomial (leading coefficient unity) of smallest degree such that $\psi_F(F) = 0$.*

Recall from the Cayley-Hamilton theorem [GANTMAKHER, 1959] that $\chi_F(F) = 0$. Recall also that ψ_F is a factor in any polynomial which annihilates F, so that $\psi_F | \chi_F$ (read "ψ_F divides χ_F"). Then we have

(3.8) **Corollary.** *If, given F, there is a G of rank 1 such that the pair $\{F, G\}$ is completely controllable, then $\chi_F = \psi_F$.*

In the Victorian era one used to say that such a matrix F was *nonderogatory*. In modern terminology we say that the linear map $F: X \to X$ is *cyclic*, which means that there is a vector g such that the set $\{g, Fg, \ldots\}$ generates X; this is clearly the same as our controllability criterion.

Proof of Theorem (3.4). *Sufficiency.* By constancy, we may assume without loss of generality that $\tau = 0$. If Σ is not completely controllable, then by the proof of Theorem (2.16) (see Formula (2.23)) there is a vector $x \neq 0$ such that $x'\Phi(0, t)G \equiv 0$ for all $t \geq 0$. Differentiating this relation once, twice, and so on with respect to t and then letting $t = 0$ gives

$$x'G = 0, \qquad x'FG = 0, \ldots$$

So $x \neq 0$ is orthogonal to every element of the matrix

$$C = [G, \ldots, F^{n-1}G],$$

and we have a contradiction if the rank of C is n.

Necessity. Suppose rank $C < n$. Then there is a vector $q \neq 0$ in X such that

$$q'G = 0, \qquad q'FG = 0, \qquad \ldots, \qquad q'F^{n-1}G = 0.$$

By the Cayley-Hamilton theorem, we then also have $q'F^nG = 0$, and it follows by induction that $q'F^pG = 0$ for all $p \geq 0$. Recalling that the transition map of $\Sigma =$ constant, linear is given by

$$\Phi(0, t) = e^{-tF} = I - tF + \frac{t^2F^2}{2!} - \cdots,$$

we conclude that $q'e^{-tF}G \equiv 0$. The relation

$$0 = \varphi(t; \tau, x, \omega) = e^{(t-\tau)F}x + \int_\tau^t e^{(t-\sigma)F}Gu(\sigma)\, d\sigma$$

implies

$$0 = x + \int_\tau^t e^{(\tau-\sigma)F}Gu(\sigma)\, d\sigma;$$

this cannot be true if $q'x \neq 0$. So if $rank\ C < n$, there is an uncontrollable state $x \neq 0$.   □

Proof of Corollary (3.5). It suffices to show that $W(\tau, t)$ is positive definite whenever $t > \tau$ and $rank\ C = n$. Indeed, if there is an $x \neq 0$ such that

$$0 = x'W(\tau, t)x = \int_\tau^t \|G'e^{(\tau-\sigma)F}x\|^2\, d\sigma,$$

then

$$x'e^{\sigma F}G \equiv 0 \text{ on } [t - \tau, 0],$$

which contradicts $rank\ C = n$, just as in the sufficiency part of the proof of Theorem (3.4).   □

The proof of Corollary (3.6) rests on

(3.9) **Lemma.**   *If q is an integer such that*

$$rank\ [G, FG, \ldots, F^{q-1}G] = rank\ [G, FG, \ldots, F^qG] = r,$$

   then

$$rank\ [G, FG, \ldots, F^pG] = r \text{ for all } p \geq q - 1.$$

Proof. To say that

$$rank\ [G, \ldots, F^{q-1}\ G] = rank\ [G, \ldots, F^qG]$$

means that every column of the matrix F^qG is linearly dependent on the columns of the matrices $G, \ldots, F^{q-1}G$. Hence every column of $F^{q+1}G$ is linearly dependent on the columns of $FG, \ldots, F^qG$. Proceeding this way by induction completes the proof.   □

Proof of Corollary (3.6). By Lemma (3.9), the rank of the matrix C must increase by at least one as each term is added, until the maximal rank n is reached. If $rank\ G = r$, it will be sufficient to include at most $n - r$ terms $FG, \ldots, F^{n-r}G$ to see whether the maximal rank of C can reach n (of course, by the Cayley-Hamilton theorem, we never have to consider a matrix bigger than $[G, \ldots, F^{n-1}G]$). $\square$

Proof of Corollary (3.8). Since χ_F is a monic polynomial and $\psi_F | \chi_F$, all we have to show is that $\delta\chi_F = \delta\psi_F$, where δ is the degree of a polynomial. Since $\psi_F(F)G = 0$, if $\delta\psi_F < n$, we contradict the controllability criterion (3.6) with $r = 1$. $\square$

(3.10) Remark. Corollary (3.5) may be puzzling to those who take a hardboiled practical position, since it is a well-known fact that real systems cannot be controlled in an arbitrarily short time, no matter how large we take the magnitude of the input signals. This "difficulty" is not one of theory but of application. In order to control a system in an arbitrarily short time, the matrices F and G must be known to an arbitrarily high accuracy; in fact, the proof of the theorem relies on the (tacit) assumption that these matrices are *exactly* known. When we apply the theory, we must, of course, worry about the degree of agreement between the behavior of our dynamical system used as a model for the plant and the behavior of the real physical plant itself.

(3.11) Remark. We must bear in mind that Theorem (3.4) supplies only a yes/no criterion for complete controllability. In many practical situations we also want a quantitative measure of the relative ease or difficulty of controllability. In case of complete controllability, we may look for this purpose at the absolute values of the $n \times n$ determinants of $[G, \ldots, F^{n-1}G]$. A less *ad-hoc* approach might be to look at the energy of the control signal ω necessary to send a given state to zero. This leads to notions of optimal control, and the natural controllability criterion turns out to be Theorem (2.16). See [KALMAN, HO, and NARENDRA, 1963] for a detailed examination of this problem.

(3.12) Remark. The deeper significance of Theorem (3.4) is that it relates structural properties of Σ (as mirrored in the algebraic properties of the *pair* of matrices $\{F, G\}$) to the question of controlling Σ.

The proofs of the next two corollaries follow the pattern already established and are left as exercises for the reader:

(3.13) Corollary. *For a constant linear system,*

$$\dim \text{range } W(\tau, t) = \text{rank } [G, FG, \ldots, F^{n-1}G]$$

whenever $t > \tau$.

(3.14) Corollary. *The controllable states of a constant linear system form an F-invariant subspace, which is the smallest such subspace containing all column vectors of the matrix G.*

In view of Corollary (3.13), the controllability theorem (2.16) and the reachability theorem (2.24) coincide for constant systems. This, plus the possibility of shifting initial times at will, plus (3.5), allow us to make the following claim:

(3.15) **Theorem.** *In a constant linear system* Σ *every event is controllable if and only if it is reachable. If* $t_2 > t_1$ *and* Σ *is completely controllable, any event* (t_1, x_1) *in* Σ *may be carried to any other event* (t_2, x_2) *by a suitable choice of the input* ω.

Up to this point we have regarded the state as an abstract quantity, that is, an abstract vector in a finite-dimensional vector space. In other words, the preceding results did not depend on a specific choice of the coordinate system in which the abstract state vectors are displayed, although it was convenient to state our criteria in the language of matrices rather than in the language of abstract linear transformations. There are situations, however, in which special choice of the basis leads to interesting results. In a way, we surrender generality (coordinate-free results) for applied-mathematical convenience (the matrices we have to work with have fewer elements). We urge the reader to consult [HALMOS, 1958] for an excellent discussion of the "coordinate-free" versus the "coordinate-dependent" point of view.

The analysis and results of Sections 2.2 *and* 2.3 *continue to hold,* mutatis mutandis, *for discrete-time linear systems.* Nonconstant discrete-time systems are of little interest, because the only tool for studying such systems consists of various types of smoothness assumptions with respect to the time dependence of the matrices F, G, and H. The constant discrete-time case is of considerable importance, however. Let us record here, without proofs, the two main facts:

(3.16) **Theorem.** *A finite-dimensional, discrete-time, linear, constant system* Σ *is equivalent to the system of equations*

(3.17)
$$x(t + 1) = Fx(t) + Gu(t),$$
$$y(t) = Hx(t),$$

where F, G, H are constant matrices and t is an integer.

This can be deduced in the same way as Theorem (2.8); of course, the reader who is not especially interested in axiomatics can take (3.17) as the definition of a *linear discrete-time* (finite-dimensional, constant) *system* (as before, we shall suppress the adjectives "finite-dimensional" and "constant").

(3.18)   **Theorem.**   *A discrete-time linear system* Σ *of dimension* n *is completely reachable if and only if*

$$\operatorname{rank} C = \operatorname{rank} [G, \ldots, F^{n-r}G] = n,$$

where $r = \operatorname{rank} G$.

(3.19)   **Remark.**   This criterion corresponds to Corollary (3.6), but since Theorem (3.15) does not hold for discrete-time systems when $det\ F = 0$, we must state (3.18) as a criterion for reachability, not controllability.   It is still true, in complete generality, that

$$\text{controllable states} \supset \text{reachable states}.$$

(3.20)   **Remark.**   It is most remarkable that the algebraic form of Theorem (3.18) is identical with that of Corollary (3.6), even though the meaning of the matrices F and G in (3.2) and (3.17) is quite different. (As an exercise, compute a discrete-time system corresponding to (3.2) by letting t take only integer values in Equation (2.11).)   It is tempting to extrapolate and guess that the foundations of linear system analysis are deeply algebraic, irrespective of whether time is taken to be continuous and discrete.   This idea is discussed in Chapter 6, and it is developed in a definitive form in Chapter 10.   As a matter of fact, Theorems (3.4) to (3.6) and (3.18) are the principal reasons for the introduction of modules into linear system theory, which is the theme of Chapter 10.

2.4   Coordinate changes and canonical forms

Now we want to develop explicit formulas for expressing the effect of a change of basis in X_Σ on the parameters defining Σ.

We shall view a state vector x as an n-tuple of numbers which are the coordinates of the (abstract) vector x with respect to some fixed basis in the state space X.   Correspondingly, we must view F and G as matrices which represent the abstract linear transformations F and G with respect to the given basis in X.   We are not interested in change of basis in U, since U is usually explicitly given by physical considerations.   But there is no "natural" basis for X, since almost always the state is to be regarded as an abstract quantity, without concrete physical meaning.   (Perhaps the only case where the state *does* have direct physical significance is classical mechanics, where the elements of the state vector consist of the positions and momenta of particles.)

A change of coordinates is a nonsingular linear transformation $A : x \mapsto \hat{x} = Ax$ of the (numerical) state vectors.   In matrix notation we write

$$(4.1) \qquad \begin{bmatrix} \hat{x}_1 \\ \cdots \\ \hat{x}_n \end{bmatrix} = \begin{bmatrix} a_{11} & \cdots & a_{1n} \\ \cdots & \cdots & \cdots \\ a_{n1} & \cdots & a_{nn} \end{bmatrix} \begin{bmatrix} x_1 \\ \cdots \\ x_n \end{bmatrix}, \qquad \det A \neq 0.$$

The transformation A induces a transformation of the matrices F and G, which is given by

$$(4.2) \qquad \begin{aligned} \hat{G} &= AG, \\ \hat{F}A &= AF. \end{aligned}$$

Let us consider the following problem. We are given two pairs of matrices, $\{F, G\}$ and $\{\hat{F}, \hat{G}\}$, which correspond to the same completely controllable system Σ, but with respect to different bases in X. How can we explicitly determine the matrix A? Iterating (4.2), we find

$$\hat{G} = AG, \qquad \hat{F}\hat{G} = AFA^{-1}AG = AFG, \ldots ;$$

so that, by (4.2) and (3.6),

$$(4.3) \qquad [\hat{G}, \hat{F}\hat{G}, \ldots, \hat{F}^{n-r}\hat{G}] = A[G, FG, \ldots, F^{n-r}G],$$

where $r = rank\ G$. If $m = 1$ (that is, if the matrix G consists of a single column, which means that the system has a single input terminal), then $0 \leq r \leq m = 1$; at the same time, $r > 0$, because otherwise $G = 0$ and Σ is surely not completely controllable. So $r = 1$. In this special case the matrices in (4.3) are square. They are nonsingular, by (3.6), since Σ is known to be completely controllable. Thus

$$(4.4) \qquad A = [\hat{g}, \hat{F}\hat{g}, \ldots, \hat{F}^{n-1}\hat{g}][g, Fg, \ldots, F^{n-1}g]^{-1},$$

where we have written g for an $n \times 1$ matrix G.

In the general case $r > 1$ this formula does not make sense, because a nonsquare matrix has no inverse. So we form $C = [G, \ldots, F^{n-r}G]$ and $D = CC'$. Then D is $n \times n$, and it is nonsingular in view of condition (3.6) (compare this construction with the definition of W in Theorem (2.16)). Thus $AC = \hat{C}$ implies $ACC' = AD = \hat{C}C'$ and $A = \hat{C}C'D^{-1}$. Note that $C'(CC')^{-1}$ is a kind of generalized inverse for C.

To sum up, we have

(4.5) **Theorem.** *If a completely controllable n-dimensional linear system Σ is represented with respect to different bases in X by the pairs of matrices $\{F, G\}$ and $\{\hat{F}, \hat{G}\}$, then the nonsingular matrix A, giving the change of coordinates $\hat{x} = Ax$ between the two bases, can be computed as*

$$A = [\hat{G}, \ldots, \hat{F}^{n-r}\hat{G}][G, \ldots, F^{n-r}G]'$$
$$\times \{[G, \ldots, F^{n-r}G][G, \ldots, F^{n-r}G]'\}^{-1},$$

where $r = \operatorname{rank} \hat{G} = \operatorname{rank} AG = \operatorname{rank} G$.

(4.6) Corollary. *The matrix A is unique.*

Proof. If we take the coordinate-free point of view, the result is immediate: two different bases of a finite-dimensional vector space are related by a unique linear map.

We can easily also give a concrete proof. Let A and B be two matrices taking $\{F, G\}$ to $\{\hat{F}, \hat{G}\}$. Applying Theorem (4.2) twice gives $B^{-1}AG = \hat{G}$ and $\hat{F}B^{-1}A = B^{-1}AF$. By Theorem (4.5), $B^{-1}A = I$, or $A = B$. ☐

(4.7) Remark. The study of square matrices under the transformation $F \mapsto AFA^{-1}$ is an old topic in linear algebra. What makes our results interesting and new is that they concern a matrix *pair* $\{F, g\}$. The assumption of complete controllability is essential. If, for instance, $G = 0$, then the transformation A is no longer unique; the equation $FA = AF$ for fixed F admits as solutions all matrices $A = \pi(F)$, where π is a polynomial in one variable with real coefficients. It can be shown [JACOBSON, 1951, sec. 3.15, corollary] that $\pi(F)$ are the *only* matrices commuting with F if and only if F belongs to a completely controllable pair $\{F, g\}$, that is, if and only if F belongs to a system which can be controlled completely through a single input terminal, or (Corollary (3.8)) if and only if F is cyclic.

Let us now avail ourselves of the possibility of simplifying the controllability criterion (3.6) with a special choice of the basis. Let G consist of a single column g.

(4.8) Theorem. *Let us write the characteristic polynomial of F as*
$$\chi_F(z) = z^n + \alpha_1 z^{n-1} + \cdots + \alpha_n.$$
Then the set of vectors

$$e_1 = F^{n-1}g + \alpha_1 F^{n-2}g + \cdots + \alpha_{n-1}g,$$
$$\cdots$$
$$e_{n-1} = Fg + \alpha_1 g,$$
$$e_n = g,$$

form a basis for X if and only if $\{F, g\}$ is completely controllable. In this basis, $\{F, g\}$ have the matrix representation

$$(4.9) \quad F = \begin{bmatrix} 0 & 1 & 0 & \cdots & 0 \\ 0 & 0 & 1 & \cdots & 0 \\ \cdots & \cdots & \cdots & \cdots & \cdots \\ 0 & 0 & 0 & \cdots & 1 \\ -\alpha_n & -\alpha_{n-1} & -\alpha_{n-2} & \cdots & -\alpha_1 \end{bmatrix}, \quad g = \begin{bmatrix} 0 \\ \cdots \\ 1 \end{bmatrix}.$$

Proof. The set of vectors $\{e_1, \ldots, e_n\}$ is a triangular linear combination of the vectors $\{g, \ldots, F^{n-1}g\}$. So $\{e_1, \ldots, e_n\}$ form a basis for X, because the same is true for $\{g, \ldots, F^{n-1}g\}$, by (3.6), with $r = 1$. That F has the given form may be seen immediately by computing $Fe_n, \ldots, Fe_2$; in particular, $Fe_1 = -\alpha_n e_n$ follows from $\chi_F(F) = 0$ (Cayley-Hamilton theorem).   □

(4.10) **Definition.** *We call* (4.9) *the* **control canonical form** *of* $\{F, g\}$.

(4.11) **Remark.** Since χ determines the matrix F in (4.9), in Victorian English F was known as the *companion matrix* of the polynomial χ. However, this terminology is not satisfactory, since several other matrices can be used equally well as "companions" of χ; for instance, the matrix

$$\begin{bmatrix} 0 & 0 & \cdots & 0 & -\alpha_n \\ 1 & 0 & \cdots & 0 & -\alpha_{n-1} \\ 0 & 1 & \cdots & 0 & -\alpha_{n-2} \\ \cdot & \cdot & \cdot & \cdot & \cdot \\ 0 & 0 & \cdots & 1 & -\alpha_1 \end{bmatrix},$$

which will be important later. For the moment, let us merely note the fact that F in (4.9) is uniquely determined by χ_F and the special basis $\{e_i\}$.

(4.12) **Exercise.** Verify the fact that $det\ (zI - F) = \chi_F(z)$; in other words, that χ_F is indeed the characteristic polynomial of the (companion) matrix F in (4.9).

(4.13) **Remark.** The genesis of the control canonical form is closely related to the well-known trick of replacing the single nth-order differential equation

(4.14) $$\frac{d^n y}{dt^n} + \alpha_1 \frac{d^{n-1}y}{dt^{n-1}} + \cdots + \alpha_n y = u(t),$$

(where u and y are scalars and α_i are constants), by the system of n first-order equations

(4.15)
$$\frac{dx_1}{dt} = x_2,$$
$$\frac{dx_2}{dt} = x_3,$$
$$\cdots$$
$$\frac{dx_n}{dt} = -\alpha_n x_1 - \cdots - \alpha_1 x_n + u(t).$$

The matrix description of (4.15) is the pair $\{F, g\}$ given by (4.9). The passage from (4.14) to (4.15) is accomplished by introducing the state variables via the identities

$$(4.16) \qquad x_i = \frac{d^{i-1}y}{dt^{i-1}} \qquad i = 1, \ldots, n.$$

The conventional treatment of Equations (4.15), as just reviewed, is deficient. Our treatment here is much sharper in at least three points (often overlooked or misexplained):

1. It is not clear that (4.14) is a completely controllable system (this question was first discussed in depth in [KALMAN, HO, and NARENDRA, 1963]).

2. It is sometimes believed that Formula (4.16) is the *only* way to convert Equation (4.14) into the "state-variable" form. This is false. The system Σ, which is the minimal realization of the input/output function $u(\cdot) \mapsto y(\cdot)$ defined by the differential equation

$$(4.17) \qquad \sum_{i=0}^{n} \alpha_i \frac{d^i y}{dt^i} = \sum_{i=0}^{m} \beta_i \frac{d^i u}{dt^i}, \qquad m < n, \qquad \alpha_0 = 1,$$

is completely controllable, and therefore this system, too, admits the canonical form (4.15). The difference between (4.14) and (4.17) is then expressed through the form of H. In the first case

$$y = x_1, \qquad H = [1 \ 0 \ \cdots \ 0];$$

in the second case (see, for example, [KALMAN, 1963c, sec. 8])

$$y = \sum_{i=0}^{n-1} \beta_i x_{i+1}, \qquad H = [\beta_0 \ \beta_1 \ \cdots \ \beta_{n-1}].$$

3. Formula (4.16), which describes the state variables as derivatives of the output, works nicely for the system (4.14) but it is not valid in general. The same does not work for Equation (4.17). In short, what is important here is the control canonical form, not the special formula (4.16).

(4.18) **Remark.** As we shall see in the next section, Theorem (4.8) provides a convenient bridge between modern control theory and the classical problem of constructing a regulator. Things are much less simple when G has more than one column, that is, when Σ has more than one input terminal. In this general case there is no simple canonical form similar to (4.9). It is interesting to note that the corresponding problem in classical control theory (the construction of regulators with multiple input terminals) has never been satisfactorily solved; in fact, there are no *general* algorithms for the design of such systems. From the point of view of the modern theory, however, the canonical form (4.9) is merely a matter of convenience. The basic issue is the question of complete controllability, which for a given system is true or false independently of the existence of any special canonical form.

2.5   The concept of a control law

We are ready to apply our abstract considerations concerning controllability to the (mathematical) problem of constructing a control system. There are many ways of doing this, because there are many possible control systems which would accomplish more or less the same task. As emphasized before, we shall confine our discussion to the simplest kind of regulators. Within this constraint we shall show that the procedures of modern control theory give results which are compatible with classical control theory, but which are also simpler to derive and easier to interpret.

The critical concept, that of "feedback," is formalized as follows:

(5.1)   **Definition.** *Consider an arbitrary dynamical system Σ. A* **control law** *is a map $k: T \times X \to U$ which assigns to the state $x(t)$ at time t the value $u(t) = k(t, x(t))$ of the input at that time; that is, the value of the input at any moment depends only on the state $x(t)$ at that moment as well as, possibly, on t. Other parameters of Σ may, however, affect the specific definition of the function k.*

The principle that the inputs should be computed from the state was enunciated and emphasized by Richard Bellman in the mid-1950s. *This is the fundamental idea of control theory.* It is, in fact, a scientific explanation of the great invention known as "feedback," which is the foundation of control engineering.

Our claim that the state description of the plant is the natural framework for formulating and solving problems in control will be amply justified by the results which will follow. The main point is, of course, that the state incorporates all information necessary to determine the control action to be taken, since (by the definition of a dynamical system) the future evolution of the plant is completely determined by its present state and the future inputs.

(5.2)   **Remark.** It is not obvious that a given control law k defines an input function $\omega: T \to U$ which belongs to the input space Ω_Σ of Σ. The trouble is that ω is determined by k only implicitly via the relation

$$\omega: t \mapsto u(t) = k(t, \varphi(t; \tau, x, \omega)).$$

It is not obvious that a solution ω exists at all, since it is not clear what conditions on k guarantee that ω is smooth enough to belong to the class Ω_Σ (which is what we need to be allowed to substitute ω into φ). So the choice of k is not free, but is indirectly constrained by the nature of Ω_Σ.

Fortunately, this vexing technical difficulty disappears in the most important special case:

Σ = finite-dimensional, linear, smooth.

Ω_Σ = continuous functions on the reals.

k = linear in x and continuous in t.

By the last assumption, we write

$$k(t, x) = K(t)x,$$

where $K\colon T \to \{m \times n \text{ matrices}\}$. Using the control law $u(t)$ in (2.9) gives the free equation

$$\frac{dx}{dt} = [F(t) - G(t)K(t)]x,$$

which has a well-determined unique solution for each event (τ, x) since F, G, and K have been assumed to be continuous in t. Moreover, the transition matrix of this equation is also continuous in t; that is, $\Phi(\cdot, \tau)x \in C^0(T \to X)$. Since

$$\omega\colon t \longmapsto u(t) = K(t)\Phi(t, \tau)x,$$

$\omega \in C^0$, and so $\omega \in \Omega_\Sigma$ for *every* choice of K.

Under the preceding assumptions the construction of a control law is a fairly immediate consequence of Theorem (2.16). The intuitive argument is this: since Formula (2.20) gives the correct input value as a function of the state alone at $\sigma = \tau$, we let $u(\sigma)$ be determined according to the same rule at each σ. So we try to define a linear control law by

$$(5.3) \qquad u(\sigma) = -G'(\sigma)W^\#(\sigma, t)x(\sigma) = -K(\sigma)x(\sigma), \qquad \sigma \in [\tau, t],$$

where $W^\#$, the *pseudo-inverse* of W, is any solution X of the matrix equation $WXW = W$ such that the map $\sigma \longmapsto X(\sigma, t)$ is piecewise continuous in σ for each t.

The existence of $W^\#$ is not quite trivial (in fact, this was overlooked in [KALMAN, HO, and NARENDRA, 1963]). We proceed to define $W^\#$ as follows:

1. We fix τ and t and then divide the interval $[\tau, t]$ into the disjoint union of subintervals such that *rank* $W(\cdot, t)$ = constant. Since the rank of $W(\sigma, t)$ increases monotonically as $\sigma \to \tau$, and since the rank can take only finitely many values, there are only a finite number of such intervals. It is easy to see also that each interval is closed on the left and is open on the right (except at $\sigma = t$).

2. On the first interval, say $[\sigma_1, t]$, we define $W^{\#}(\cdot, t)$ as follows. If $X = Y \oplus Z$ (orthogonal direct sum) such that W is nonsingular on Y and zero on Z, we let $W^{\#}|Y = W^{-1}$ and $W^{\#}|Z = I$.†

3. We then proceed similarly and define $W^{\#}$ on the entire set $[\tau, t]$. Thus $W^{\#}$ is piecewise continuous.

Notice that $\|W^{\#}(\sigma, t)\| \to \infty$ as $\sigma \to \sigma_i$ from the left. This mathematical difficulty will "disappear" after we multiply $K(\sigma)$ by $x(\sigma)$ (see Equations (5.5) and (5.6) below). The practical difficulty, that K has unbounded components, cannot be ignored, however.

Let us try to prove now that Formula (5.3) fulfills our expectations:

(5.4) Theorem. *In a finite-dimensional, linear, smooth system Σ the input determined by the control law* (5.3) *carries every event* (τ, x) *into* $(t, 0)$ *if and only if* $x \in$ range $W(\tau, t)$.

Proof. The "only-if" part has been proved already (in Theorem (2.16)). To prove the "if" part we must show that the input ω_* generated by rule (5.3) is identical with the input ω_x defined via (2.20). That is,

$$(5.5) \qquad u_*(\sigma) = -G'(\sigma)W^{\#}(\sigma, t)x(\sigma)$$

and

$$(5.6) \qquad u_x(\sigma) = -G'(\sigma)\Phi'(\tau, \sigma)z_x$$

must be the same for all $\sigma \in [\tau, t]$. Substituting (5.6) into (2.11) (and using (2.21) several times) gives the transition function explicitly as

$$x(\sigma) = \varphi(\sigma; \tau, x, \omega_x) = \Phi(\sigma, \tau)[W(\tau, t) - W(\tau, \sigma)]z_x,$$
$$= W(\sigma, t)\Phi'(\tau, \sigma)z_x.$$

Substituting this expression into (5.5), we see that $u_*(\sigma) \equiv u_x(\sigma)$ if

$$(5.7) \qquad G'(\sigma)w = G'(\sigma)W^{\#}(\sigma, t)W(\sigma, t)w$$

for all σ and w (actually, $w = \Phi'(\tau, \sigma)z_x$). To prove (5.7), we compute

$$(5.8) \qquad \int_{\sigma}^{t} \|G'(\rho)\Phi'(\sigma, \rho)[I - W^{\#}(\sigma, t)W(\sigma, t)]w\|^2 \, d\rho = 0.$$

This follows easily from the definition of W and the relation $WW^{\#}W = W$. By continuity, the integrand vanishes identically on $[\sigma, t]$, and so in particular also at $\rho = \sigma$, which proves (5.7).

So (5.6) is a solution of (2.4) in both cases: when ω is given explicitly by (2.20) and when it is given implicitly by (5.3). Moreover, everything we said holds for arbitrary (τ, x). $\square$

The identity (5.8) is of fundamental importance; it is the so-called "pseudo-inverse lemma" [KALMAN, 1963b, app.].

---

† Here $A|X$ means "A restricted to X."

(5.9) **Remark.** The control law (5.3) is seldom used in practice because $K(\sigma) \to \infty$ as $\sigma \to t$. This is due to the overly stringent boundary condition imposed by the requirement that $x(t) = 0$. A thorough analysis of the situation requires the calculus of variations and properly belongs to optimal control theory.

For us the most important implication of Theorem (5.4) is just this: *The possibility of constructing an arbitrarily good control law is limited only by the controllability properties of the plant.* If the plant is finite dimensional, linear, smooth, and *constant*, there is no limitation whatever once complete controllability is present. This leads to a rather surprising result:

(5.10) **Theorem.** *Let the pair $\{F, g\}$ be completely controllable, and let $\theta(z) = z^n + \beta_1 z^{n-1} + \cdots + \beta_n$ be an arbitrary polynomial, where $n = \dim F$. Then there is a vector k such that $\chi_{F-gk'} = \theta$.*

As far as the authors know, this theorem was first obtained in 1959 by J. E. Bertram, who used "root-locus" methods. In 1961 R. W. Bass independently formulated and proved the theorem in unpublished lecture notes; his proof was based on linear algebra. The theorem was rediscovered independently many times since then. The present proof follows [KALMAN, 1963*d*, main lemma] and is a good deal simpler than some others which have been published.

Proof. Since the characteristic polynomial is independent of the choice of coordinates, we may take any convenient basis for computing k. We choose the control canonical form of $\{F, g\}$ given by (4.9). With respect to this basis, we let $k' = (k_1, \ldots, k_n)$, where

$$\beta_i = \alpha_i + k_{n-i+1}, \qquad i = 1, \ldots, n.$$

Because of the special form of g, the matrix $F - gk'$ is a companion matrix whose bottom row is $(-\beta_n, \ldots, -\beta_1)$. So $\chi_{F-gk'} = \theta$. $\square$

(5.11) **Interpretation.** The theorem says that in a single-input, completely-controllable, constant system Σ the characteristic roots can be chosen at will if we allow feedback. This shows that feedback is enormously important; the dynamics of a plant may be altered by means of feedback in a completely arbitrary fashion, subject only to the requirement that we have an *accurate* representation of the physical behavior of the plant as a linear constant system. It is not necessary to build a plant (such as an airplane or a nuclear reactor) so that it has inherently good dynamical characteristics; the desired characteristics can be achieved later, artificially, by means of feedback. As a result, one has great freedom in the design and physical construction of the plant.

Since there is no good theory of nonlinear controllability at present, it is impossible to say what *global* limitations on controllability are implied by nonlinear elements in the plant. In other words, the sweeping potentialities of feedback mentioned above can be realized in general only in the *local* (linear) case.

(5.12) **Remark.** As pointed out already in Section 2.1, the deeper theory of control requires that we *compute* the feedback coefficients $\beta_1, \ldots, \beta_n$ so that they are optimal according to some specific performance criterion. The relevant theory is presented in Section 3.5. The end result (Corollary (5.57) of Chapter 3) is that under reasonable assumptions optimization on the interval $[0, \infty)$ leads to a strictly stable control law; that is,

(5.13) $$\operatorname{Re} \lambda_i[F - gk'] < 0 \qquad \text{for all } i.†$$

To simplify the remaining discussion, we shall simply *assume* from now on that our control law satisfies (5.13). Actually, this is not a bad assumption. Very often in classical control theory $\chi_{F-gk'}$ is selected according to intuitive physical considerations, subject, of course, to the stability constraint (5.13). In surprisingly many cases the resulting systems are actually optimal in the sense of the modern theory [KALMAN, 1964].

2.6   State determination

Our definition of a control law implicitly assumed that we know what the state of the plant is at each instant, in other words, that all the internal variables of the plant can be read out as outputs. In most practical cases we should not expect this to be true. Indeed, we should always think of the state as an abstract quantity which represents inaccessible variables inside the plant. This is the first difficulty we must come to grips with in trying to apply Bellman's recipe: *control is a function of state.*

The natural solution is this: in addition to the control law, the regulator must contain another component whose task is *state determination*. From the definition of a dynamical system, we see that we need two different kinds of information to determine the state of the plant:

    1. Knowledge of the structure of the plant—its transition map, its readout map, etc.,

    2. Knowledge of the actual inputs and outputs of the plant.

---

† We write $\lambda_i[A]$ for the eigenvalues of a square matrix A.

In this section we shall seek a data-processing scheme which converts these two types of data into a good estimate of the unknown present state of the plant.

(6.1) Remark. In present-day engineering practice it is customary to assume that most data of the first type are given *a priori* (for instance, supplied by the manufacturer of the plant), and that data of the second type are available through real-time measurements on the plant while it is in actual operation. When data of the first type are not available and must be somehow inferred from input/output data (the problem of *identification*), we are faced with an *adaptive control problem*. The theory of adaptive systems is much talked about, but very little has been accomplished. In the nonadaptive control problem (when data on the plant structure are given) dynamical properties of the plant are assumed to be exactly known, and it remains "only" to determine the instantaneous state. This is relatively easy, for structural data represent a very large amount of information, stemming from centuries of work in physics and chemistry. A machine which could provide adaptive control for arbitrary plants could also replace human beings in scientific experimentation and model building. We regard adaptive control as a problem for the future and shall not discuss it further here.

Returning to the problem of state determination, we shall find, surprisingly, that it is the "dual" of the problem of control. This is the DUALITY PRINCIPLE of control and estimation announced in [KALMAN, 1960*b*].

In the finite-dimensional linear case the definition of "duality" involves the usual notions: the dual of a vector space, the adjoint of a linear map, etc. Still, a decent axiomatic development of duality would unfortunately require a higher level of abstraction than is possible without violating the elementary character of this chapter. We shall be content merely to get to the main results as quickly as possible. We do so with no sacrifice of rigor, but the reader may not find the definitions as well motivated as before.

We shall distinguish between two kinds of state-determination problems:

1. The *observation problem*, where the present state $x(\tau)$ is to be determined from the knowledge of the future outputs $\{y(\sigma): \sigma \geq \tau\}$,

2. The *reconstruction problem*, where the present state $x(\tau)$ is to be determined from the knowledge of the past outputs $\{y(\sigma): \sigma \leq \tau\}$.

We always assume that the functions φ, η, and ω are known for the system in question. In the first case we observe future effects of the present state and try to unravel the underlying cause. In the second case we attempt to reconstruct the present state, but without complete knowledge of the actual state transitions.

Let us introduce some basic definitions which provide abstract necessary and sufficient conditions for the solvability of these two problems. We start with the definitions for general systems.

(6.2) **Definition.** *Two events, (τ, x_1) and (τ, x_2), of a dynamical system Σ belong to the same* **observation class** *(are* **indistinguishable in the future***) iff*

$$\eta(t, \varphi(t; \tau, x_1, \omega)) = \eta(t, \varphi(t; \tau, x_2, \omega))$$

for all $t \geq \tau$ and all ω.

The complementary concept is

(6.3) **Definition.** *Two events, (τ, x_1) and (τ, x_2), of a dynamical system Σ belong to the same* **reconstruction class** *(are* **indistinguishable in the past***) iff*

$$\eta(\sigma, \varphi(\sigma; \tau, x_1, \omega)) = \eta(t, \varphi(t; \tau, x_2, \omega))$$

for all $\sigma \leq \tau$ and all ω.

Note that here we talk about the *set*

$$\varphi(\sigma; \tau, x, \omega) = \{\bar{x}; \varphi(\tau; \sigma, \bar{x}, \omega) = x\}.$$

These definitions are unnecessarily elaborate in the linear case. By linearity of $\varphi(t; \tau, \cdot, \cdot)$ and $\eta(t, \cdot)$, we have

$$\begin{aligned} \eta(t, \varphi(t; \tau, x, \omega)) &= \eta(t, \varphi(t; \tau, x, 0) + \varphi(t; \tau, 0, \omega)), \\ &= \eta(t, \varphi(t; \tau, x, 0)) + \eta(t, \varphi(t; \tau, 0, \omega)). \end{aligned}$$

Since the second term depends only on η, φ, and ω, it cancels when we compare

$$\begin{aligned} \eta(t, \varphi(t; \tau, x_1, \omega)) &- \eta(t, \varphi(t; \tau, x_2, \omega)) \\ &= \eta(t, \varphi(t; \tau, x_1, 0)) - \eta(t; \varphi(t; \tau, x_2, 0)), \\ &= \eta(t, \varphi(t; \tau, x_1 - x_2, 0)). \end{aligned}$$

In the last step we have again used the linearity of φ on $X \times \Omega$.

So we can rephrase our definitions as follows:

(6.4) **Definition.** *An event (τ, x) of a linear dynamical system Σ is* **unobservable** *iff it belongs to the observation class of $(\tau, 0)$; that is, iff*

$$\eta(t; \tau, x, 0) = 0$$

for all $t \geq \tau$.

(6.5) **Definition.** *An event* (τ, x) *of a linear dynamical system* Σ *is* **unconstructible** (*or* **unreconstructible**)† *iff it belongs to the reconstruction class of* $(\tau, 0)$; *that is, iff*

$$\eta(\sigma; \tau, x, 0) = 0$$

for all $\sigma \leq \tau$.

We shall now give a fairly explicit characterization of unconstructible events in a *linear* system [KALMAN, 1960a–b].

(6.6) **Theorem.** *In a finite-dimensional, smooth, linear dynamical system* Σ *an event* (τ, x) *is unconstructible if and only if* $x \in$ kernel $M(s, \tau)$ *for all* $s \leq \tau$, *where*

$$M(s, \tau) = \int_s^\tau \Phi'(\sigma, \tau) H'(\sigma) H(\sigma) \Phi(\sigma, \tau) \, d\sigma.$$

This theorem is the natural counterpart of (2.16). However, the proof is much simpler; here we are not asked to describe a (constructive) procedure for finding x from knowledge of $y(\sigma) = H(\sigma)\Phi(\sigma, \tau)x$, $\sigma \leq \tau$, but merely to characterize the case where this cannot be done.   Later we shall show that an event which is *not* unconstructible can be actually constructed (reconstructed); in other words, we shall give an explicit recipe for computing x from $\{y(\sigma): s \leq \sigma \leq \tau\}$.

Proof. Necessity. The formula

$$x'M(s, \tau)x = \int_s^\tau \|H(\sigma)\Phi(\sigma, \tau)x\|^2 \, d\sigma$$

shows that if (τ, x) is unconstructible, then $x'M(s, \tau)x = 0$ for all s. Since M is symmetric and nonnegative definite, $M = N'N$.   Then $x'Mx = \|Nx\|^2 = 0$, which implies $Nx = 0$, and so $Mx = N'Nx = 0$.

Sufficiency. If $x \in kernel$ M, then $x'Mx = 0$, and the same formula shows that $H(\sigma)\Phi(\sigma, \tau)x = 0$ for all $\sigma \leq \tau$.   □
The "dual" result is

(6.7) **Theorem.** *In a finite-dimensional, linear, smooth, dynamical system* Σ *an event* (τ, x) *is unobservable if and only if* $x \in$ kernel $\hat{M}(\tau, t)$ *for all* $t \geq \tau$, *where*

$$\hat{M}(\tau, t) = \int_\tau^t \Phi'(\sigma, \tau) H'(\sigma) H(\sigma) \Phi(\sigma, \tau) \, d\sigma.$$

There is an obvious similarity between the definition of W (Theorem (2.16)) and the definition of M and that of $\hat{W}$ (Theorem (2.24)) and $\hat{M}$. That is, in some sense constructibility is the natural counterpart of controllability, and observability is the natural counterpart of reachability.

† L. Weiss has suggested the terminology "unidentifiable" in place of "unconstructible."   It may turn out to be a better word than ours.

The simplest way of making this clear is to convert the integrand of W into the integrand of M. The appropriate transformations are, for *fixed* τ and arbitrary real α,

$$G(\tau + \alpha) \to H'(\tau - \alpha),$$
(6.8) $$\Phi(\tau, \tau + \alpha) \to \Phi'(\tau - \alpha, \tau),$$
$$F(\tau + \alpha) \to F'(\tau - \alpha).$$

In other words, we take the mirror image of the graph of each function $G(\cdot)$, $H(\cdot)$, and $F(\cdot)$ about the point $t = \tau$ on the time axis, and then take transposes of each matrix. In case of controllability and constructibility the parameter α is nonnegative; in case of reachability and observability α is nonpositive.

For constant systems these transformations are simply

$$G \to H',$$
(6.9) $$e^{-tF} \to e^{-tF'},$$
$$F \to F'.$$

The duality relations (6.8) and (6.9) are clearly one-to-one; in fact, the inverse transformations are given by

(6.10) $$\begin{aligned} H(\tau - \alpha) &\to G'(\tau + \alpha), \\ F(\tau - \alpha) &\to F'(\tau + \alpha), \end{aligned}$$

and

(6.11) $$\begin{aligned} H &\to G', \\ F &\to F'. \end{aligned}$$

With the aid of these relations we can give criteria for observability and constructibility in terms of reachability and controllability, and then use the results of Sections 2.2 and 2.3 to get explicit conditions. In view of Theorem (6.7), it is convenient to use the following definitions:

Σ *is completely observable at time* τ iff no event (τ, x) of Σ is unobservable except $(\tau, 0)$ and

Σ *is completely constructible at time* τ iff no event (τ, x) of Σ is unconstructible except $(\tau, 0)$.

By (6.10) we have immediately

(6.12) **Proposition.** *The pair of matrix functions*

$$t \mapsto F(t), \qquad t \mapsto H(t)$$

defines a system Σ *which is completely observable at time* τ *if and only if the pair of matrix functions*

$$t \mapsto F'(2\tau - t) = F^*(t), \qquad t \mapsto H'(2\tau - t) = G^*(t)$$

defines a system Σ^* *which is completely reachable at time* τ.

We leave it to the reader to formulate further results of this sort. Throughout the rest of this chapter the word "linear" will be used as an abbreviation for "finite dimensional, linear, smooth, and constant." In view of (6.9), we can rephrase Corollary (3.13) as follows:

(6.13) **Proposition.** *In an n-dimensional linear system* Σ

$$\dim \ker M(\tau, t)$$
$$= n - \dim \text{ range } M(\tau, t),$$
$$= n - \text{rank } [H', F'H', \ldots, F'^{n-1}H'] \qquad \textit{for all } t > \tau.$$

Either from this statement or from Proposition (6.12) we get

(6.14) **Proposition.** *A pair of constant matrices* $\{F, H\}$ *corresponds to a completely constructible system if and only if the pair* $\{F', H'\}$ *corresponds to a completely controllable system.*

To make the story complete, let us note also the dual of Theorem (3.4):

(6.15) **Theorem.** *A linear system* Σ *of dimension n is completely constructible if and only if*

$$\text{rank } D = \text{rank } [H', F'H', \ldots, (F')^{n-1}H'] = n.$$

Here D is an $n \times pn$ matrix made up of the columns of the matrices $H', F'H', \ldots, (F')^{n-1}H'$.

Now we wish to give a scheme for reconstructing the present state of a completely constructible system from past observations of values of its output. If we had to do this by strict duality with our discussion of controllability in Section 2.2, we would first try to describe a scheme for constructing $x(\tau)$ from output data over the *finite interval* $[s, \tau]$. Instead, we prefer to go directly to the dual of Theorem (5.10) and Remark (5.12) utilizing a notion due to R. W. Bass (first formulated in 1963 in an unpublished report):

(6.16) **Definition.** *A linear system* $\hat{\Sigma}$ *is an* **asymptotic state estimator** *for a linear system* Σ *iff*

$$\tilde{x}(t) = x(t) - \hat{x}(t) \to 0 \qquad \textit{with } t \to \infty.$$

More precisely, $\hat{\Sigma}$ *is given by the system of equations*

$$\frac{d\hat{x}}{dt} = F_\Sigma(t)\hat{x} + L(t)[y(t) - H_\Sigma(t)\hat{x}] + G_\Sigma(t)u(t),$$
$$\hat{y}(t) := \hat{x}(t).$$

The subscript Σ designates matrices belonging to Σ.

We shall not develop the theory of asymptotic state estimators per se, but shall pass immediately to the most important special case.

(6.17)   **Theorem.**   *A linear single-output $(p = 1)$ system Σ has an asymptotic state estimator if and only if it is completely constructible.*

Proof.   We shall do more than required; we shall construct $\hat{\Sigma}$ explicitly, as follows. Let us write h' for the $1 \times n$ matrix H. Assuming that (F, h') is completely constructible, we know from Proposition (6.14) that (F', h) is completely controllable. We adopt that special basis in X in which the matrices (F', h) are displayed in the control canonical form (4.9). Then

$$(6.18) \quad F = \begin{bmatrix} 0 & 0 & 0 & \cdots & 0 & -\alpha_n \\ 1 & 0 & 0 & \cdots & 0 & -\alpha_{n-1} \\ 0 & 1 & 0 & \cdots & 0 & -\alpha_{n-2} \\ \cdot & \cdot & \cdot & \cdots & \cdot & \cdots \\ 0 & 0 & 0 & \cdots & 1 & -\alpha_1 \end{bmatrix}, \quad h' = [0 \; \cdots \; 1].$$

We let $\theta(z) = z^n + \beta_1 z^{n-1} + \cdots + \beta_n$ be an arbitrary monic polynomial whose zeros have negative real parts. If we then define a column vector l with components

$$l_{n-i+1} = \beta_i - \alpha_i, \qquad i = 1, \ldots, n,$$

it follows (from Theorem (5.10) or by inspection) that

$$(6.19) \qquad\qquad \chi_{F-lh'} = \theta.$$

Now if

$$\frac{dx}{dt} = Fx + Gu(t),$$

and

$$\frac{d\hat{x}}{dt} = F\hat{x} + l[y(t) - h'\hat{x}] + Gu(t)$$

are the equations of Σ and $\hat{\Sigma}$, it is easy to see that $\tilde{x} = x - \hat{x}$ obeys the differential equation

$$(6.20) \qquad\qquad \frac{d\tilde{x}}{dt} = (F - lh')\tilde{x},$$

subject to the initial condition

$$(6.21) \qquad\qquad \tilde{x}(s) = x(s),$$

where s is the time when the observation of the output of Σ starts. In effect, we have assumed by (6.21) that $\hat{x}(s) = 0$, because initially we know nothing about the state of Σ. Since, by (6.19) and the assumption on θ, the eigenvalues of $F - lh'$ in equation (6.20) have negative real parts, $\bar{x}(t) \to 0$ for every value of $x(s)$. This proves the "if" part of the theorem.

The "only-if" part is immediate, from the definition of constructibility. □

Recalling that the polynomial θ was completely arbitrary except for the requirement that its zeros have negative real part, we also have

(6.22) **Corollary.** *The error dynamics (6.20) of the asymptotic state estimator can be chosen arbitrarily, subject only to the condition that all solutions of this equation approach zero with $t \to \infty$; that is,*

$$\text{Re } \lambda_i[F - lh'] < 0, \qquad i = 1, \ldots, n.$$

We shall call (6.18) the *state-construction canonical form*.

As an illustration of these results let us now solve a typical problem in linear system theory.

(6.23) **Definition.** *Let $\hat{\Omega} = \{continuous \ functions \ T \to \mathbf{R}\}$. A linear single-input $(m = 1)$ system $\hat{\Sigma}$ is an* **asymptotic differentiator** *of order $n - 1$ iff we have, for all $\hat{u}(\cdot) \in \hat{\Omega}$,*

$$\hat{x}_i(t) = \frac{d^i \hat{u}(t)}{dt^i} + \hat{e}_i(t), \qquad i = 0, 1, \ldots, n - 1,$$

and each $\hat{e}_i(t) \to 0$ with $t \to \infty$.

(6.24) **Theorem.** *If $\hat{\Omega} = \{all \ polynomials \ in \ t \ of \ degree \leq n - 1 \ with \ coefficients \ in \ \mathbf{R}\}$, then there exists an asymptotic differentiator $\hat{\Sigma}$ of order $n - 1$ (and dimension n).*

Proof. The inputs $\hat{\omega}$ may be regarded as the outputs $y(t) = \hat{u}(t)$ of the system Σ defined by

$$F = \begin{bmatrix} 0 & \cdots & 0 & 0 \\ 1 & \cdots & 0 & 0 \\ \cdots & & \cdots \\ 0 & \cdots & 1 & 0 \end{bmatrix}, \qquad H = h' = [0 \ \cdots \ 1], \qquad G = 0.$$

In fact, a simple computation shows that

$$e^{Ft} = \sum_0^\infty \frac{t^k F^k}{k!} = \begin{bmatrix} 1 & \cdots & 0 & 0 \\ t & \cdot & 0 & 0 \\ \cdots & \cdot & \cdots \\ \frac{t^{n-1}}{(n-1)!} & \cdots & t & 1 \end{bmatrix}.$$

So

(6.25) $$y(t) = h'e^{Ft}x = \sum_0^{n-1} \frac{t^k}{k!} x_{n-k-1}$$

is a polynomial. In other words, any polynomial of degree $<n$ can be obtained as the output $y(\cdot)$ of the system Σ by appropriate choice of the initial state $x \in X_\Sigma$. Since Σ is completely constructible, we can construct a state estimator, as in Theorem (6.17). Then (6.24) shows that estimating the state of Σ is the same as estimating derivatives of $y(\cdot)$, that is, those of $\hat{\omega}$. $\quad\square$

(6.26) **Comment.** This example shows the importance of assumptions concerning $\hat{\Omega}$. Indeed, our construction works nicely only because polynomials of degree $<n$ form an n-dimensional vector space.

(6.27) **Remark.** Just as in Theorem (5.10) and Remark (5.12), the rational choice of the coefficients $\beta_1, \ldots, \beta_n$ requires a criterion of optimality. If the signal $y(t)$ is contaminated by noise, the "best" choice of the β_k is determined via the Kolomogorov-Wiener filtering theory, in the Kalman-Bucy formulation [KALMAN and BUCY, 1961]. The result will depend on the spectrum of the noise and also on the dynamics of Σ (see Section 3.6). In the linear case this theory is the exact dual of optimal control theory. In the nonlinear case no one has yet been able to give an effective method of solution.

(6.28) **Remark.** Note that in the above example $dim\ X_\Sigma = dim\ X_{\hat{\Sigma}}$. This is a direct consequence of linearity. In the nonlinear case $dim\ X_{\hat{\Sigma}} = \infty$ even if $dim\ X_\Sigma$ is finite.

There is an interesting alternative to Theorem (6.17). Suppose we choose state variables in Σ in such a way that

$$x_{n-p+1}(t) = y_1(t), \ldots, x_n(t) = y_p(t).$$

If the output variables $y_1(t), \ldots, y_p(t)$ are "clean" (that is, exactly known), then we can identify them with the state variables $x_{n-p+1}(t), \ldots,$ $x_n(t)$, so that only the state variables $x_1(t), \ldots, x_{n-p}(t)$ remain to be determined. (Note that this procedure is invalid if $y(t)$ is contaminated with even a small amount of noise; then the full machinery of the Kalman-Bucy theory must be used. However, if $y(t)$ contains no noise at all, then we have a singular case of this theory which requires special discussion. This is the reason for considering this case separately.)

A naïve but suggestive way of setting up a state estimator $\hat{\Sigma}$ would be to write

$$\hat{\Sigma}: \begin{cases} \dfrac{d\hat{x}_i}{dt} = \displaystyle\sum_{j=1}^{n-p} f_{ij}\, \hat{x}_j + \sum_{j=1}^{p} f_{i,n-p+j} y_j(t), & i = 1, \ldots, n-p; \\ \hat{x}_{n-p+i}(t) = y_i(t), & i = 1, \ldots, p. \end{cases}$$

However, $\hat{\Sigma}$ may be an unacceptable estimator, because there is no guarantee that the $(n - p) \times (n - p)$ upper left-hand submatrix

$$\hat{F} = [f_{ij}], \qquad \begin{array}{l} i = 1, \ldots, n - p \\ j = 1, \ldots, n - p \end{array}$$

of F will be stable (have eigenvalues with negative real parts). It was [LUENBERGER, 1964] who first noticed that *even in this case there exist asymptotic state estimators whose eigenvalues can be chosen arbitrarily.*

We shall prove this important result in a way quite different from the original proof of Luenberger. The main ideas are:

1. We observe that, even though the eigenvalues of a matrix F are coordinate free (that is, the eigenvalues of F are the same as those of BFB^{-1}, with B an arbitrary nonsingular matrix), the *same does not necessarily hold for submatrices of F.*

2. We use the condition of complete constructibility to choose the state variables in such a way that submatrix $\hat{F}$, which governs the dynamics of $\hat{\Sigma}$, has any desired set of eigenvalues.

For simplicity, we shall consider only the case $p = 1$. We have then

(6.29) **Lemma.** *If the pair $\{F, h'\}$ is completely constructible, there is a linear transformation B, given by the matrix*

$$B = \begin{bmatrix} I & \begin{matrix} \beta_{n-1} \\ \cdots \\ \beta_1 \end{matrix} \\ \hline 0 & 1 \end{bmatrix},$$

which will take the state-construction canonical form of $\{F, h'\}$ into the modified canonical form

(6.30)
$$F = \begin{bmatrix} \bar{F} & \vdots & ? \\ \hline ? & \vdots & ? \end{bmatrix} = \begin{bmatrix} 0 & & 0 & -\beta_{n-1} & \\ 1 & \cdot & 0 & -\beta_{n-2} & ? \\ & \cdot & \cdot & & \\ \cdots & & \cdot & \cdots & \\ & \cdot & & & \\ 0 & \cdots & 1 & -\beta_1 & \\ \hline & ? & & & ? \end{bmatrix},$$

$$h' = [0 \ \cdots \ 0 \ 1],$$

where ? denotes elements of F which are of no interest in the applications of the lemma.

Proof. Apply the canonical form (6.18) and verify (6.30). □

Since $\bar{F}$ is a companion matrix, it follows, as in Section 2.4, that

$$\chi_{\bar{F}}(z) = z^{n-1} + \beta_1 z^{n-2} + \cdots + \beta_{n-1};$$

in other words, we have Luenberger's

(6.31)   **Theorem.**   *The eigenvalues of $\bar{F}$ can be picked at will by suitable choice of the transformation B.*

We shall give an example of the Luenberger estimator in the next section.

(6.32)   **Remark.**   Of course, it is possible to dualize this result back to the control case.   The reader should carefully verify the truth of the following statement: *If (F, g) is completely controllable, we can choose state variables in such a way that the subset of state variables $\{x_1, \ldots, x_{n-1}\}$ forms a subsystem with arbitrarily prescribed internal dynamics.*

2.7   Construction of the regulator

Let us review what has been accomplished so far.

If the plant is completely controllable, we can give control laws which take any initial state to zero in a finite interval of time (Theorem (5.4)); or, if we insist on a constant control law for a constant plant, we can alter the dynamics arbitrarily (Theorem (5.10)).

If the plant is completely constructible, then, at least in the constant case, we can give a state-reconstruction scheme (Theorem 6.15)) which turns out to be the dual of the control scheme of Theorem (5.10).

Two problems remain:

1. Under what circumstances can we be sure that the plant will indeed have the critical properties of complete controllability and complete constructibility?

2. How do we combine the solution of the control problem (with the state variables assumed known) and the solution of the state-construction problem into the solution of the originally posed regulator problem?

A full discussion of the first problem would be too great a digression here.   Let us note two important results, however, which show that the conditions of complete controllability and complete constructibility are entirely natural for the problem at hand.

(7.1)   **Remark.**   According to the "canonical decomposition theorem" for linear dynamical systems (see [KALMAN, 1962a, 1963c; WEISS and KALMAN, 1965]), any such system may be regarded as the direct sum (in a certain precise sense) of four parts, of which the first is completely reachable and observable, the second is completely reachable but unobservable, and so forth.   Moreover, all causal paths from input to output pass through the first part and only through that part.   Hence it is impossible to include any of the other three parts of the plant in a closed loop, which shows that only the first part is relevant as far as control is concerned. This situation is very much a consequence of linearity, as is discussed further in Section 6.3.

(7.2) Remark. It has always been assumed in classical control theory (without firm logical justification) that the plant is given in input/output form (Definition (1.8) of Chapter 1). Although a result like Theorem (5.10) was never proved classically, its truth was generally taken for granted. Why?

It turns out (and this is one of the fundamentally new results of modern system theory) that, *over some interval any smooth, linear, zero-state input/output function is realizable as a smooth, linear, dynamical system Σ (see Theorem (2.8)) which is completely reachable and completely observable.* See Section 10.13 for the full technical statement.

Now, if the plant is constant, then Theorem (3.15) and its dual show that reachability is equivalent to controllability and observability is equivalent to constructibility. In other words, *if the plant is assumed to be constant and regarded as the minimal realization of its zero-state input/output function, then it is always completely controllable and completely constructible.* This subtle accident explains the success of the classical theory in spite of its naïve and shaky logical foundations. By the same token, the *lack of success* of the classical theory in dealing with nonconstant linear plants is also explained, because in the nonconstant case reachability does not always imply controllability nor does observability imply constructibility. We refer again to Section 10.13 for further discussion.

Let us turn now to the second problem; we must combine our preceding results into an explicit specification of the regulator for a linear constant plant.

At each moment τ of time the task of the regulator is to emit a control signal $u(\tau)$ based on all the knowledge available, that is, on all the past values of the output signal $\{y(\sigma): \sigma \leq \tau\}$. It would be desirable to have, if possible,

$$(7.3) \qquad\qquad u(\tau) = -Kx(\tau);$$

but of course, this formula cannot be implemented because in general $x(\tau)$ is unknown and its exact value cannot be inferred from $\{y(\sigma): \sigma \leq \tau\}$. So it is plausible to replace (7.3) by

$$(7.4) \qquad\qquad u(\tau) = -K\hat{x}(\tau),$$

where $\hat{x}(\tau)$ is the *estimate of the state constructed from* $\{y(\sigma): \sigma \leq \tau\}$ according to the theory of Section 2.6.

The passage from (7.3) to (7.4) can be rigorously justified only with the help of stochastic optimal control theory. (It must be proved that "no information is lost" through this seemingly arbitrary step.)   The following elementary observation provides a partial justification for (7.4), as well as some intuitive insight:

(7.5)   **Proposition.**   *Consider the equations of the plant, state estimator (6.16), and control law (7.4):*

$$\frac{dx}{dt} = Fx + Gu(t),$$

(7.6) $$y(t) = Hx(t),$$

$$\frac{d\hat{x}}{dt} = F\hat{x} + L[y(t) - H\hat{x}] + Gu(t),$$

$$u(t) = -K\hat{x}(t).$$

Then the characteristic polynomial χ_{overall} *of (7.6) satisfies*

$$\chi_{\text{overall}} = \chi_{\text{control}} \times \chi_{\text{state est.}} = \chi_{F-GK} \times \chi_{F-LH}.$$

Proof. χ_{overall} is the characteristic polynomial of the matrix

$$F_{\text{overall}} = \begin{bmatrix} F & -GK \\ LH & F - LH - GK \end{bmatrix}.$$

The desired conclusion is not immediately evident by inspection of F_{overall} but follows easily after a change of variables.   We replace the pair $(x, \hat{x})$ by the pair $(x, \tilde{x})$.   This transformation is linear and one-to-one, so it does not affect χ_{overall}.   In terms of the new coordinates, relations (7.6) become

$$\frac{dx}{dt} = (F - GK)x - GK\tilde{x},$$

$$\frac{d\tilde{x}}{dt} = (F - LH)\tilde{x}.$$

Since this is a triangular system (the second equation "drives" the first), the desired conclusion is immediate.   □

We can now summarize the entire investigation in the following way:

(7.7)   **Theorem.**   *Consider a linear plant Σ which is completely controllable as well as completely constructible.   Pick a matrix K yielding a stable control law, that is, so that χ_{F-GK} is a stable polynomial. Similarly, pick a matrix L yielding a stable state estimator, so that χ_{F-LH} is a stable polynomial.   Define the regulator as the system comprising the state estimator and control law (7.4).   The equations of the overall system (plant plus regulator) are given by (7.6).*

Then the overall (closed-loop) system is stable.

Moreover, the dynamical behavior of the overall system is the direct sum of the dynamics of the regulator loop (the matrix F − GK) and the dynamics of the estimation loop (the matrix F − LH).

This shows that, as long as the "performance" of the system is measured solely in terms of its stability properties, the design of a regulator is an almost immediate consequence of the controllability and constructibility properties of the plant. No further mathematical, conceptual, or physical considerations enter *as long as the plant is accurately representable as a finite-dimensional, linear, constant, smooth dynamical system.* Remember that control theory does not deal with the real world, but only with mathematical models of it.

Exactly the same result as Theorem (7.7) holds also when the Luenberger estimator is used. The precise formulation and proof is left to the reader.

We conclude with a detailed example of the computation of the regulator:

(7.8) **Example.** Consider the simplest plant of interest in control theory, which has the transfer function $1/s^2$ (two cascaded integrators). It is known from the general theory of linear systems (see Chapter 10) that this transfer function has a minimal "realization" in terms of the standard linear model (3.2) of Section 2.3, which is two-dimensional. By Remark (7.2), this minimal realization is completely controllable as well as completely constructible.

For the plant $1/s^2$ the minimal realization can be written down by inspection. It is given by the matrices

$$F = \begin{bmatrix} 0 & 1 \\ 0 & 0 \end{bmatrix}, \qquad g = \begin{bmatrix} 0 \\ 1 \end{bmatrix}, \qquad h = \begin{bmatrix} 1 \\ 0 \end{bmatrix}.$$

If we take

$$\chi_{\text{control}}(s) = s^2 + \alpha_1 s + \alpha_2,$$

then

$$k = \begin{bmatrix} \alpha_2 \\ \alpha_1 \end{bmatrix}.$$

Let us determine a state estimator of the Luenberger type. First we must transform coordinates to the constructibility canonical form. In this very simple case the transformation is merely an interchange of the indices of the state variables: $(x_1, x_2) \to (x_2, x_1)$. Hence the matrices F, g, h, and k become

$$F = \begin{bmatrix} 0 & 0 \\ 1 & 0 \end{bmatrix}, \qquad g = \begin{bmatrix} 1 \\ 0 \end{bmatrix}, \qquad h = \begin{bmatrix} 0 \\ 1 \end{bmatrix}, \qquad k = \begin{bmatrix} \alpha_1 \\ \alpha_2 \end{bmatrix}.$$

As in the proof of Lemma (6.29), we apply to these matrices the linear transformation

$$B = \begin{bmatrix} 1 & -\beta \\ 0 & 1 \end{bmatrix}, \qquad B^{-1} = \begin{bmatrix} 1 & \beta \\ 0 & 1 \end{bmatrix}$$

and obtain

$$F_{\text{new}} = \begin{bmatrix} -\beta & -\beta^2 \\ 1 & \beta \end{bmatrix}, \qquad g_{\text{new}} = g = \begin{bmatrix} 1 \\ 0 \end{bmatrix},$$

$$h_{\text{new}} = h = \begin{bmatrix} 0 \\ 1 \end{bmatrix}, \qquad k_{\text{new}} = \begin{bmatrix} \alpha_1 \\ \alpha_1\beta + \alpha_2 \end{bmatrix}.$$

We can now write down the "wiring diagram" of the regulator by inspection. It is shown in Figure 2.1. Notice that it is quite complicated.

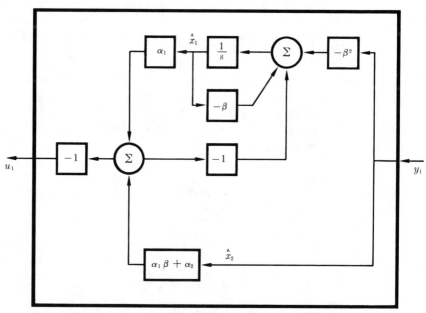

Figure 2.1   Detailed block diagram for the plant $1/s^2$.

If this problem had been solved by classical control theory, the controller would have been specified via its transfer function. To compare results we must therefore compute the transfer function of the system shown. The result, after some computation, is

$$\frac{u_1(s)}{y_1(s)} = T(s) = \frac{(\alpha_1\beta + \alpha_2)(s + \alpha_1) + \alpha_2\beta}{s + \alpha_1 + \beta},$$

$$= (\alpha_1\beta + \alpha_2)\frac{s + \alpha_1 + \alpha_2\beta/(\alpha_1\beta + \alpha_2)}{s + \alpha_1 + \beta}.$$

It is well known that this transfer function should represent a lead network, namely,

$$T(s) = \gamma \frac{s + \delta}{s + \epsilon}, \qquad \gamma, \, \delta, \, \epsilon > 0, \, \delta < \epsilon.$$

Since α_1, α_2, and β are all positive, the result of the new theory is in agreement with the result of the classical theory.

We can make the connection between the two approaches somewhat more explicit by writing

$$\chi_{\text{control}}(s) = s^2 + 2\xi\omega s + \omega^2,$$

where $2\xi\omega = \alpha_1$ and $\omega^2 = \alpha_2$. Taking ξ, the *damping ratio*, equal to $1/\sqrt{2}$, which is the optimum value suggested by practical experience, we get

(7.9)
$$\gamma = \omega(\sqrt{2}\,\beta + \omega),$$
$$\delta = \sqrt{2}\,\omega + \frac{\omega\beta}{(\sqrt{2}\,\beta + \omega)} = \omega\left(\frac{3\beta + \sqrt{2}\,\omega}{\sqrt{2}\,\beta + \omega}\right),$$
$$\epsilon = \beta + \sqrt{2}\,\omega.$$

The classical design procedure consists in making the ratio ϵ/δ as large as possible in regard to high-frequency noise in the output of the plant; then γ is adjusted so as to give suitable stability for the overall system. In our case

(7.10)
$$\frac{\epsilon}{\delta} = \frac{(\beta/\omega)^2 + (3/\sqrt{2})(\beta/\omega) + 1}{(3/\sqrt{2})(\beta/\omega) + 1}.$$

If β is large (no noise in the output), then ϵ/δ can be very large. As the desired bandwidth ω of the internal dynamics of the system approaches the bandwidth of the noise as measured by β, $\epsilon/\delta \to 1.4$. This is regarded as far too small a value of ϵ/δ for such a regulator to be of practical benefit.

It is remarkable that classical control theory has never been able to give explicit formulas such as (7.9) and (7.10) even for this very simple example. The conclusions of the previous paragraph are usually obtained by lengthy and fuzzy arguments which purport to capitalize on the engineer's physical intuition. The reasons for the difficulties encountered by the classical theory are not hard to find. The desired regulator is a complicated function of the primary design parameters ξ, ω, and β. Even after elimination of ξ, the formulas for γ, δ, and ϵ are much too intricate to be derivable simply from physical "intuition." At this point classical control theory usually resorts to graphical calculations and practical rules, in other words, to a very primitive kind of applied mathematics.

(7.11) **Remark.** Classical control theory could never really cope with large-scale systems (n large, more than one control variable, etc.) because the formulas analogous to (7.10) are much too complex.　Even in modern control theory we do not attempt to apply the present technique when the plant is much more complicated than $1/s^2$.　There is no need to, because the only applied-mathematical problem remaining is the numerical determination of K and L; this is easily accomplished with the powerful algorithms provided by optimal control theory.　In this context, see especially Sections 5.3 to 5.5.

(7.12)　**Remark.**　Even in this simple case, the "canonical wiring diagram" of the regulator shown in Figure 2.1 is much too complicated to be intuitively understandable.　Here the superiority of modern control theory is especially striking; we obtained the wiring diagram step by step, without any preconceived notion of what it might look like when we were finished.

　　It is tempting to speculate about the implications of this result for problems in biology, especially the theory of the brain in the higher animals.　It could be true that it is hopeless to try to understand brain functions solely on the basis of anatomy (wiring diagrams).　Perhaps the problem will become relatively transparent only after we have developed a theory (vaguely analogous to the present one) powerful enough to give us the main features of the anatomy.

PART TWO

Optimal control theory

P. L. Falb

3   Basic optimal control theory

The underlying motivation for modern system theory is the system design problem, which is, in essence, the statement of a task to be performed by either an existing physical process or a physical process which is to be developed. The task statement includes the specification of various goals and objectives, and the ultimate system design is, of course, subject to a variety of constraints, such as time, money, and physical laws. The usual procedure for solving a system design problem entails, first and foremost, the construction of an adequate mathematical model of the physical process; this is generally a very difficult matter. Assuming that such a mathematical model has been developed, the designer then proceeds to produce a "pencil-and-paper" design for the model; this design is then translated back into the real world, implemented, and tested. Control problems are a particular class of system design problems.

Any control problem contains four essential elements:

1. A system to be controlled.
2. A desired output or objective.
3. A set of admissible controls (or control inputs).
4. A measure of the cost or effectiveness of control actions.

The translation of these intuitive elements into a viable and useful mathematical framework leads to what is currently known as optimal control theory and to a statement of the control problem. Optimal control theory is the subject of this portion of the book. We shall usually omit the qualification "optimal," since we are not concerned with control theory in the old-fashioned sense. In other words, we are interested in the scientific foundations of control, not merely engineering know-how. Of course, the scientific methods we shall examine are often of great practical importance.

We begin, in the next section, with a statement of the abstract control problem. The notion of a smooth dynamical system is then defined. Next the particular class of control problems for which we shall develop some theoretical results are delineated. In Section 3.4 we study the Hamilton-Jacobi theory with a view toward establishing some useful sufficient conditions. Having done this, we apply the sufficiency theorem to the case of a linear system and a quadratic cost functional, obtaining some familiar results involving Riccati equations. We then apply these results to the filtering problem and obtain the Kalman filter. We turn our attention to necessary conditions for optimality in Chapter 4 and discuss the variational approach in Section 4.1. In Section 4.2 the Pontryagin maximum principle is stated, with a brief heuristic argument given for it. Since use of the necessary conditions requires knowledge of the existence of optimal controls, this question is briefly examined in Section 4.3. Chapter 5 on control system design includes several sections on numerical techniques in control theory.

Although we shall be dealing with fairly general spaces (such as Banach spaces), the reader may, if this frightens him, assume that everything is finite dimensional. This assumption will not change any of the proofs. However, since there are interesting systems which are infinite dimensional (for example, distributed-parameter systems), we have thought it wise to be reasonably general in our approach.

3.1   The abstract control problem

We begin with the translation of the essential elements of the control problem into precise mathematical language. We assume that the system to be controlled is adequately described by the dynamical system

$$\Sigma = (T, U, \Omega, X, Y, \varphi, \eta)$$

with transition function $\varphi(t; \tau, x, u(\cdot))$ and output function $\eta(t, x)$. We let S_0 be a given subset of $T \times X \times Y$,

$$S_0 \subset T \times X \times Y,$$

and we let Γ be a given subset of Ω,

$$\Gamma \subset \Omega.$$

We call S_0 the (output) *target set* and Γ the *set of admissible controls*. We now have

(1.1) **Definition.** *A control function $u(\cdot)$* **transfers** *the event (t_0, x_0) to S_0 if the set*

$$\{(t, \varphi(t; t_0, x_0, u(\cdot)), \eta(t, \varphi(t; t_0, x_0, u(\cdot)))) : t \geq t_0\}$$

meets S_0. If $u(\cdot)$ transfers (t_0, x_0) to S_0 and if t_1 is the time of first meeting, then t_1 is called the **transfer time.**

Let $M(t, x, y, \varphi, u, \eta)$ be a given real-valued function defined on $T \times X \times Y \times (X^T) \times \Omega \times (Y^T)$. If $u(\cdot)$ is a control which transfers (t_0, x_0) to S_0 with transfer time t_1 (or $t_1(u)$), then

$$M(t_1, x_1, y_1, \varphi_{(t_0, t_1]}(\cdot; t_0, x_0, u(\cdot)), u(\cdot), \eta_{(t_0, t_1]}(\cdot, \varphi(\cdot; t_0, x_0, u(\cdot)))),$$

where $x_1 = \varphi(t_1; t_0, x_0, u(\cdot))$ and $y_1 = \eta(t_1, x_1)$, is a well-defined real number which we shall denote by $J(t_0, x_0, u(\cdot), t_1, x_1)$. We call $J(t_0, x_0, u(\cdot),$ $t_1, x_1)$ the *cost of the control $u(\cdot)$ relative to the initial event (t_0, x_0)*. We now have

(1.2) **Definition.** *An* **abstract control problem** *is the composite concept consisting of a dynamical system Σ, a target set S_0, a set Γ of admissible controls, a subset I of $T \times X$ (called the set of initial events), a cost-of-control functional J, and the statement: "Determine for each initial event (t_0, x_0) an admissible control $u(\cdot)$ which transfers (t_0, x_0) to S_0 and which in so doing minimizes $J(t_0, x_0, u(\cdot), t_1, x_1)$, where t_1 is the transfer time and x_1 is the transfer point."*

Of course, very few substantive statements can be made about an abstract control problem in view of the generality of the definition. Consequently, we shall make additional assumptions, particularly with regard to smoothness, which will enable us to develop some useful theorems.

3.2 Smooth dynamical systems

We shall consider only continuous-time differential dynamical systems in this portion of the book. In other words, we shall require that the time set T of our dynamical systems be an open subinterval (T_1, T_2) (which may be of the form $(-\infty, T_2)$, etc.) of the real line **R** and that the transition function φ correspond to the (unique) solutions of a differential equation. We shall also assume that the various spaces X, Y, U, and Ω have both an algebraic and a topological structure; for example, X may be a Banach space. Moreover, we shall demand that the functions with which we deal are sufficiently smooth, that is, that they satisfy suitable continuity and differentiability requirements. Dynamical systems which satisfy these various conditions (soon to be made more precise) shall be called *smooth*.

(2.1) **Definition.** *A dynamical system* $\Sigma = \langle T,\ U,\ \Omega,\ X,\ Y,\ \varphi,\ \eta \rangle$ *is called* **smooth** *if the following conditions are satisfied:*

(a) $T = (T_1,\ T_2)$ *is an open subinterval of* **R**.

(b) U, X, *and* Y *are subsets of Banach spaces* $\mathcal{B}_U$, $\mathcal{B}_X$, *and* $\mathcal{B}_Y$, *respectively.*

(c) *The set of control functions* Ω† *satisfies the following conditions:*

   (1) *Every element* $u(\cdot)$ *of* Ω *is measurable and bounded.*

   (2) *If* $u(\cdot) \in \Omega$, *then* $u(t) \in U$ *for all* t *in* $(T_1,\ T_2)$.

   (3) *If* $u(\cdot) \in \Omega$, $w \in U$, *and* $[t,\ t'] \subset (T_1,\ T_2)$, *then the function* $\hat{u}(\cdot)$ *given by*

$$\hat{u}(s) = \begin{cases} w, & s \in [t,\ t'], \\ u(s), & s \notin [t,\ t'], \end{cases}$$

   is in Ω.

   (4) *If* $[t_1,\ t_2] \subset (T_1,\ T_2)$, $t_1 \neq t_2$, *and* $v(\cdot)$ *is a function from* $[t_1,\ t_2]$ *into* $\mathcal{B}_U$ *satisfying* (1) *and* (2), *then there is a* $u(\cdot) \in \Omega$ *such that* $u_{[t_1,t_2]}(\cdot) = v(\cdot)$; *or*

$$u(t) = v(t) \text{ for } t \in [t_1,\ t_2].$$

(d) *The transition function* φ *satisfies the following conditions:*

   (1) φ *is continuous in all of its arguments.*

   (2) *Given any* t_0 *in* $(T_1,\ T_2)$, *any* x_0 *in* X, *and any* $u(\cdot)$ *in* Ω, $\varphi(t;\ t_0,\ x_0,\ u(\cdot))$ *is the unique solution of a differential equation*

$$\frac{dx(t)}{dt} = f(x(t),\ u(t),\ t) \ddagger$$

   satisfying the initial condition

$$x(t_0) = x_0$$

   and is defined for all t *in* $[t_0,\ T_2)$; *moreover,* $\varphi(t;\ t_0,\ x_0,\ u(\cdot)) \in X$

† Some typical sets Ω might be the function spaces L^1, L^2, or L^∞; however, the function space C^∞ of infinitely differentiable functions does not satisfy condition (3) and so is not a suitable Ω.

‡ $dx(t)/dt$ is an element of $\mathcal{L}(\mathbf{R},\ \mathcal{B}_X)$, which we identify with $\mathcal{B}_X$. Thus f is a map of $\mathcal{B}_X \times \mathcal{B}_U \times (T,\ T_2)$ into $\mathcal{B}_X$ (see [DIEUDONNE, 1960]).

for t in $[t_0, T_2)$. *The function* $f(x, u, t)$ *will be called the* **generator** *of the dynamical system* Σ.

(e) *The output function* η *is continuous in all of its arguments.*

From now on in this portion of the book, we shall use the term "dynamical system" in place of the term "smooth dynamical system."

3.3 The standard control problem

We are now going to delineate the version of the control problem for which we shall develop some theory. We shall call this version of the control problem the *standard control problem*.

Suppose that Σ is our (smooth) dynamical system and that f is the generator of Σ. If S_0 is a given subset of $(T_1, T_2) \times X \times Y$ (our output-target set), then we shall say that Σ is *constructible relative to* S_0 if $\eta(t, x) \in \pi_Y(S_0)$ implies that $(t, x, \eta(t, x)) \in S_0$, where π_Y is the projection of $(T_1, T_2) \times X \times Y$ on Y.

(3.1) Remark. Let $S = \eta^{-1}(\pi_Y(S_0))$ and suppose that Σ is constructible relative to S_0. Then a control $u(\cdot)$ will transfer (t_0, x_0) to S_0 if and only if the set

$$(3.2) \qquad \{(t, \varphi(t; t_0, x_0, u(\cdot))): t \geq t_0\}$$

meets S.

For the standard control problem we shall assume that the target set S is a given subset of the event space $(T_1, T_2) \times X$, and we shall say that a control $u(\cdot)$ transfers (t_0, x_0) to S if the set

$$(3.3) \qquad \{(t, \varphi(t; t_0, x_0, u(\cdot))): t \geq t_0\}$$

meets S. We let $\Gamma \subset \Omega$ be a given set of admissible controls, and we let $I \subset (T_1, T_2) \times X$ be a given set of initial events. All that remains to specify the standard control problem is a description of the cost functional J. To that end we let K be a real-valued function defined on S, and we let L be a continuous real-valued function defined on $X \times U \times (T_1, T_2)$ with the property that if $[t_1, t_2]$ is a closed bounded subinterval of (T_1, T_2), if $x(\cdot)$ is a continuous function from $[t_1, t_2]$ into X, and if $u(\cdot) \in \Omega$, then $L(x(\cdot), u(\cdot), \cdot)$ is integrable on $[t_1, t_2]$. It follows that if $u(\cdot)$ is a control which transfers (t_0, x_0) to S with t_1 as transfer time, then

$$K(t_1, \varphi(t_1; t_0, x_0, u(\cdot)))$$

and

$$\int_{t_0}^{t_1} L(\varphi(t; t_0, x_0, u(\cdot)), u(t), t) \, dt$$

are well-defined real numbers. We call their sum the cost of the control

$u(\cdot)$, and we write $J(t_0, x_0, u(\cdot))$ for this cost; that is,

$$(3.4) \qquad J(t_0, x_0, u(\cdot)) = K(t_1, x_1) + \int_{t_0}^{t_1} L(\varphi(t; t_0, x_0, u(\cdot)), u(t), t)\, dt,$$

where $x_1 = \varphi(t_1; t_0, x_0, u(\cdot))$ is the transfer point. The term $K(t_1, x_1)$ is called the *terminal cost*, and the integral term in (3.4) is called the *cost along the trajectory*. To formalize, we have

(3.5) **Definition.** *A* **standard control problem** *is the composite concept consisting of a smooth dynamical system* Σ, *a target set* S *contained in the event space, a set* Γ *of admissible controls, a set* I *of initial events, a cost-of-control functional* $J(t_0, x_0, u(\cdot))$ *defined by functions* K *and* L *as in (3.4), and the statement: "Determine for each event* (t_0, x_0) *in* I *the control* $u(\cdot)$ *in* Γ *which transfers* (t_0, x_0) *to* S *and which in so doing minimizes* $J(t_0, x_0, u(\cdot))$."

We shall use the term "control problem" in place of the term "standard control problem" from now on in this portion of the book. We shall also speak of solutions of a control problem as *optimal controls*.

(3.6) **Example.** Let $T = (-\infty, \infty) = \mathbf{R}$ be the entire real line; let $X = \mathbf{R}_n$ (euclidean n-space); let U be a bounded subset of $\mathbf{R}_m$—for example, $U = \{u: \|u\| \le 1\}$—and let Ω be the set of all measurable functions $u(\cdot)$ from $\mathbf{R}$ into $\mathbf{R}_m$ such that $u(t) \in U$ almost everywhere; let f be a function from $\mathbf{R}_n \times \mathbf{R}_m \times \mathbf{R}$ into $\mathbf{R}_n$ such that $f, \nabla_x f, \nabla_t f \in C(\mathbf{R}_n \times \bar{U} \times \mathbf{R}) \cap \mathcal{B}(\mathbf{R}_n \times \bar{U} \times \mathbf{R})$,[†] where $\bar{U}$ is the closure of U in $\mathbf{R}_m$, and let $\varphi(t; t_0, x_0, u(\cdot))$ denote the (unique) solution of the differential equation

$$\frac{dx(t)}{dt} = f(x(t), u(t), t)$$

satisfying the initial condition

$$x(t_0) = x_0.$$

Let $Y = \mathbf{R}_n$ and $\eta(t, x) = x$. Then the dynamical system $\Sigma = \langle \mathbf{R}, U, \mathbf{R}_n, \mathbf{R}_n, \varphi, \eta \rangle$ is smooth with generator f. Let S be a given subset of $\mathbf{R} \times \mathbf{R}_n$—for example, $S = \mathbf{R} \times \{0\}$; let Γ be a subset of Ω—for example, $\Gamma = \Omega$; let I be a given subset of $\mathbf{R} \times \mathbf{R}_n$—for example, $I = \{0\} \times \{x: \|x\| \ge 1\}$; let $K(t, x) = 0$ and let $L(x, u, t)$ be a suitably smooth real-valued function—for example, $L(x, u, t) = \frac{1}{2}\langle x, Q(t)x \rangle + \frac{1}{2}\langle u, R(t)u \rangle$, where $Q(\cdot)$, and $R(\cdot)$ are matrix-valued functions. Then, if $u(\cdot) \in \Omega$ transfers (t_0, x_0) to S with transfer time t_1, the cost of the control $u(\cdot)$, $J(t_0, x_0, u(\cdot))$, is given by

$$J(t_0, x_0, u(\cdot)) = \int_{t_0}^{t_1} L(\varphi(t; t_0, x_0, u(\cdot)), u(t), t)\, dt.$$

[†] If $f = f(x, u, t)$, then $\nabla_x f$ is the gradient of f with respect to x and $\nabla_t f = \partial f/\partial t$. $\mathcal{B}(\mathbf{R}_n \times \bar{U} \times \mathbf{R})$ is set of all bounded functions from $\mathbf{R}_n \times \bar{U} \times \mathbf{R}$ into $\mathbf{R}_n$.

For example, if $(0, x) \in \{0\} \times \{x: \|x\| \geq 1\}$ and if $u(\cdot)$ transfers $(0, x)$ to $\mathbf{R} \times \{0\}$ in time t_1, then

$$J(t_0, x_0, u(\cdot)) = \tfrac{1}{2} \int_{t_0}^{t_1} \{ \langle \varphi(t; t_0, x_0, u(\cdot)), Q(t)\varphi(t; t_0, x_0, u(\cdot)) \rangle + \langle u(t), R(t)u(t) \rangle \} \, dt.$$

The statement of the control problem is just as in Definition (3.5).

We observe that a control problem may be viewed as equivalent to a problem of minimizing a functional defined on a subset of a normed linear space. We also note that, given a control problem, two questions immediately come to mind:

1. Do there exist solutions of the problem; that is, do optimal controls exist?

2. If there are optimal controls, then how can they be determined?†

The development of answers to these questions is the basic subject matter of control theory.

Various approaches have been used in control theory and may be broadly classified into the following four categories:

1. Hamilton-Jacobi theory, the derivation and use of a basic set of sufficiency conditions depending on the solution of a partial differential equation (see Section 3.4).

2. Development of necessary conditions, such as the Pontryagin maximum principle (see Section 4.1).

3. Functional-analysis methods (see [FALB, 1967; KRANC and SARACHIK, 1963; KULIKOWSKI, 1959; NEUSTADT, 1965]).

4. Numerical solutions (see [COLLATZ, 1964; TODD, 1962; WITSENHAUSEN, 1965] and Chapter 5).

The numerical-methods approach is quite different in kind from the other three approaches. We shall examine the Hamilton-Jacobi theory in the next section and the role of necessary conditions in Chapter 4. We shall not examine functional-analysis methods, as these require somewhat more sophisticated mathematical knowledge.‡ We shall only comment on numerical techniques, as these too are a more specialized topic.

The various approaches used in control theory may be compared in terms of the following criteria: (1) generality of applicability, (2) sharpness of results, (3) ease of practical implementation, and (4) physical relevance of assumptions made. These criteria of comparison should be kept in mind during subsequent discussions.

---

† That these are, in fact, quite different questions may be seen by considering, for example, a continuous real-valued function defined on a compact subset of $\mathbf{R}_n$. Such a function always has a minimum, but it is something else again to determine the point(s) at which this minimum occurs.

‡ But see the last section of Chapter 4.

3.4   Hamilton-Jacobi theory

Let us suppose that we are faced with a standard control problem. We propose to develop a basic set of sufficiency conditions for this problem using a lemma of Carathéodory in a manner suggested by [KALMAN, 1963a]. To begin with, we introduce the following notation:

(4.1)   **Notation.**   If $\mathcal{B}$ is a Banach space, then $\mathcal{B}^*$ denotes the dual space of $\mathcal{B}$ (that is, $\mathcal{B}^*$ is the set of continuous linear functionals on $\mathcal{B}$), and if $\lambda \in \mathcal{B}^*$ and $b \in \mathcal{B}$, then $\langle \lambda, b \rangle$ denotes the operation of λ on b.†

(4.2)   **Notation.**   If $\mathcal{B}_1$ and $\mathcal{B}_2$ are Banach spaces and Λ is a continuous linear transformation from $\mathcal{B}_1$ into $\mathcal{B}_2$ [that is, $\Lambda \in \mathcal{L}(\mathcal{B}_1, \mathcal{B}_2)$], then the adjoint of Λ is denoted by Λ^*. We recall that Λ^* is a continuous linear transformation from $\mathcal{B}_2^*$ into $\mathcal{B}_1^*$ (that is, $\Lambda^* \in \mathcal{L}(\mathcal{B}_2^*, \mathcal{B}_1^*)$) which is defined by $\Lambda^* \lambda_2 = \lambda_2 \Lambda$ for $\lambda_2 \in \mathcal{B}_2^*$, so that

(4.3) $\langle \Lambda^* \lambda_2, b_1 \rangle = \langle \lambda_2, \Lambda b_1 \rangle$

for $\lambda_2 \in \mathcal{B}_2^*$ and $b_1 \in \mathcal{B}_1$. In the finite-dimensional case, where Λ may be viewed as a matrix, Λ^* is simply the transpose of Λ.

(4.4)   **Definition.**   *Let $\mathcal{R} \subset (T_1, T_2) \times X$. If $(t_0, x_0) \in \mathcal{R}$ and $\hat{u}(\cdot) \in \Omega$, then we say that $\hat{u}(\cdot)$ $\mathcal{R}$-transfers (t_0, x_0) to S if $\hat{u}(\cdot)$ transfers (t_0, x_0) to S along a path lying entirely in $\mathcal{R}$, that is, if $(t, \varphi(t; t_0, x_0, \hat{u}(\cdot))) \in \mathcal{R}$ for $t \geq t_0$. An admissible control $u^0(\cdot)$ is called **optimal relative to** $\mathcal{R}$ and a given initial event (t_0, x_0) in $\mathcal{R}$ (a) if $u^0(\cdot)$ $\mathcal{R}$-transfers (t_0, x_0) to S and (b) if $u^1(\cdot)$ is any element of Ω which $\mathcal{R}$-transfers (t_0, x_0) to S, then $J(t_0, x_0, u^0(\cdot)) \leq J(t_0, x_0, u^1(\cdot))$.*

With these preliminaries out of the way, we can now state and prove the basic Carathéodory lemma.

(4.5)   **Lemma.**   *Suppose that $K(t, x) = 0$ for $(t, x) \in S$ and that for each event (t, x) in a region $\mathcal{R}$ with $\mathcal{R} \subset (T_1, T_2) \times X$ the function $L(x, u, t)$ has, as a function of u, zero as its unique absolute minimum with respect to all u in U at $u^0(t, x)$, that is,*

(4.6) $0 = L(x, u^0(t, x), t) < L(x, u, t)$

for all $u \in U$, $u \neq u^0(t, x)$. Let $\hat{u}(\cdot) \in \Omega$ be such that
(a) $\hat{u}(\cdot)$ $\mathcal{R}$-transfers (t_0, x_0) to S with t_1 as transfer time.
(b) $\hat{u}(t) = u^0(t, \hat{x}(t))$ for (almost) all t in $[t_0, t_1]$, where for convenience $\hat{x}(t) = \varphi(t; t_0, x_0, \hat{u}(\cdot))$.

† Although this notation is consistent with our use of $\langle \ , \ \rangle$ for the scalar product on $\mathbf{R}_n$, the reader should be wary of confusing the two.

Then $\hat{u}(\cdot)$ is optimal relative to $\mathfrak{R}$ and (t_0, x_0).

Proof. We observe that

$$J(t_0, x_0, \hat{u}(\cdot)) = K(t_1, \hat{x}(t_1)) + \int_{t_0}^{t_1} L(\hat{x}(t), u^0(t, \hat{x}(t)), t)\, dt = 0,$$

since $K(t, x) = 0$ on S and the integrand is zero by virtue of (4.6). Now, if $u^1(\cdot)$ is an element of Ω which $\mathfrak{R}$-transfers (t_0, x_0) to S with t_2 as transfer time, then

$$\begin{aligned}
J(t_0, x_0, u^1(\cdot)) &= K(t_2, x^1(t_2)) + \int_{t_0}^{t_2} L(x^1(t), u^1(t), t)\, dt, \\
&\geq \int_{t_0}^{t_2} L(x^1(t), u^0(t, x^1(t)), t)\, dt, \\
&= 0,
\end{aligned}$$

since $K(t, x) = 0$ on S and (4.6) holds (note that we have set

$$x^1(t) = \varphi(t; t_0, x_0, u^1(\cdot))).$$

The lemma follows. $\qquad\square$

We observe that $\hat{u}(\cdot)$ need not be an optimal control, since there may be controls $\bar{u}(\cdot)$ in Ω which transfer (t_0, x_0) to S along paths that do not lie in $\mathfrak{R}$ but for which the cost is smaller than 0; that is,

$$J(t_0, x_0, \bar{u}(\cdot)) < J(t_0, x_0, \hat{u}(\cdot)) = 0.$$

Now, in order to apply this lemma we must introduce the hamiltonian functional and the notion of regularity (or normality). In a very loose way, the hamiltonian functional may be used to measure the difference between the costs of different controls and so may be viewed as representing the differential (or gradient) of the cost functional. This will become clearer in the sequel. To be more precise, we have

(4.7) **Definition.** *The real-valued function H defined on $X \times \mathfrak{B}_X^* \times U \times (T_1, T_2)$ by the relation*

(4.8) $$H(x, \lambda, u, t) = L(x, u, t) + \langle \lambda, f(x, u, t) \rangle,$$

*where f is the generator of the dynamical system Σ, is called the **hamiltonian** (or hamiltonian functional) of the control problem.*

(4.9) **Definition.** *Let $\mathfrak{R} \subset (T_1, T_2) \times X$. If for each $(t, x) \in \mathfrak{R}$ and $\lambda \in \mathfrak{B}_X^*$ the function $H(x, \lambda, u, t)$ has, as a function of u, a unique absolute minimum with respect to all u in U, at $u^0(t, x, \lambda)$, say—that is,*

(4.10) $$H(x, \lambda, u^0(t, x, \lambda), t) < H(x, \lambda, u, t)$$

*for all $u \in U$, $u \neq u^0(t, x, \lambda)$—then we say that H is **regular** (normal)* *relative to* $\mathfrak{R}$. *In that case the mapping* $u^0(t, x, \lambda)$ *of* $\mathfrak{R} \times \mathfrak{B}_X^*$ *into* U *is called the* H-**minimal control** *relative to* $\mathfrak{R}$.†

We recall (see [DIEUDONNE, 1960]) that if $V(t, x)$ is a real-valued function defined on $(T_1, T_2) \times X$, then $\partial V/\partial t$ is a continuous linear transformation of $\mathbf{R}$ into $\mathbf{R}$ and $\partial V/\partial x$ is a continuous linear transformation of $\mathfrak{B}_X$ into $\mathbf{R}$, that is, an element of $\mathfrak{B}_X^*$. Thus we have

(4.11) **Definition.** *If* H *is regular relative to* $\mathfrak{R}$, *then the partial differential equation*

$$(4.12) \qquad \frac{\partial V(t, x)}{\partial t} + H\left(x, \frac{\partial V(t, x)}{\partial x}, u^0\left(t, x, \frac{\partial V(t, x)}{\partial x}\right), t\right) = 0$$

is called the **Hamilton-Jacobi equation of the control problem** *(relative to* $\mathfrak{R}$).

Observe now that if $V(t, x)$ and $x(t)$ are continuously differentiable, then $V(t, x(t))$ is continuously differentiable and

$$(4.13) \qquad \frac{dV(t, x(t))}{dt} = \frac{\partial V(t, x(t))}{\partial t} + \left\langle \frac{\partial V(t, x(t))}{\partial x}, \frac{dx(t)}{dt} \right\rangle$$

(see [DIEUDONNE, 1960]). We shall use this observation in the proof of the following sufficiency theorem:

(4.14) **Theorem.** *Let* $\mathfrak{R} \subset (T_1, T_2) \times X$. *Suppose that* H *is regular relative to* $\mathfrak{R}$ *and that the transition function* φ *is continuously differentiable on* $\mathfrak{R}$ *(that is, on an open set containing the projection of* $\mathfrak{R}$ *on* (T_1, T_2)*). Let* $\hat{u}(\cdot) \in \Omega$ *be such that*
  (a) $\hat{u}(\cdot)$ $\mathfrak{R}$-*transfers* (t_0, x_0) *to* S *with* $\hat{t}$ *as transfer time.*
  (b) *There is a continuously differentiable solution* $V(t, x)$ *of the Hamilton-Jacobi equation on* $\mathfrak{R}$ *satisfying the boundary condition*

$$(4.15) \qquad\qquad V(t, x) = K(t, x)$$

for $(t, x) \in \mathfrak{R} \cap S$ *such that*

$$(4.16) \qquad \hat{u}(t) = u^0\left(t, \hat{x}(t), \frac{\partial V(t, \hat{x}(t))}{\partial x}\right)$$

for (almost) all t *in* $[t_0, t_1]$, *where for convenience*

$$\hat{x}(t) = \varphi(t; t_0, x_0, \hat{u}(\cdot)).$$

Then $\hat{u}(\cdot)$ *is optimal relative to* $\mathfrak{R}$ *and* (t_0, x_0).

Proof. The proof of the theorem is a direct application of the Carathéo-

---

† The mapping $u^0(t, x, \lambda)$ is, of course, not an actual admissible control.

dory lemma. First we introduce the function $\hat{L}(x, u, t)$, defined by

$$\hat{L}(x, u, t) = \frac{\partial V(t, x)}{\partial t} + H\left(x, \frac{\partial V(t, x)}{\partial x}, u, t\right),$$

$$= \frac{\partial V(t, x)}{\partial t} + L(x, u, t) + \left\langle \frac{\partial V(t, x)}{\partial x}, f(x, u, t) \right\rangle.$$

We claim that $\hat{L}$ satisfies the hypotheses of the lemma. To verify this claim we make the following observations: $\partial V(t, x)/\partial t$ does not depend on u; since H is regular relative to $\mathfrak{R}$, $H(x, \partial V(t, x)/\partial x, u, t)$ has, as a function of u, a unique absolute minimum over U at $u^0(t, x, \partial V(t, x)/\partial x)$, which implies that $\hat{L}(x, u, t)$ has, as a function of u, a unique absolute minimum at $\hat{u}^0(t, x) = u^0(t, x, \partial V(t, x)/\partial x)$ for $(t, x) \in \mathfrak{R}$; V satisfies (4.12) on $\mathfrak{R}$. Let $\hat{K}(t, x) = 0$ and let $\hat{J}(t_0, x_0, u(\cdot))$ be the standard-cost functional relative to $\hat{K}$ and $\hat{L}$. We suppose that $u^1(\cdot)$ $\mathfrak{R}$-transfers (t_0, x_0) to S with t^1 as transfer time. Then it follows from the lemma that

(4.17)
$$\begin{aligned}
0 &= \hat{J}(t_0, x_0, \hat{u}(\cdot)), \\
&= \int_{t_0}^{\hat{t}} \hat{L}(\hat{x}(t), \hat{u}(t), t)\, dt, \\
&\leq \hat{J}(t_0, x_0, u^1(\cdot)), \\
&= \int_{t_0}^{t^1} \hat{L}(x^1(t), u^1(t), t)\, dt.
\end{aligned}$$

But in view of (4.13), we have

(4.18)
$$\begin{aligned}
\frac{d}{dt}[V(t, \hat{x}(t))] + L(\hat{x}(t), \hat{u}(t), t) &= \hat{L}(\hat{x}(t), \hat{u}(t), t), \\
\frac{d}{dt}[V(t, x^1(t))] + L(x^1(t), u^1(t), t) &= \hat{L}(x^1(t), u^1(t), t).
\end{aligned}$$

However, (4.17) and (4.18) and the boundary condition (4.15) together imply that

$$\begin{aligned}
0 &= K(\hat{t}, \hat{x}(\hat{t})) + \int_{t_0} L(\hat{x}(t), \hat{u}(t), t)\, dt - V(t_0, x_0), \\
&= J(t_0, x_0, \hat{u}(\cdot)) - V(t_0, x_0), \\
&\leq K(t^1, x^1(t^1)) + \int_{t_0}^{t^1} L(x^1(t), u^1(t), t)\, dt - V(t_0, x_0), \\
&\leq J(t_0, x_0, u^1(\cdot)) - V(t_0, x_0),
\end{aligned}$$

and the theorem follows. $\qquad\square$

(4.19) **Corollary.** *Under the hypotheses of the theorem the solution $V(t, x)$ of the Hamilton-Jacobi equation represents the optimum cost relative to $\mathfrak{R}$ in the sense that*

$$V(t_0, x_0) = J(t_0, x_0, \hat{u}(\cdot)).$$

Moreover, if $t \in [t_0, \hat{t}]$, *then* $\hat{u}(\cdot)$ *is optimal relative to* $\Re$ *and* $(t, \hat{x}(t))$, *and*

$$V(t, \hat{x}(t)) = J(t, \hat{x}(t), \hat{u}(\cdot)).$$

Now let us make some comments concerning the sufficiency theorem. We observe first of all that it is essentially a local result in that everything is relative to the region $\Re$ of the event space. Of course, if $\Re = (T_1, T_2) \times X$, then the theorem represents a sufficient condition for global optimality. However, this requires global regularity of the hamiltonian, which is often not the case. Furthermore, even when there is global regularity, it may not be possible to obtain a global solution of the Hamilton-Jacobi equation. In practice these difficulties are often circumvented by "splitting" the event space into suitable regions in which the requisite hypotheses are satisfied (see [ATHANS and FALB, 1966; BOLTYANSKII, 1966]). This partitioning process is frequently used as a check on the optimality of controls obtained by other means (for example, through use of the necessary conditions).

In essence, we are demanding that the equation of the system be integrable when the H-minimal control evaluated with respect to the optimum cost is substituted into the system equation. In other words, we require that

$$(4.20) \qquad \frac{dx(t)}{dt} = f\left(x(t), u^0 \left(t, x(t), \frac{\partial V(t, x(t))}{\partial x}, t \right) \right)$$

have a solution satisfying the initial condition $x(t_0) = x_0$ which lies entirely in $\Re$ and meets the target set S. Equation (4.20) is often used as a basis for numerical approaches to the solution of a control problem. As a consequence of these ideas, we also have

(4.21) **Corollary.** *Let* $\Re \subset (T_1, T_2) \times X$. *Suppose that* H *is regular relative to* $\Re$ *and that there is a continuously differentiable solution* $V(t, x)$ *of the Hamilton-Jacobi equation on* $\Re$ *satisfying the boundary condition* $V(t, x) = K(t, x)$ *for* $(t, x) \in \Re \cap S$. *If the equation*

$$(4.22) \qquad \frac{dx(t)}{dt} = f\left(x(t), u^0 \left(t, x(t), \frac{\partial V(t, x(t))}{\partial x}, t \right) \right)$$

has for each initial pair (t_0, x_0) *in* $\Re$ *a continuously differentiable solution lying entirely in* $\Re$ *and meeting* S, *then there is an optimal control relative to* $\Re$ *for every* (t_0, x_0) *in* $\Re$.

This corollary is frequently used in practice.

We observe now that the requirement of continuous differentiability is a bit too strong; in point of fact, it is sufficient that (4.13) be satisfied almost everywhere along trajectories of the system which lie entirely in $\Re$. This will be true, for example, if the partial derivatives of $V(t, x)$ are regulated (rather than continuous)† on ℛ and φ is continuously differentiable on ℛ.

The sufficiency theorem and the Hamilton-Jacobi equation can be interpreted in an intuitively appealing geometric way. The function $V(t, x)$ may be viewed as determining a "surface" in $(T_1, T_2) \times X \times \mathbf{R}$ (see Figure 3.1) called the cost surface. The Hamilton-Jacobi equation

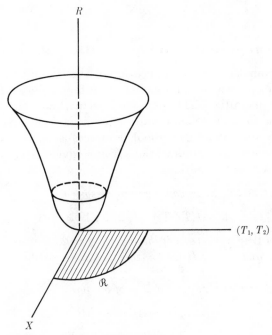

Figure 3.1

represents the equation of the curve of steepest descent consistent with the contraints along the cost surface; in other words, the Hamilton-Jacobi equation describes the evolution of the optimum cost, and the hamiltonian may be viewed as the time rate of change of the optimum cost. To put it still another way, there is at each point P of the cost surface a set of "directions" whose "projections" on the event space generate the "cone" of admissible trajectory directions‡ at the projection of the point P on the event space, and the "direction" satisfying the Hamilton-Jacobi

---

† A function is *regulated* if it has one-sided limits at every point of its domain. In other words, the only discontinuities of the function are simple "jumps" (see [Dieudonne, 1960]).

‡ That is, the set of vectors of the form $(1, f(x, u, t))$, $u \in U$, viewed as attached to the point (t, x).

equation "projects" onto the direction of the optimum trajectory. A curve whose tangent direction satisfies the Hamilton-Jacobi equation will thus project onto a portion of an optimal trajectory. The relevance of these ideas will become clearer after our discussion of necessary conditions in Chapter 4.

We apply the results developed here to the case of a linear system and a quadratic cost criterion in the next section.

3.5   Linear systems and quadratic criterion

A very important and useful application of the Hamilton-Jacobi theory is the situation in which our dynamical system is linear and our cost functional is quadratic. This problem is of particular interest because of its relation to conventional control methods (see [KALMAN, 1963a, 1964; WILLIS and BROCKETT, 1965]) and because of its relation to the question of correcting small deviations from a given path. As usual, we begin with a series of definitions:

(5.1)   **Definition.**   *The smooth dynamical system*

$$\Sigma = \langle (T_1, T_2), U, \Omega, X, Y, \varphi, \eta \rangle$$

is called a **linear system** *if the following conditions are satisfied:*

(a) *X, Y, and $\mathfrak{B}_U$ are Hilbert spaces with inner products $\langle \ , \ \rangle_X, \langle \ , \ \rangle_Y$, and $\langle \ , \ \rangle_U$, respectively.†*

(b) *The generator $f(x, u, t)$ of Σ is of the form*

$$f(x, u, t) = A(t)x + B(t)u,$$

where $A(t) \in \mathcal{L}(X, X)$, $B(t) \in \mathcal{L}(\mathfrak{B}_U, X)$,   and   the   mappings $t \to A(t)$ and $t \to B(t)$ are continuous (or regulated) on (T_1, T_2).

(c) *The output function $\eta(t, x)$ is of the form*

$$\eta(t, x) = C(t)x,$$

where $C(t) \in \mathcal{L}(X, Y)$ and the mapping $t \to C(t)$ is regulated on (T_1, T_2).

We recall that if $\mathcal{3C}$ is a Hilbert space with [,] as inner product, then $\mathcal{3C}^*$ can be identified with $\mathcal{3C}$.‡ In particular, this means that if $\xi \in \mathcal{L}(\mathcal{3C}, \mathcal{3C})$, then we may view the adjoint ξ^* of ξ as also belonging to $\mathcal{L}(\mathcal{3C}, \mathcal{3C})$. We say that a selfadjoint (or symmetric) element ξ of $\mathcal{L}(\mathcal{3C}, \mathcal{3C})$

---

† We shall frequently omit the subscripts X, Y, and U, as it will be clear from the context which inner product is intended.

‡ We shall make this identification for all the Hilbert spaces that we consider in the sequel.

is *positive* if $[h, \xi h] \geq 0$ for all h in $\mathfrak{IC}$ and is *positive definite* if $[h, \xi h] > 0$ for all h in $\mathfrak{IC}$ with $h \neq 0$ and ξ is invertible.† Finally, if ξ is positive, then we shall frequently use the notation $\| \ \|_\xi^2$ in place of $[\cdot, \xi \cdot]$; that is, if $h \in \mathfrak{IC}$, then

$$\|h\|_\xi^2 = [h, \xi h].$$

In the finite-dimensional case we may view ξ as a (symmetric) matrix, and positivity or positive definiteness corresponds to the familiar analogous matrix notion.

We now have

(5.2) **Definition.** *Let $z(t)$ be a regulated mapping of (T_1, T_2) into Y. Let φ be a positive element of $\mathfrak{L}(Y, Y)$, let $\rho(t)$ be a positive element of $\mathfrak{L}(Y, Y)$ such that the mapping $t \to \rho(t)$ is regulated on (T_1, T_2), and let $\sigma(t)$ be a positive definite element of $\mathfrak{L}(\mathfrak{B}_U, \mathfrak{B}_U)$ such that the mapping $t \to \sigma(t)$ is continuous on (T_1, T_2). If $K(t, x)$ is given by*

$$K(t, x) = \tfrac{1}{2}\|z(t) - C(t)x\|_\varphi^2$$

and if $L(x, u, t)$ is given by

$$L(x, u, t) = \tfrac{1}{2}[\|z(t) - C(t)x\|_{\rho(t)}^2 + \|u\|_{\sigma(t)}^2),$$

then the cost functional $J(t_0, x_0, u(\cdot))$ determined by K and L is called a **quadratic cost functional** *(for the linear system of Definition (5.1)).*

Now let us suppose that Σ is our given linear system, that K and L are our terminal- and trajectory-cost terms, respectively, that t_0 is a given element of (T_1, T_2), that x_0 is a given element of X, that t_1 is a given element of (t_0, T_2) (that is, $t_1 > t_0$), and that $S = \{t_1\} \times X$ is our target set.

Let us further assume that

1. $U = \mathfrak{B}_U$, so that there are no magnitude constraints on our admissible controls.

2. Ω is the set of all bounded regulated functions‡ from (T_1, T_2) into U.

Since the time t_1 is fixed and the terminal state is free, every $u(\cdot) \in \Omega$ will transfer (t_0, x_0) to S. Thus, if $u(\cdot) \in \Omega$, then we have

(5.3) $$J(t_0, x_0, u(\cdot)) = \tfrac{1}{2}\|z(t_1) - C(t_1)x_u(t_1)\|_\varphi^2$$
$$+ \tfrac{1}{2}\int_{t_0}^{t_1} [\|z(t) - C(t)x_u(t)\|_{\rho(t)}^2 + \|u(t)\|_{\sigma(t)}^2]\,dt,$$

where $x_u(t)$ is the solution of our system equation starting from x_0 at t_0 and

---

† This second condition is actually redundant.

‡ See [DIEUDONNE, 1960] and recall that *regulated* means having one-sided limits.

generated by the control $u(\cdot)$; that is, $x_u(t) = \varphi(t; t_0, x_0, u(\cdot))$, and our control problem is to determine the $u(\cdot)$ which minimizes $J(t_0, x_0, u(\cdot))$.

We shall soon solve this control problem using the Hamilton-Jacobi theory (cf. [KALMAN, 1963a]). However, before we develop the solution we shall examine the physical basis for this particular problem and then describe the path to the formal result.

If we view $z(t)$ as representing a desired output signal and if we set $y_u(t) = C(t)x_u(t)$, then the difference

$$e_u(t) = z(t) - y_u(t)$$

may be viewed as the error associated with the control signal $u(t)$. We want to choose our control signal in such a way as to make this error small without requiring excessive expenditures of control energy. We observe that the cost functional J of (5.3), which may be written as

$$(5.4) \qquad J(t_0, x_0, u(\cdot)) = \tfrac{1}{2}\|e_u(t_1)\|_\varphi^2 + \tfrac{1}{2}\int_{t_0}^{t_1} \|e_u(t)\|_{\rho(t)}^2 \, dt$$

$$+ \tfrac{1}{2}\int_{t_0}^{t_1} \|u(t)\|_{\sigma(t)}^2 \, dt,$$

is designed with this objective in mind. J is always greater than or equal to zero, large errors are weighed more heavily than small errors, and there is a term which takes control "energy" into account (the last term in (5.4)). Moreover, the optimal control law, as we shall see, leads to a linear feedback controller which is desirable from a practical standpoint.

Development of the solution to our control problem begins with a consideration of a simpler version, called the *state-regulator problem*, in which $C(t) = I$ (the identity) and $z(t) = 0$ for all t. The key step is the introduction of the Riccati equation, with a concomitant examination of the necessary properties of its solution. We next treat the so-called *output-regulator problem*, in which $z(t) = 0$ for all t. Then we consider the generally posed problem and obtain its solution. We conclude the section with an account of regulator problems for time-invariant systems over an infinite interval.

So, let us turn our attention to the state-regulator problem. In other words, let us assume that $X = Y$ and that $C(t) = I$ and $z(t) = 0$ for all t. Then

$$K(t, x) = \tfrac{1}{2}\|x\|_\varphi^2,$$
$$L(x, u, t) = \tfrac{1}{2}\|x\|_{\rho(t)}^2 + \tfrac{1}{2}\|u\|_{\sigma(t)}^2,$$

and, noting that X^* is identified with X, we see that the hamiltonian H is given by

$$(5.5) \qquad H(x, \lambda, u, t) = L(x, u, t) + \langle \lambda, f(x, u, t)\rangle,$$
$$= \tfrac{1}{2}\|x\|_{\rho(t)}^2 + \tfrac{1}{2}\|u\|_{\sigma(t)}^2 + \langle \lambda, A(t)x + B(t)u\rangle.$$

We now have

(5.6) **Proposition.** *Let* $\mathfrak{R} = (T_1, T_2) \times X$, *let* $(\hat{t}, \hat{x})$ *be any element of* $\mathfrak{R}$, *and let* $\hat{\lambda}$ *be any element of* $X(= X^*)$. *Then the function* $H(\hat{x}, \hat{\lambda}, u, \hat{t})$ *of* u *has a unique absolute minimum at* $u^0(\hat{t}, \hat{x}, \hat{\lambda})$, *where*

(5.7) $$u^0(\hat{t}, \hat{x}, \hat{\lambda}) = -\sigma^{-1}(t)B^*(\hat{t})\hat{\lambda}.$$

In other words, H is regular relative to $(T_1, T_2) \times X$, *and*

$$u^0(t, x, \lambda) = -\sigma^{-1}(t)B^*(t)\lambda$$

is the corresponding H-minimal control.

Proof.   The terms in $H(\hat{x}, \hat{\lambda}, u, \hat{t})$ which depend on u are

(5.8) $$\tfrac{1}{2}\langle u, \acute{\sigma}u \rangle + \langle \hat{B}^*\hat{\lambda}, u \rangle,$$

where $\acute{\sigma} = \sigma(\hat{t})$ and $\hat{B}^* = B^*(\hat{t})$.   Now H will have a unique absolute minimum at $u^0(\hat{t}, \hat{x}, \hat{\lambda})$ if and only if (5.8) has a unique absolute minimum at $u^0(\hat{t}, \hat{x}, \hat{\lambda})$.   However, (5.8) will have a unique absolute minimum at $u^0(\hat{t}, \hat{x}, \hat{\lambda})$ if and only if the expression

(5.9) $$\langle u, \acute{\sigma}u \rangle + 2\langle \hat{B}^*\hat{\lambda}, u \rangle + \langle \hat{B}^*\hat{\lambda}, \acute{\sigma}^{-1}\hat{B}^*\hat{\lambda} \rangle$$

has a unique absolute minimum at $u^0(\hat{t}, \hat{x}, \hat{\lambda})$, as $\acute{\sigma}$ is positive definite. But

$$\langle u, \acute{\sigma}u \rangle + 2\langle \hat{B}^*\hat{\lambda}, u \rangle + \langle \hat{B}^*\hat{\lambda}, \acute{\sigma}^{-1}\hat{B}^*\hat{\lambda} \rangle$$
$$= \langle u + \acute{\sigma}^{-1}\hat{B}^*\hat{\lambda}, \acute{\sigma}(u + \acute{\sigma}^{-1}\hat{B}^*\hat{\lambda}) \rangle = \| u + \acute{\sigma}^{-1}\hat{B}\hat{\lambda} \|_{\acute{\sigma}}^2,$$

and thus it follows from the fact that $\acute{\sigma}$ is positive definite that (5.9) has a unique absolute minimum at $u = u^0(\hat{t}, \hat{x}, \hat{\lambda})$.

Now we observe that if $\pi(\cdot)$ is a continuously differentiable mapping of (T_1, T_2) into $\mathfrak{L}(X, X)$, then the function $W(t, x)$ defined by

(5.10) $$W(t, x) = \tfrac{1}{2}\langle x, \pi(t)x \rangle$$

is continuously differentiable on $(T_1, T_2) \times X$.   It follows that if $u(\cdot)$ is any element of Ω and $x_u(\cdot)$ is a trajectory of our system, then

$$\frac{dW(t, x_u(t))}{dt} = \frac{\partial W(t, x_u(t))}{\partial t} + \left\langle \frac{\partial W(t, x_u(t))}{\partial x}, \frac{dx_u(t)}{dt} \right\rangle$$

almost everywhere on (T_1, T_2).   In view of the remarks immediately after Corollary (4.21), it is apparent that if we can find a solution of the Hamilton-Jacobi equation of the form (5.10), then we can apply Theorem (4.14) to obtain an optimal control for our problem.   We begin with the following lemma:

(5.11)   **Lemma.**   *If $\pi(t)$ is a solution of the Riccati differential equation*

(5.12) $\dot{\pi}(t) = -\pi^*(t)A(t) - A^*(t)\pi(t) + \pi^*(t)S(t)\pi(t) - \rho(t),$†

where

(5.13) $S(t) = B(t)\sigma^{-1}(t)B^*(t),$

then the function $W(t, x)$ defined by (5.10) is a solution of the Hamilton-Jacobi equation.

Proof. We have

(5.14) $\dfrac{\partial W(t, x)}{\partial t} = \dfrac{1}{2}\langle x, \dot{\pi}(t)x\rangle,\qquad \dfrac{\partial W(t, x)}{\partial x} = \pi(t)x,$

and so, from Equations (5.14) and (5.5), the left-hand side of (4.12) becomes

$$\frac{1}{2}\langle x, \dot{\pi}(t)x\rangle + \frac{1}{2}\langle x, \rho(t)x\rangle + \frac{1}{2}\langle B^*(t)\pi(t)x, \sigma^{-1}(t)B^*(t)\pi(t)x\rangle$$
$$+ \langle \pi(t)x, A(t)x - S(t)\pi(t)x\rangle,$$
$$= \frac{1}{2}\langle x, \dot{\pi}(t)x\rangle + \frac{1}{2}\langle x, \rho(t)x\rangle$$
$$- \frac{1}{2}\langle x, \pi^*(t)S(t)\pi(t)x\rangle + \frac{1}{2}\langle x, \pi^*(t)A(t)x\rangle + \frac{1}{2}\langle x, A^*(t)\pi(t)x\rangle.$$

The lemma follows immediately.   ☐

We can now see that a consideration of the differential equation (5.12) may prove fruitful.   If we set

(5.15) $g(t, \pi) = -\pi^*A(t) - A^*(t)\pi + \pi^*S(t)\pi - \rho(t),$

then $g(t, \pi)$ is a locally lipschitzian mapping of $(T_1, T_2) \times \mathcal{L}(X, X)$ into $\mathcal{L}(X, X)$ which is regulated in t and continuous in π.   This implies that if t_1 is a given element of (T_1, T_2), then there is an open interval about t_1, say (r_1, s_1), in which (5.12) has a *unique* solution $\pi_1(t)$ satisfying the condition

(5.16) $\pi_1(t) = \phi$

(see [DIEUDONNE, 1960]).   We observe that since $\phi = \phi^*$ and

$$g(t, \pi^*) = -\pi A(t) - A^*(t)\pi^* + \pi S(t)\pi^* - \rho(t),$$

$\pi_1^*(t)$ is also a solution of (5.12) satisfying (5.16).   It follows that $\pi_1(t)$ is selfadjoint.   We now have

(5.17) **Lemma.** *Let t be an element of (r_1, t_1) and let x be any element of X.   Then there is an optimal control $u^0(\cdot)$ relative to (t, x), and the cost of this optimal control, $J(t, x, u^0(\cdot))$, is given by*

(5.18) $J(t, x, u^0(\cdot)) = \frac{1}{2}\langle x, \pi_1(t)x\rangle.$

† $\dot{\pi}(t)$ is a shorthand for $d\pi(t)/dt.$

Proof. The lemma is an immediate consequence of Proposition (5.6), Lemma (5.11), and Theorem (4.14). □

(5.19) **Corollary.** *If $t \in (r_1, t_1)$, then $\pi_1(t)$ is positive.*

Proof. Let x be any element of X. Then

$$
\begin{aligned}
\langle x, \pi_1(t)x \rangle &= 2J(t, x, u^0(\cdot)), \\
&= \|x^0(t_1)\|_\phi^2 + \int_t^{t_1} [\|x^0(s)\|_{\rho(s)}^2 + \|u^0(s)\|_{\sigma(s)}^2]\, ds, \\
&\geq 0,
\end{aligned}
$$

where $x^0(\cdot)$ is the trajectory corresponding to $u^0(\cdot)$, in view of our assumptions on ϕ, $\rho(\cdot)$, and $\sigma(\cdot)$.

(5.20) **Corollary.** *The mapping $t \to g(t, \pi_1(t))$ is bounded on $(r_1, t_1]$ (if $r_1 > T_1$).*

Proof. In view of our assumptions on the mappings $t \to A(t)$, $t \to B(t)$, and $t \to \rho(t)$, it will be enough to show that the mapping t into $\pi_1(t)$ is bounded. Let us suppose for the moment that there is a positive π such that $\pi - \pi_1(t)$ is positive for *all* t in $(r_1, t_1]$. Then we claim that $\|\pi\| \geq \|\pi_1(t)\|$ for all t in $(r_1, t_1]$. To verify this claim we observe that for $x \in X$

$$
\begin{aligned}
\langle x, \{\pi - \pi_1(t)\}\, x \rangle &= \langle x, \pi x \rangle - \langle x, \pi_1(t)x \rangle, \\
&= \langle \sqrt{\pi}\, x, \sqrt{\pi}\, x \rangle - \langle \sqrt{\pi_1(t)}\, x, \sqrt{\pi_1(t)}\, x \rangle, \\
&\geq 0,
\end{aligned}
$$

where $\sqrt{\pi}$ and $\sqrt{\pi_1(t)}$ are the positive square roots of π and $\pi_1(t)$, respectively.[†] It follows that $\|\sqrt{\pi}\| \geq \|\sqrt{\pi_1(t)}\|$, and hence that

$$
\begin{aligned}
\|\pi\| &= \|\sqrt{\pi} \cdot \sqrt{\pi}\| = \|\sqrt{\pi}\|^2 \geq \|\sqrt{\pi_1(t)}\|^2 \\
&= \|\sqrt{\pi_1(t)} \cdot \sqrt{\pi_1(t)}\| = \|\pi_1(t)\|,
\end{aligned}
$$

since $\|\xi\xi^*\| = \|\xi\|^2$ for all ξ in $\mathcal{L}(X, X)$. Thus it will suffice to exhibit a suitable π.

We let $\Phi(r, s)$ denote the transition map (or resolvent or fundamental linear map) of our system; (see [DIEUDONNE, 1960] and Section 2.2). Then there is a $c > 0$ such that

$$
\|\Phi(r, s)\|^2 \|\rho(r)\| \leq c \text{ for all } (r, s) \text{ in } [r_1, t_1] \times [r_1, t_1],
$$

and

$$
\|\Phi(t_1, s)\|^2 \cdot \|\phi\| \leq c \text{ for all } s \text{ in } [r_1, t_1].
$$

Let us set $\pi = \dfrac{c}{2}(1 + t_1 - r_1)I$. Then π is a positive-definite element

---

[†] The existence of these roots is a well-known consequence of the spectral theorem (see [DUNFORD and SCHWARTZ, 1958]).

of $\mathcal{L}(X, X)$, and we claim that

$$\langle x, \pi x \rangle \geq \langle x, \pi_1(t)x \rangle$$

for all x in X and t in $(r_1, t_1]$.

To verify this claim we observe first of all that if $x \in X$ and $t \in (r_1, t_1]$, then the solution of our system starting from x at t generated by the control $0(\cdot) = 0$ is given by $\Phi(r, t)x$. It follows that

$$2J(t, x, 0(\cdot)) = \|\Phi(t_1, t)x\|_\phi^2 + \int_t^{t_1} \|\Phi(r, t)x\|_{\rho(r)}^2 \, dr,$$

$$\leq [\|\Phi(t_1, t)\|^2 \cdot \|\phi\| + \int_t^{t_1} \|\Phi(r, t)\|^2 \cdot \|\rho(r)\| \, dr] \cdot \|x\|^2,$$

$$\leq c(1 + t_1 - r_1) \cdot \|x\|^2,$$

and hence that

$$J(t, x, 0(\cdot)) \leq \left\langle x, \frac{c}{2}(1 + t_1 - r_1)Ix \right\rangle = \langle x, \pi x \rangle.$$

But for t in (r_1, t_1), Lemma (5.17) implies that

$$\langle x, \pi_1(t)x \rangle = J(t, x, u^0(\cdot)) \leq J(t, x, 0(\cdot)) \leq \langle x, \pi x \rangle.$$

Thus the corollary is established.   □

Combining Corollary (5.20) with 10.5.5 of [DIEUDONNE, 1960], we deduce that (5.12) has a solution $\pi(t)$ satisfying (5.16) which is defined and unique on $(T_1, t_1]$. Our deduction is based on the following argument: Let $\hat{s} = \mathrm{glb}\ \{s : s \in (T_1, t_1]$ and there is a positive solution $\pi_s(\cdot)$ of (5.12) defined on $[s, t_1]$ and satisfying (5.16)$\}$. If $\hat{s}$ were greater than T_1, then there would be a positive solution $\pi_1(\cdot)$ of (5.12) defined on $(\hat{s}, t_1]$, satisfying (5.16), and by the argument of the proof of Corollary (5.20), *bounded* on $(\hat{s}, t_1]$. It would follow that there was a solution of (5.12) satisfying (5.16) which was defined on $[\hat{s} - \epsilon, t_1]$ for some $\epsilon > 0$. Taking ϵ small enough to ensure that $\hat{s} - \epsilon > T_1$, we obtain a contradiction, and so $\hat{s} = T_1$ (or, if $T_1 = -\infty$, then the set of numbers s is not bounded below). Thus we have established the following result:

(5.21) **Theorem.** *For any t_0 in $(T_1, t_1]$ and any x_0 in X, the state-regulator problem has a solution. The optimal control is given as a function of the event by*

(5.22) $$u^0(t, x) = -\sigma^{-1}(t)B^*(t)\pi(t)x,$$

where $\pi(\cdot)$ is the unique solution of the Riccati equation satisfying the condition $\pi(t_1) = \phi$. The optimal trajectory is the solution of the differential equation

(5.23) $$\frac{dx(t)}{dt} = [A(t) - S(t)\pi(t)]x(t),$$

starting at x_0 at t_0. Finally, the minimum cost is given by

(5.24) $$J(t_0,\, x_0,\, u^0(\cdot)) = \tfrac{1}{2}\langle x_0,\, \pi(t_0)x_0\rangle.$$

We observe that the optimal control law leads to a time-varying feedback system in which the solution of the Riccati equation acts as a gain (see Figure 3.2).

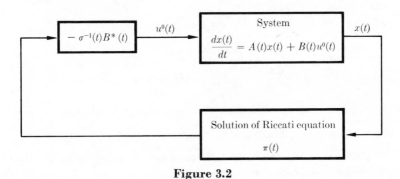

Figure 3.2

Now let us consider the output-regulator problem. In other words, let us assume that $z(t) = 0$ for all t. Then

$$K(t,\, x) = \tfrac{1}{2}\|C(t)x\|_\phi^2,$$
$$L(x,\, u,\, t) = \tfrac{1}{2}\|C(t)x\|_{\rho(t)}^2 + \tfrac{1}{2}\|u\|_{\sigma(t)}^2.$$

Suppose that we fix t_1 and that we set

$$\hat{\phi} = C^*(t_1)\phi C(t_1),$$
$$\hat{\rho}(t) = C^*(t)\rho(t)C(t),$$
$$\hat{K}(t,\, x) = \tfrac{1}{2}\|x\|_{\hat{\phi}}^2,$$
$$\hat{L}(x,\, u,\, t) = \tfrac{1}{2}\|x\|_{\hat{\rho}(t)}^2 + \tfrac{1}{2}\|u\|_{\sigma(t)}^2.$$

Then we could apply our result, Theorem (5.21), to the state-regulator problem defined by $\hat{K}$ and $\hat{L}$ in order to solve the output-regulator problem if we could show that $\hat{\phi}$ and $\hat{\rho}(t)$ were positive. But this is indeed the case in view of

(5.25) **Lemma.** *Let α be a positive element of $\mathcal{L}(X,\, X)$ and let β be any element of $\mathcal{L}(X,\, X)$. Then $\beta^*\alpha\beta$ is positive.*

Proof. Observe that $\langle x,\, \beta^*\alpha\beta x\rangle = \langle \beta x,\, \alpha\beta x\rangle$. $\qquad\square$
Thus we are led to

(5.26) **Theorem.** *For any t_0 in $(T_1,\, t_1]$ and any x_0 in X the output-*

regulator problem has a solution. An optimal control is given by

(5.27) $$u^0(t, x) = -\sigma^{-1}(t)B^*(t)\hat{\pi}(t)x,$$

where $\hat{\pi}(t)$ is the unique solution of the Riccati equation

(5.28) $$\frac{d\hat{\pi}(t)}{dt} = -\hat{\pi}^*(t)A(t) - A^*(t)\hat{\pi}(t) + \hat{\pi}^*(t)S(t)\hat{\pi}(t) - \hat{\rho}(t)$$

satisfying the condition

(5.29) $$\hat{\pi}(t_1) = \hat{\phi}.$$

The optimal trajectory is the solution of

(5.30) $$\frac{dx(t)}{dt} = [A(t) - S(t)\hat{\pi}(t)]x(t),$$

starting at x_0 at t_0. Finally, the minimum cost is given by

(5.31) $$J(t_0, x_0, u^0(\cdot)) = \tfrac{1}{2}\langle x_0, \hat{\pi}(t_0)x_0\rangle.$$

We observe that the only difference between Theorems (5.21) and (5.26) is the replacement of $\pi(t)$ by $\hat{\pi}(t)$, ϕ by $\hat{\phi}$, and $\rho(t)$ by $\hat{\rho}(t)$. We note also that the optimal control is a function of the *state* rather than of the output; this is because (loosely speaking) the state summarizes all the information needed to predict the future output, given knowledge of the control.

Now let us examine the general quadratic cost functional determined by the K and L, respectively, of Definition (5.2). We observe that the hamiltonian is now given by

(5.32) $$H(x, \lambda, u, t) = \tfrac{1}{2}\|z(t) - C(t)x\|^2_{\rho(t)} + \tfrac{1}{2}\|u\|^2_\sigma$$
$$+ \langle\lambda, A(t)x + B(t)u\rangle.$$

Since the u dependence of (5.32) and (5.5) is the same, we deduce that (5.32) is regular relative to $(T_1, T_2) \times X$ and that the H-minimal control is given by

(5.33) $$u^0(t, x, \lambda) = -\sigma^{-1}(t)B^*(t)\lambda.$$

However, in view of the dependence of our present hamiltonian on $z(t)$, we can see that (5.10) will not define a solution of the Hamilton-Jacobi equation. So let us consider the function $Z(t, x)$ defined by

(5.34) $$Z(t, x) = \tfrac{1}{2}\langle x, \pi(t)x\rangle - \langle x, \xi(t)\rangle + \varphi(t),$$

where $\pi(\cdot)$ is a continuously differentiable mapping into $\mathcal{L}(X, X)$, $\xi(\cdot)$ is a continuously differentiable mapping into X, and $\varphi(\cdot)$ is a continuously differentiable mapping into the reals $\mathbf{R}$. We have

(5.35) **Lemma.** *If $\pi(t)$ is a solution of the Riccati equation*

(5.36) $\qquad \dot{\pi}(t) = -\pi^*(t)A(t) - A^*(t)\pi(t) + \pi^*(t)S(t)\pi(t) - C^*(t)\rho(t)C(t),$

if $\xi(t)$ is a solution of the differential equation

(5.37) $\qquad \dot{\xi}(t) = -[A(t) - S(t)\pi(t)]^*\xi(t) - C^*(t)\rho(t)z(t),$

and if $\varphi(t)$ is a solution of the differential equation

(5.38) $\qquad \dot{\varphi}(t) = \frac{1}{2}[\|z(t)\|^2_{\rho(t)} - \langle \xi(t), S(t)\xi(t)\rangle],$

where $S(t) = B(t)\sigma^{-1}(t)B^*(t)$, then $Z(t, x)$ is a solution of the Hamilton-Jacobi equation.

The proof is a direct calculation, which is left to the reader.

In view of our previous work, we observe that (5.36) has a unique solution $\tilde{\pi}(t)$ on $(T_1, t_1]$ satisfying the condition

(5.39) $\qquad\qquad \tilde{\pi}(t_1) = C^*(t_1)\phi C(t_1).$

Since (5.37) is linear, it has a unique solution $\xi(t)$ on $(T_1, t_1]$ satisfying the condition

(5.40) $\qquad\qquad \xi(t_1) = C^*(t_1)\phi z(t_1)$

(see [DIEUDONNE, 1960]). Similarly, (5.38) has a unique solution $\bar{\varphi}(t)$ on $(T_1, t_1]$ satisfying the condition

(5.41) $\qquad\qquad \bar{\varphi}(t_1) = \langle z(t_1), \pi(t_1)z(t_1)\rangle.$

Thus, in view of the remarks immediately following Corollary 5.21, we can apply Theorem 5.14 to obtain the following result, which exhibits the solution of our problem:

(5.42) **Theorem.** *For any t_0 in $(T_1, t_1]$ and any x_0 in X, the linear-system quadratic cost problem has a solution. An optimal control is given by*

(5.43) $\qquad\qquad u^0(t, x) = -\sigma^{-1}(t)B^*(t)[\tilde{\pi}(t)x - \xi(t)].$

The optimal trajectory is the solution of

(5.44) $\qquad\qquad \dfrac{dx(t)}{dt} = [A(t) - S(t)\tilde{\pi}(t)]x(t) + S(t)\xi(t),$

starting at x_0 at t_0. Finally, the minimum cost is given by

(5.45) $\qquad J(t_0, x_0, u^0(\cdot)) = \frac{1}{2}\langle x_0, \tilde{\pi}(t_0)x_0\rangle - \langle x_0, \xi(t_0)\rangle + \bar{\varphi}(t_0).$

We observe that the Riccati equation for $\tilde{\pi}(\cdot)$ does not depend on $z(\cdot)$, and so the feedback structure of this optimal system is identical to the feedback structure of the optimal output-regulator system. The major distinction between the two optimal systems is the "forcing" function $\xi(t)$, which tends to "correct" the regulating features of the general optimal system. Note also that determination of $\xi(t)$ requires knowledge of $z(s)$ for $t \leq s \leq t_1$. In other words, the *present* optimal control requires knowledge of the *future* desired output. This requires predictors for the realization of the optimal system.†

We now turn to a consideration of the state-regulator problem for a time-invariant system over an infinite interval. In other words, we suppose that $A(t) = A$, $B(t) = B$, $\rho(t) = \rho$, and $\sigma(t) = \sigma$, that $X = Y$ and $C(t) = I$, that $z(t) = 0$, that $\phi = 0$, and that $T_1 = -\infty$ and $T_2 = \infty$. Thus our system equation is

$$(5.46) \qquad \dot{x}(t) = Ax(t) + Bu(t),$$

and our cost functional is of the form

$$(5.47) \qquad \tilde{J}(t_0, x_0, u(\cdot)) = \tfrac{1}{2} \int_{t_0}^{\infty} [\|x_u(t)\|_\rho^2 + \|u(t)\|_\sigma^2]\, dt.$$

However, in view of the time invariance of our system and of ρ and σ, we can readily see (using a time translation) that we may assume that $t_0 = 0$, and so we shall consider the cost functional

$$(5.48) \qquad J(x_0, u(\cdot)) = \tfrac{1}{2} \int_0^{\infty} [\|x_u(t)\|_\rho^2 + \|u(t)\|_\sigma^2]\, dt.$$

We now have

(5.49) **Lemma.** *Let $J_T(x_0, u(\cdot))$ be the cost functional defined by*

$$(5.50) \qquad J_T(x_0, u(\cdot)) = \tfrac{1}{2} \int_0^T [\|x_u(t)\|_\rho^2 + \|u(t)\|_\sigma^2]\, dt.$$

Let $\pi(\cdot\,; T)$ denote the solution of the Riccati equation corresponding to this cost functional and let $\tilde{\pi}(T) = \pi(0; T)$. Suppose that
(a) $\lim_{T \to \infty} \tilde{\pi}(T) = \hat{\pi}$ *exists.*
(b) $\hat{\pi}$ *satisfies the equation*

$$(5.51) \qquad -\hat{\pi}A - A^*\hat{\pi} + \hat{\pi}S\hat{\pi} - \rho = 0.$$

Then the infinite-interval state-regulator problem has a solution for any x_0 in X. The optimal control is given by

$$(5.52) \qquad u^0(x) = -\sigma^{-1}B^*\hat{\pi}x;$$

the optimal trajectory is the solution of the differential equation

$$(5.53) \qquad \frac{dx(t)}{dt} = (A - S\hat{\pi})x(t),$$

† These results in a slightly less general form first appeared in [KALMAN, 1963a]; many interesting examples can be found in [ATHANS and FALB, 1966], and part of the development appears in [FALB and KLEINMAN, 1966].

starting from x_0 at $t = 0$; and finally, the minimum cost is given by

(5.54) $$J(x_0, u^0(\cdot)) = \tfrac{1}{2}\langle x_0, \hat{\pi}x_0 \rangle.$$

Proof. We first assert that

(5.55) $$J(x_0, u^0(\cdot)) = \lim_{T \to \infty} J_T(x_0, u^0(\cdot)) = \tfrac{1}{2}\|x_0\|_{\hat{\pi}}^2.$$

To verify this assertion we note that the problem with cost functional $J^T(x_0, u(\cdot))$ given by

$$J^T(x_0, u(\cdot)) = \tfrac{1}{2}\|x_u(T)\|_{\hat{\pi}}^2 + J_T(x_0, u(\cdot))$$

has a solution; moreover, in view of (5.51), we can see that $u^0(\cdot)$ is the optimal control for this problem and that

(5.56) $$\tfrac{1}{2}\|x_0\|_{\hat{\pi}}^2 = J^T(x_0, u^0(\cdot)) = \tfrac{1}{2}\|x_{u^0}(T)\|_{\hat{\pi}}^2 + J_T(x_0, u^0(\cdot)).$$

It follows that

$$\tfrac{1}{2}\|x_0\|_{\hat{\pi}}^2 \geq J_T(x_0, u^0(\cdot)) \geq \tfrac{1}{2}\|x_0\|_{\hat{\pi}(T)}^2$$

and hence, by condition (a), that (5.55) holds. Now let $u(\cdot)$ be a given control and suppose that

$$J(x_0, u^0(\cdot)) - J(x_0, u(\cdot)) \geq \delta > 0.$$

However, if T is large enough, then

$$J(x_0, u^0(\cdot)) = \tfrac{1}{2}\|x_0\|_{\hat{\pi}(T)}^2 + \frac{\delta}{2},$$
$$\geq J_T(x_0, u(\cdot)) + \delta,$$

which is a contradiction. The lemma follows immediately.   □

(5.57) **Corollary.** *Under the hypotheses of the Lemma* (5.49), *the system* (5.53) *is strictly $\hat{\pi}$-stable,[†] and in particular, if $\hat{\pi}$ is positive definite, then the system* (5.53) *is strictly stable.*

Proof. Apply Equations (5.55) and (5.56).   □

We remark that if the dynamical system is finite dimensional (that is, if X and U are finite dimensional), then the conditions of Lemma (5.49) are satisfied if the system is completely controllable. For more details see [KALMAN, 1960a]. The question of the limiting behavior of solutions of the Riccati equation in the general case is still open.[‡]

[†] *Strictly $\hat{\pi}$-stable* means that $\| \cdot \|_{\hat{\pi}}$ goes to zero along the trajectories of the system.

[‡] The proof of [KALMAN, 1960a] does not apply, since an infinite-dimensional system can be completely controllable without a bound on the times required to drive the states to the origin.

3.6   The Kalman-Bucy filter

We examine the question of determining the "best" linear filter, in an expected-squared-error sense, for a "signal" generated by a linear stochastic differential equation on a Hilbert space.   Our treatment, which is based on [KALMAN and BUCY, 1961] and [FALB, 1967], is essentially heuristic; a fully rigorous development would entail concepts beyond the scope of this book.   However, modulo the use of delta functions, the arguments are accurate in the finite-dimensional case, and full details for the more general Hilbert space case can be found in [FALB, 1967].   The material in this section requires somewhat more mathematical sophistication and knowledge than the other portions of the book relating to control theory.   In particular, the reader should have some familiarity with measure theory, integration theory for Banach space–valued functions (see [DUNFORD and SCHWARTZ, 1958]), and the theory of stochastic processes (see [DOOB, 1953]).

Let $(\Omega, \mathcal{P}, \mu)$ be a probability space, with $\mathcal{P}$ as Borel field and μ as measure.   If X is a Banach space and $x(\cdot)$ is a measurable function from Ω into X, † then we shall call $x(\cdot)$ a *random variable*.   If $x(\cdot)$ is an integrable random variable, then we say that $x(\cdot)$ has an expected value (or mean) which we denote by $E\{x(\cdot)\}$,

$$E\{x(\cdot)\} = \int_{\Omega} x(\omega) \, d\mu.$$

We observe that $E\{x(\cdot)\}$ is also an element of the Banach space X.   We shall assume from now on that the random variables which we consider have expected values, and we shall often delete the explicit ω dependence of these random variables.

We need several additional notions in order to develop the Kalman-Bucy filter.   In particular, we need to define such concepts as stochastic process, covariance, orthogonal increments, and stochastic integral.   Loosely speaking, a stochastic process is a parametrized family of random variables; more precisely,

(6.1)   **Definition.**   *Let T be a real interval and let X be a Banach space. Then a function $x(t, \omega)$ from $T \times \Omega$ into X, which is measurable in the pair (t, ω) (using Lebesgue measure on T), is called a **stochastic process**.‡*

---

† Recall that a function $x(\cdot)$ defined on Ω with values in X is *measurable* if $x(\cdot)$ is essentially separably valued, and if $x^{-1}(\mathcal{O}) \in \mathcal{P}$ for each open set $\mathcal{O}$ in X (see [DUNFORD and SCHWARTZ, 1958]).

‡ This definition is somewhat more restrictive than the usual one (cf. [DOOB, 1953]) but is adequate for our purposes.

We often write $x(t)$ in place of $x(t, \omega)$ when discussing stochastic processes.

Let us now suppose that H is a Hilbert space and that h_1 and h_2 are elements of H. We then can define an element $h_1 \circ h_2$ of $\mathfrak{L}(H, H)$ by setting

(6.2) $$(h_1 \circ h_2)h = h_1\langle h_2, h\rangle$$

where $\langle\ ,\ \rangle$ is the inner product on H. We observe that if $H = \mathbf{R}_n$ (that is, H is finite dimensional), then $h_1 \circ h_2$ can be identified with the matrix $h_1 h_2'$. Moreover we have

(6.3) **Proposition.** *Let H be a Hilbert space and let ψ be the mapping of $H \oplus H$ into $\mathfrak{L}(H, H)$ defined by*

(6.4) $$\psi(h_1, h_2) = h_1 \circ h_2.$$

Then ψ has the following properties:
(a) ψ is linear in both h_1 and h_2.
(b) ψ is continuous, since

(6.5)† $$\|\psi(h_1, h_2)\| = \|h_1\| \cdot \|h_2\|.$$

(c) $\psi(h_1, h_2)^* = \psi(h_2, h_1) = h_2 \circ h_1$, and hence $\psi(h_1, h_1)$ is symmetric.
(d) If h is an element of H, then

(6.6) $$\langle h, \psi(h_1, h_2)h\rangle = \langle h, h_1\rangle \cdot \langle h, h_2\rangle.$$

Since H^* can be identified with H, this can be written as

$$h^*(h_1 \circ h_2)h = (h^* h_1) \cdot (h^* h_2),$$

where h^* is h viewed as an element of H^*.

Proof. Properties (a), (c), and (d) are obvious. As for (b), we simply note that

$$\|\psi(h_1, h_2)\| = \sup_{\|h\| \leq 1} \|(h_1 \circ h_2)h\| = \sup_{\|h\| \leq 1} \|h_1\langle h_2, h\rangle\|,$$
$$= \|h_1\| \sup_{\|h\| \leq 1} |\langle h_2, h\rangle| = \|h_1\| \cdot \|h_2\|$$

(see, for example, [KANTOROVICH and AKILOV, 1964]). □

Proposition (6.3) leads us to the following:

(6.7) **Proposition.** *Let $x(\cdot)$ and $y(\cdot)$ be H-valued random variables. Then $x(\cdot) \circ y(\cdot)$ is an $\mathfrak{L}(H, H)$-valued measurable function.*

Proof. Since $x(\cdot)$ and $y(\cdot)$ are essentially separably valued, it is clear that $x(\cdot) \circ y(\cdot)$ is essentially separably valued, so it will be enough to

---

† Observe that this also implies that the linear transformation $h_1 \circ h_2 = \psi(h_1, h_2)$ is actually an element of $\mathfrak{L}(H, H)$.

prove that if $\mathcal{O}$ is open in $\mathfrak{L}(H, H)$, then $[x(\cdot) \circ y(\cdot)]^{-1} (\mathcal{O})$ is in $\mathcal{P}$ (see [DUNFORD and SCHWARTZ, 1958]). Now,

$$\{\omega: \psi(x(\omega), y(\omega)) \in \mathcal{O}\} = \{\omega: (x(\omega), y(\omega)) \in \psi^{-1}(\mathcal{O})\}.$$

But ψ is continuous, and so $\psi^{-1}(\mathcal{O})$ is open in $H \oplus H$. Thus

$$\psi^{-1}(\mathcal{O}) = \cup(\mathcal{O}_i \times \tilde{\mathcal{O}}_i),$$

where the $\mathcal{O}_i$ and $\tilde{\mathcal{O}}_i$ are open in H. It follows that

$$\{\omega: (x(\omega), y(\omega)) \in \psi^{-1}(\mathcal{O})\} = \cup[\{\omega: x(\omega) \in \mathcal{O}_i\} \cap \{\omega: y(\omega) \in \tilde{\mathcal{O}}_i\}]$$
$$= \cup x^{-1}(\mathcal{O}_i) \cap y^{-1}(\tilde{\mathcal{O}}_i).$$

Since $x(\cdot)$ and $y(\cdot)$ are measurable, the proposition follows. $\qquad\square$
Thus we have

(6.8) **Definition.** *Let x and y be H-valued random variables.† Then the **covariance** of x and y, in symbols* cov $[x, y]$, *is the element (if it exists) of $\mathfrak{L}(H, H)$ given by*

(6.9)   cov $[x, y] = E\{x \circ y\} - E\{x\} \circ E\{y\}.$

If cov $[x, y] = 0$, *then we say that x and y are **uncorrelated**.*

(6.10) **Definition.** *Let $x(t)$ be an H-valued random process. Then $x(t)$ is called a **process with orthogonal increments** if*
   (a) $E\{\|[x(t) - x(s)] \circ [x(t) - x(s)]\|\} < \infty$ *for t, $s \in T$.*
   (b) $E\{[x(t_2) - x(s_2)] \circ [x(t_1) - x(s_1)]\} = 0$ *for $s_1 < t_1 \leq s_2 < t_2$.*
   *If $E\{[x(t) - x(s)] \circ [x(t) - x(s)]\}$ depends only on $t - s$, then $x(t)$ is said to have (**wide-sense**) **stationary increments** (cf. [DOOB, 1953]).*

We now prove a proposition about processes with orthogonal increments.

(6.11) **Proposition.** *Let $x(t)$ be a process with orthogonal increments and let $\lambda(t)$ be given by*

(6.12) $\lambda(t) = \begin{cases} E\{[x(t) - x(t_0)] \circ [x(t) - x(t_0)]\}, & t \geq t_0, \\ -E\{[x(t) - x(t_0)] \circ [x(t) - x(t_0)]\}, & t < t_0. \end{cases}$

   Then

(6.13) $E\{[x(t) - x(s)] \circ [x(t) - x(s)]\} = \lambda(t) - \lambda(s), \qquad s < t,$

   and $\lambda(t) - \lambda(s)$ is a positive element of $\mathfrak{L}(H, H)$ (in this sense $\lambda(\cdot)$ is a monotone-nondecreasing function).

   † Note that we are deleting the ω dependence in accordance with our earlier remarks.

Proof. A simple calculation, which is left to the reader, verifies (6.13). In view of Proposition (6.3c), $\lambda(t) - \lambda(s)$ is symmetric. Now, if $h \in H$, then

$$
\begin{aligned}
\langle [\lambda(t) - \lambda(s)]h, h \rangle &= \langle E\{[x(t) - x(s)] \circ [x(t) - x(s)]\}h, h \rangle, \\
&= \langle E\{[x(t) - x(s)]\langle [x(t) - x(s)], h \rangle\}, h \rangle, \\
&= E\{\langle [x(t) - x(s)], h \rangle \cdot \langle [x(t) - x(s)], h \rangle\}, \\
&\geq 0,
\end{aligned}
$$

so that $\lambda(t) - \lambda(s)$ is positive. $\square$

Now, if $x(t)$ is a process with orthogonal increments, then $\lambda(t)$ can be used to define an $\mathcal{L}(H, H)$-valued measure, and it would be possible to develop a stochastic integral based on this measure [Doob, 1953]. However, we shall be considerably less general and shall deal with processes analogous to Wiener processes. More precisely, we have

(6.14) **Definition.** *Let $w(t)$ be an H-valued random process with identically zero means (that is, $E\{w(t)\} = 0$ for all t). If $w(t)$ has orthogonal increments, if $w(t)$ is continuous almost everywhere (with respect to ω), if*

(6.15) $E\{[w(t) - w(s)] \circ [w(t) - w(s)]\} = |t - s|W,$

where W is a positive definite element of $\mathcal{L}(H, H)$ with $Wh_\alpha = \lambda_\alpha h_\alpha$ for an orthonormal basis of H (that is, W is "diagonal"), and if

(6.16) $E\{\langle w(t_2) - w(s_2), S[w(t_1) - w(s_1)]\rangle\} = 0$

whenever $s_1 < t_1 \leq s_2 < t_2$ and $S \in \mathcal{L}(H, H)$,[†] then $w(t)$ is called a **Wiener process.**

We now introduce a stochastic integral based on Definition (6.14); this will enable us to discuss stochastic differential equations and to state the filtering problem. Let us suppose that $S(t)$ is a step function from T into $\mathcal{L}(H, H)$ and that $w(t)$ is a Wiener process. In other words, there are $t_0 < t_1 < \cdots < t_n$ in T and $S_j, j = 1, \ldots, n$, in $\mathcal{L}(H, H)$ such that

$$
S(t) = \begin{cases}
0, & t < t_0, \\
S_j, & t_{j-1} \leq t < t_j, \quad j = 1, \ldots, n, \\
0, & t \geq t_n.
\end{cases}
$$

Then we define the *(stochastic) integral of $S(t)$ with respect to $dw(t)$* by setting

$$
\int S(t)\, dw(t) = \sum_{j=1}^{n} S_j[w(t_j) - w(t_{j-1})].
$$

---

† Equation (6.16) is an "independence" condition which is redundant in the finite-dimensional case.

We observe that this integral is an H-valued random variable with zero mean.† We now have

(6.17) **Lemma.** *Let $R(t)$ and $S(t)$ be $\mathfrak{L}(H, H)$-valued step functions. Then*

(6.18) $E\{[\int R(t)\, dw(t)] \circ [\int S(t)\, dw(t)]\} = \int R(t) W S^*(t)\, dt,$

and hence

(6.19) $E\{\|\int R(t)\, dw(t)\| \cdot \|\int S(t)\, dw(t)\|\} \leq \int \|R(t)\| \, \|W\|_{\mathrm{tr}} \, \|S(t)\| \, dt.$

The proof of Lemma (6.17) is a simple calculation which is left to the reader.

With Lemma (6.17) we can define a stochastic integral in a manner analogous to that used in [DOOB, 1953]. In other words, if $\Phi(t)$ is an element of $L^2(T, \mathfrak{L}(H, H))$, so that

$$\int \|\Phi(t)\|^2\, dt < \infty,$$

then [DUNFORD and SCHWARTZ, 1958] $\Phi(t)$ is a limit of step functions $S_m(t)$; that is,

(6.20) $\lim_{m \to \infty} \int \|\Phi(t) - S_m(t)\|^2\, dt = 0,$

and we can define $\int \Phi(t)\, dw(t)$ by setting

$$\int \Phi(t)\, dw(t) = \lim_{m \to \infty} \int S_m(t)\, dw(t);$$

that is,

(6.21) $\lim_{m \to \infty} E\{\|\int \Phi(t)\, dw(t) - \int S_m(t)\, dw(t)\|^2\} = 0.$

We note that (6.21) follows from (6.20) by virtue of the inequality (6.19). Moreover, $\int \Phi(t)\, dw(t)$ is also an H-valued random variable with zero mean (see [DUNFORD and SCHWARTZ, 1958]).

Now we observe that if $T = [t_0, t_1]$ (or $[t_0, \infty)$), then

(6.22) $x(t) = \int_{t_0}^{t} \Phi(s)\, dw(s), \qquad t \in [t_0, t_1]$

is an H-valued random process. With some suitable definitions (cf. [FALB, 1967]), $x(t)$ is a martingale. Although we shall not pursue this matter here, we shall show that $x(t)$ is a "wide-sense" martingale:

(6.23) **Proposition.** *Let $x(t)$ be the stochastic integral of (6.22). Then $x(t)$ is an element of $L^2(\Omega, H)$ and $x(t_2) - x(s_2)$ is orthogonal to $x(t)$*

---

† Of course, the integral is to be viewed in an almost-everywhere sense, but we shall slur over this point here.

for $t \leq s_2 < t_2$ *(in the $L^2(\Omega, H)$ sense); that is,*

(6.24) $E\{\langle x(t_2) - x(s_2), x(t)\rangle\} = 0.$

In other words, $x(t)$ is a wide-sense martingale (see [DOOB, 1953]*).*

Proof. In view of inequality (6.19), it is clear that $x(t) \in L^2(\Omega, H)$. Now we observe that (6.24) is true for step functions by virtue of (6.16) and the fact that

$$x(t_2) - x(s_2) = \int_{s_2}^{t_2} \Phi(s) \, dw(s).$$

If $S_m(s)$ is a sequence of step functions which converges to $\Phi(s)$, then

$$\left| E\left\{\left\langle \int_{s_2}^{t_2} \Phi(s) \, dw(s), \int_{t_0}^{t} \Phi(s) \, dw(s)\right\rangle - \left\langle \int_{s_2}^{t_2} S_m(s) \, dw(s), \int_{t_0}^{t} S_m(s) \, dw(s)\right\rangle\right\} \right|$$

$$\leq E\left\{\left| \left\langle \int_{s_2}^{t_2} [\Phi(s) - S_m(s)] \, dw(s), \int_{t_0}^{t} \Phi(s) \, dw(s)\right\rangle\right.\right.$$

$$\left.\left. + \left\langle \int_{s_2}^{t_2} S_m(s) \, dw(s), \int_{t_0}^{t} [\Phi(s) - S_m(s)] \, dw(s)\right\rangle\right|\right\}$$

$$\leq E\left\{\left\| \int_{s_2}^{t_2} [\Phi(s) - S_m(s)] \, dw(s) \right\| \cdot \left\| x(t) \right\|\right.$$

$$\left. + \left\| \int_{s_2}^{t_2} S_m(s) \, dw(s) \right\| \cdot \left\| x(t) - x_m(t) \right\|\right\}^{\dagger}$$

$$\leq E\{\|x(t_2) - x_m(t_2)\| \cdot \|x(t)\|\} + E\{\|x(s_2) - x_m(s_2)\| \cdot \|x(t)\|\}$$

$$+ E\{\|x_m(t_2) - x_m(s_2)\| \cdot \|x(t) - x_m(t)\|\}.$$

Applying Hölder's inequality (see [DUNFORD and SCHWARTZ, 1958]), we deduce that

$$\lim_{m \to \infty} E\{\langle x(t_2) - x(s_2), x(t)\rangle - \langle x_m(t_2) - x_m(s_2), x_m(t)\rangle\} = 0,$$

and hence, since $E\{\langle x_m(t_2) - x_m(s_2), x_m(t)\rangle\} = 0$ for all m, that (6.24) holds.   □

(6.25) **Corollary.** $E\{\|x(s)\|^2\} \leq E\{\|x(t)\|^2\}$ *if $s \leq t$.*

For a proof of the corollary see [DOOB, 1953, p. 165].

Now, we observe that if $S(t)$ is a step function, then

$$x(t) = \int_{t_0}^{t} S(s) \, dw(s)$$

is continuous in t (almost everywhere in ω). In the finite-dimensional case it is then reasonably easy to prove that the stochastic integral (6.22) is also continuous in t (almost everywhere in ω) [DOOB, 1953]. To avoid considerable complexity, we shall suppose from now on that

$\dagger x_m(t) = \int_{t_0}^{t} S_m(s) \, dw(s)$

the functions $\Phi(t)$ which we integrate stochastically are such that $x(t) = \int_{t_0}^{t} \Phi(s) \, dw(s)$ is continuous in t (almost everywhere in ω). We shall call such functions $\Phi(t)$ *s-integrable*.† If we suppose that $x(t_0)$ is an H-valued random variable with $E\{\|x(t_0)\|^2\} < \infty$, that $A(t)$ is a regulated function from T into $\mathfrak{L}(H, H)$, and that $M(t)$ is an s-integrable function from T into $\mathfrak{L}(H, H)$, then we can consider the following (stochastic) integral equation:

$$(6.26) \qquad x(t) = x(t_0) + \int_{t_0}^{t} A(s)x(s) \, ds + \int_{t_0}^{t} M(s) \, dw(s),$$

which we often write in the form

$$(6.27) \qquad\qquad dx = A(t)x \, dt + M(t) \, dw,$$

or, more intuitively, in the form

$$(6.28) \qquad\qquad \dot{x} = A(t)x + M(t)\xi(t),$$

where $\xi(t)$ is "white noise." If $M(t)$ is of the form $B(t)\sigma(t)$, then we speak of $u(t) = \sigma(t)\xi(t)$ as "white noise with covariance, cov $[u(t), u(\tau)]$, given by $\sigma(t)\sigma^*(t) \, \delta(t - \tau)$, where $\delta(\cdot)$ is the delta function"; that is,

$$(6.29) \qquad\qquad \text{cov } [u(t), u(\tau)] = \sigma(t)\sigma^*(t) \, \delta(t - \tau).$$

Of course, (6.29) is purely formal. Now, with regard to (6.26) we have

(6.30) **Theorem.** *Let $x(t_0)$ be an H-valued random variable with $E\{\|x(t_0)\|^2\} < \infty$. Let $A(t)$ be a regulated function from T into $\mathfrak{L}(H, H)$ and let $M(t)$ be an s-integrable function from T into $\mathfrak{L}(H, H)$. Let $\Phi(t, t_0)$ be the transition map of the (nonstochastic) differential equation*

$$\dot{h} = A(t)h$$

(see [DIEUDONNE, 1960] or Section 2.2) and suppose that $\Phi(t_0, s)M(s)$ is s-integrable. Then (6.26) has an (essentially unique) solution $x(t)$ which is given by

$$(6.31) \qquad x(t) = \Phi(t, t_0) \left[x(t_0) + \int_{t_0}^{t} \Phi(t_0, s)M(s) \, dw(s) \right].$$

Moreover, $E\{\|x(t)\|^2\} < \infty$.

Proof. We may suppose without loss of generality that $x(t_0) = 0$, since

$$x(t_0) + \int_{t_0}^{t} A(s)\Phi(s, t_0)x(t_0) \, ds = \Phi(t, t_0)x(t_0)$$

† Conditions for s-integrability are given in [FALB, 1968a].

(cf. [DIEUDONNE, 1960]). Thus we wish to show that

$$\int_{t_0}^{t} A(s)\Phi(s, t_0)\left[\int_{t_0}^{s}\Phi(t_0, \tau)M(\tau)\,dw(\tau)\right]ds + \int_{t_0}^{t} M(s)\,dw(s)$$
$$= \Phi(t, t_0).\int_{t_0}^{t}\Phi(t_0, s)M(s)\,dw(s).$$

Letting

$$\psi(s) = \int_{t_0}^{s}\Phi(t_0, \tau)M(\tau)\,dw(\tau)$$

and noting that $\psi(s)$ is continuous† and that $A(s)\Phi(s, t_0) = d\Phi(s, t_0)/ds$, we observe that there are partitions $t_0 = a_0 < a_1 < \cdots < a_n = t$ of $[t_0, t]$ such that

$$(6.32)\qquad \lim_{\mu\to 0}\left\|\int_{t_0}^{t}\frac{d\Phi(s, t_0)}{ds}\psi(s)\,ds - \sum_{1}^{n}[\Phi(a_i, t_0)\right.$$
$$\left. - \Phi(a_{i-1}, t_0)]\psi(\alpha_i)\right\| = 0,$$

where $a_{i-1} \le \alpha_i \le a_i$ and $\mu = \sup |a_i - a_{i-1}|$. Since

$$\sum_{1}^{n}[\Phi(a_i, t_0) - \Phi(a_{i-1}, t_0)]\psi(\alpha_i)$$
$$= \Phi(t, t_0)\psi(t) - \sum_{0}^{n-1}\Phi(a_i, t_0)[\psi(\alpha_{i+1}) - \psi(\alpha_i)],$$
$$= \Phi(t, t_0)\psi(t) - \sum_{0}^{n-1}\int_{\alpha_i}^{\alpha_{i+1}}\Phi(a_i, t_0)\Phi(t_0, s)M(s)\,dw(s),$$

and since $\|\Phi(a_i, t_0)\Phi(t_0, s) - I\| \to 0$ for each i as $\mu \to 0$, we deduce that

$$\lim_{\mu\to 0}\left\|\sum_{0}^{n-1}\int_{\alpha_i}^{\alpha_{i+1}}\Phi(a_i, t_0)\Phi(t_0, s)M(s)\,dw(s) - \int_{t_0}^{t}M(s)\,dw(s)\right\| = 0;$$

and, using (6.32), that

$$(6.33)\qquad \int_{t_0}^{t}A(s)\Phi(s, t_0)\psi(s)\,ds + \int_{t_0}^{t}m(s)\,dw(s)$$
$$= \Phi(t, t_0)\int_{t_0}^{t}\Phi(t_0, s)M(s)\,dw(s).$$

The final assertion of the theorem is obvious. □

(6.34) **Corollary.** *If $M(s)$ is a continuously differentiable function of s, then $M(s)$ is s-integrable and*

$$(6.35)\qquad \int_{t_0}^{t}M(s)\,dw(s) = M(t)w(t) - M(t_0)w(t_0) - \int_{t_0}^{t}M'(s)w(s)\,ds.$$

† From now on we omit the phrase "almost everywhere with respect to ω."

The proof of the corollary is based on an argument similar to that used in going from (6.32) to (6.33). For full details see [FALB, 1968a].

We can now state the filtering problem. We suppose that the "signal" $x(t)$ is generated by the equation

$$(6.36) \qquad dx = A(t)x\,dt + B(t)q(t)\,dw$$

and that the "observation" is generated by the equation

$$(6.37) \qquad dz = C(t)x\,dt + r(t)\,dw_1,$$

where $R(t) = r(t)w_1r^*(t)$ is positive definite, cov $[w(t), w_1(\tau)] = 0$ and $E\{\langle w(t), w_1(\tau)\rangle\} = 0$. More intuitively, we say that

$$\dot{z} = C(t)x + v(t)$$

is observed where $v(t)$ is "white noise with covariance

$$\text{cov } [v(t), v(\tau)] = R(t)\,\delta(t - \tau)."$$

Now our filtering problem can be stated as follows:

(6.38) **Filtering problem.** *Given $z(s)$ for $t_0 \le s \le t$, determine an estimate $\hat{x}(t_1|t)$ of $x(t_1)$ of the form*

$$(6.39) \qquad \hat{x}(t_1|t) = \int_{t_0}^{t} A(t_1, s)\,dz(s) \triangleq \int_{t_0}^{t} A(t_1, s)C(s)x(s)\,ds$$
$$+ \int_{t_0}^{t} A(t_1, s)r(s)\,dw_1(s),$$

where the bounded linear-transformation-valued function $A(\cdot, \cdot)$ is integrable in both arguments, with the property that

$$(6.40) \qquad E\{\langle x^*, x(t_1) - \hat{x}(t_1|t)\rangle^2\}, \qquad x^* \in H^* = H$$

is minimized for all x^*; that is, the expected squared error in estimating any linear functional of the signal is minimized.

The following theorem, which involves a Wiener-Hopf equation, gives the basic necessary and sufficient condition for $\hat{x}(t_1|t)$ to be a solution of the filtering problem.

(6.41) **Theorem (Wiener-Hopf equation).** *Let*

$$\tilde{x}(t_1|t) = x(t_1) - \hat{x}(t_1|t).$$

Then $\hat{x}(t_1|t)$ is a solution of the filtering problem if and only if

$$(6.42) \qquad E\{\tilde{x}(t_1|t) \circ [z(\sigma) - z(\tau)]\} = 0$$

for all σ and τ, $t_0 \leq \tau < \sigma \leq t$, or equivalently, if and only if

$$(6.43) \quad \text{cov } [x(t_1), z(\sigma) - z(\tau)] - \text{cov } \left[\int_{t_0}^{t} A(t_1, s) \, dz(s), z(\sigma) - z(\tau) \right] = 0$$

for all σ and τ, $t_0 \leq \tau < \sigma \leq t$.

Proof. Let x^* be an element of $H^* = H$ and let $X(x^*)$ be the space of all real random variables of the form $\langle x^*, x(\cdot) \rangle$, where $x(\cdot) \in L^2(\Omega, H)$. The inner product on $X(x^*)$ is given by $E\{\langle x^*, x(\cdot) \rangle \cdot \langle x^*, y(\cdot) \rangle\}$. We let $U(x^*)$ denote the subspace of $X(x^*)$ generated by elements of the form

$$\langle x^*, y(a) \rangle = \left\langle x^*, \int_{t_0}^{a} B(t_1, s) \, dz(s) \right\rangle, \qquad a \leq t,$$

where $B(t_1, s)$ is integrable.† By the well-known orthogonal projection lemma (see, for example, [DUNFORD and SCHWARTZ, 1958]), $\hat{x}(t_1|t)$ *will be a solution of the filtering problem if and only if $\tilde{x}(t_1|t)$ is orthogonal to $U(x^*)$ in $X(x^*)$ for every x^*.* In other words, $\hat{x}(t_1|t)$ is a solution of the filtering problem if and only if

$$E\{\langle x^*, \tilde{x}(t_1|t) \rangle \langle x^*, y(a) \rangle\} = 0$$

for all x^*, where

$$(6.44) \qquad y(a) = \int_{t_0}^{a} B(t_1, s) \, dz(s), \qquad a \leq t.$$

Let us suppose that (6.42) holds. We observe that, in view of (6.6),

$$E\{\langle x^*, \tilde{x}(t_1|t) \rangle \langle x^*, y(a) \rangle\} = x^* E\{\tilde{x}(t_1|t) \circ y(a)\} x.$$

However, $y(a)$ is given by (6.44); thus, if $B(t_1, s)$ is a step function, then

$$E\{\tilde{x}(t_1|t) \circ y(a)\} = \Sigma E\{\tilde{x}(t_1|t) \circ B_j[z(\sigma_j) - z(\tau_j)]\}$$
$$= \Sigma E\{\tilde{x}(t_1|t) \circ [z(\sigma_j) - z(\tau_j)]\} B_j^* = 0.$$

It follows (say, from Hölder's inequality) that $E\{\tilde{x}(t_1|t) \circ y(a)\} = 0$ for any $y(a)$ given by (6.44), and hence that

$$x^* E\{\tilde{x}(t_1|t) \circ y(a)\} x = 0$$

for all x^*. Therefore $\hat{x}(t_1|t)$ is a solution of the filtering problem.

On the other hand, let us suppose that $\hat{x}(t_1|t)$ is a solution of the filtering problem. Now assume that (6.42) does not hold, so that

$$(6.45) \qquad E\{\tilde{x}(t_1|t) \circ [z(\sigma) - z(\tau)]\} = \text{cov } [\tilde{x}(t_1|t), z(\sigma) - z(\tau)] \neq 0$$

† Note that $U(x^*)$ is the subspace of $X(x^*)$ generated by elements of the form $\left\langle x^*, \int_{t_0}^{t} C(t_1, s) \, dz(s) \right\rangle$ with $C(t_1, s)$ integrable, since the fact that $B(t_1, s), t_0 \leq s \leq a$, is integrable implies that $C(t_1, s) = B(t_1, s), t_0 \leq s \leq a$, and $C(t_1, s) = 0 \; a < s \leq t$, is also integrable.

for some σ and τ, $t_0 \leq \tau < \sigma \leq t$. If we define $B(t_1, s)$ by setting

$$B(t_1, s) = \begin{cases} 0, & s < \tau, \\ \text{cov } [\tilde{x}(t_1|t), z(\sigma) - z(\tau)], & \tau \leq s \leq \sigma, \\ 0, & s > \sigma, \end{cases}$$

then $B(t_1, s)$ is integrable and

$$y(t) = \int_{t_0}^{t} B(t_1, s) \, dz(s) = \text{cov } [\tilde{x}(t_1|t), z(\sigma) - z(\tau)] \cdot [z(\sigma) - z(\tau)].$$

Thus

(6.46) $E\{\langle x^*, \tilde{x}(t_1|t)\rangle\langle x^*, y(t)\rangle\}$
$$= x^* E\{\tilde{x}(t_1|t) \circ y(t)\}x,$$
$$= x^* E\{\tilde{x}(t_1|t) \circ \text{cov } [\tilde{x}(t_1|t), z(\sigma) - z(\tau)][z(\sigma) - z(\tau)]\}x,$$
$$= x^* \text{cov } [\tilde{x}(t_1|t), z(\sigma) - z(\tau) \text{ cov } [\tilde{x}(t_1|t), z(\sigma) - z(\tau)]^* x.$$

But (6.45) implies that the right-hand side of (6.46) is *not* zero for some x; this is a contradiction. The proof of the theorem is now complete. $\square$

We can use Theorem (6.41) and some properties of covariances to determine an equation for the solution of the filtering problem. We shall do this now, beginning with some lemmas.

(6.47) **Lemma.** *Let $\Phi(s)$ and $\Psi(s)$ be elements of $L^2(T, \mathcal{L}(H, H))$. Then*

(6.48) $$\text{cov } \left[\int_{t_0}^{t} \Phi(s) \, dw(s), \int_{t_0}^{t} \Psi(s) \, dw_1(s) \right] = 0$$

and

(6.49) $$\text{cov } \left[\int_{t_0}^{t} \Phi(s) \, dw(s), \int_{t_0}^{t} \Psi(s) \, dw(s) \right] = \int_{t_0}^{t} \Phi(s) W \Psi^*(s) \, ds.$$

The lemma is an easy consequence of Lemma (6.17), and a proof is given in [FALB, 1967].

(6.50) **Lemma.** *Suppose that $\text{cov } [w(t), x(t_0)] = 0$ for all t and that $K(t, s)$ is integrable with respect to $dw(s)$. Then*

(6.51) $$\text{cov } \left[\int_{t_0}^{t} K(t, s) \, dw(s), z(\sigma) - z(\tau) \right]$$
$$= \int_{t_0}^{\sigma} K(t, s) W \cdot q^*(s) B^*(s) \varphi_\sigma^*(s) \, ds,$$

where

(6.52) $$\varphi_\sigma(s) = \int_{s}^{\sigma} C(a) \Phi(a, s) \, da,$$

and $\Phi(\cdot, \cdot)$ is the transition map of the system of Theorem (6.30).

Proof. First of all we note that

$$z(\sigma) - z(\tau) = \int_\tau^\sigma C(s)x(s)\,ds + \int_\tau^\sigma r(s)\,dw_1(s),$$
$$= \int_\tau^\sigma C(s)\Phi(s, t_0)\left[x_0 + \int_{t_0}^s \Phi(t_0, a)B(a)q(a)\,dw(a)\right]ds$$
$$+ \int_\tau^\sigma r(s)\,dw_1(s).$$

Since cov $[w(t), x_0] = 0$, we observe, in view of (6.48), that

$$\text{cov}\left[\int_{t_0}^t K(t, s)\,dw(s), z(\sigma) - z(\tau)\right]$$
$$= \text{cov}\left\{\int_{t_0}^t K(t, s)\,dw(s), \int_\tau^\sigma C(s)\left[\int_{t_0}^s \Phi(s, a)B(a)q(a)\,dw(a)\right]ds\right\},$$

and hence, letting $\tilde{C}(s)$ be given by

$$(6.53) \qquad \tilde{C}(s) = \begin{cases} 0, & s > \sigma, \\ C(s), & \tau \le s \le \sigma, \\ 0, & s < \tau, \end{cases}$$

that

$$\text{cov}\left[\int_{t_0}^t K(t, s)\,dw(s), z(\sigma) - z(\tau)\right]$$
$$= \text{cov}\left\{\int_{t_0}^t K(t, s)\,dw(s), \int_{t_0}^t\left[\int_a^t \tilde{C}(s)\Phi(s, a)\,ds\right]B(a)q(a)\,dw(a)\right\},$$

where the interchange of integrations is justified by a suitable version of Fubini's theorem (see [DUNFORD and SCHWARTZ, 1958; DOOB, 1953; FALB, 1967]). It then follows from (6.49) and (6.53) that

$$\text{cov}\left[\int_{t_0}^t K(t, s)\,dw(s), z(\sigma) - z(\tau)\right] = \int_{t_0}^\sigma K(t, s)Wq^*(s)B^*(s)\varphi_\sigma^*(s)\,ds.$$

Thus the lemma is established. $\qquad\square$

Now, if $a(t)$ and $b(t)$ are random processes with cov $[a(t), b(t)] = h(t)$ (a "sure" function), then it is natural to set

$$\frac{d}{dt}\text{cov}\,[a(t), b(t)] = \frac{d}{dt}h(t) = \dot{h}(t)$$

when $\dot{h}(t)$ exists. With this in mind, we have

(6.54) Corollary. *If $K(t, s)$ is continuously differentiable with respect to t, then for $\sigma < t$*

$$(6.55) \qquad \frac{d}{dt}\text{cov}\left[\int_{t_0}^t K(t, s)\,dw(s), z(\sigma) - z(\tau)\right]$$
$$= \text{cov}\left[\int_{t_0}^t \frac{\partial K(t, s)}{\partial t}\,dw(s), z(\sigma) - z(\tau)\right].$$

(6.56)   **Corollary.**   *Suppose that* cov $[w_1(t),\ x_0] = 0$ *for all* t. *Then for* $\sigma < t$

(6.57) $\dfrac{d}{dt}$ cov $[x(t),\ z(\sigma) - z(\tau)] =$ cov $[A(t)x(t),\ z(\sigma) - z(\tau)].$

Proof.   Let $K(t,\ s) = \Phi(t,\ s)B(s)q(s)$ and apply Corollary (6.54). $\square$

We shall suppose from now on that cov $[w(t),\ x_0] = 0$ and that cov $[w_1(t),\ x_0] = 0$. We now have

(6.58)   **Lemma.**   *Suppose that $L(t,\ s)$ is integrable with respect to $dz(s)$. Then*

(6.59)   cov $\left[\displaystyle\int_{t_0}^{t} L(t,\ s)\ dz(s),\ z(\sigma) - z(\tau) \right]$

$$= \int_{t_0}^{\sigma} \psi(t,\ s)B(s)q(s)Wq^*(s)B^*(s)\varphi_\sigma^*(s)\ ds$$

$$+ \int_{\tau}^{\sigma} L(t,\ s)r(s)W_1r^*(s)\ ds$$

$$+ \int_{t_0}^{t} L(t,\ s)C(s)\Phi(s,\ t_0)\ ds \text{ cov } [x_0,\ x_0] \int_{\tau}^{\sigma} \Phi^*(s,\ t_0)C^*(s)\ ds,$$

where

(6.60) $\psi(t,\ s) = \displaystyle\int_{s}^{t} L(t,\ b)C(b)\Phi(b,\ s)\ db,$

and $\varphi_\sigma(s)$ is given by (6.52).

Proof.   Noting that

$$\int_{t_0}^{t} L(t,\ s)\ dz(s) = \int_{t_0}^{t} L(t,\ s)C(s)x(s)\ ds + \int_{t_0}^{t} L(t,\ s)r(s)\ dw_1(s),$$

$$= \int_{t_0}^{t} L(t,\ s)C(s) \left[\Phi(s,\ t_0)x_0 + \int_{t_0}^{s} \Phi(s,\ a)B(a)q(a)\ dw(a) \right] ds$$

$$+ \int_{t_0}^{t} L(t,\ s)r(s)\ dw_1(s),$$

we deduce from (6.48), the "independence" of x_0 of w and w_1, and a suitable Fubini theorem [FALB, 1967] that

cov $\left[\displaystyle\int_{t_0}^{t} L(t,\ s)\ dz(s),\ z(\sigma) - z(\tau) \right]$

$$= \text{cov} \left[\int_{t_0}^{t} \psi(t,\ s)B(s)q(s)\ dw(s),\ z(\sigma) - z(\tau) \right]$$

$$+ \int_{\tau}^{\sigma} L(t,\ s)r(s)W_1r^*(s)\ ds$$

$$+ \int_{t_0}^{t} L(t,\ s)C(s)\Phi(s,\ t_0)\ ds \text{ cov } [x_0,\ x_0] \int_{\tau}^{\sigma} \Phi^*(s,\ t_0)C^*(s)\ ds.$$

The lemma follows from Lemma (6.50). $\square$

(6.61) **Corollary.** *If $L(t, s)$ is continuously differentiable with respect to t, then for $\sigma < t$*

$$(6.62) \quad \frac{d}{dt} \operatorname{cov} \left[\int_{t_0}^{t} L(t, s) \, dz(s), z(\sigma) - z(\tau) \right]$$
$$= \operatorname{cov} \left[\int_{t_0}^{t} \frac{\partial L}{\partial t} (t, s) \, dz(s), z(\sigma) - z(\tau) \right]$$
$$+ \operatorname{cov} \left[L(t, t) C(t) x(t), z(\sigma) - z(\tau) \right].$$

The proof is a simple calculation which is left to the reader.
These lemmas and corollaries lead us to the following theorem:

(6.63) **Theorem.** *Suppose that there is a solution of the filtering problem of the form*

$$(6.64) \qquad \hat{x}(t|t) = \int_{t_0}^{t} L(t, s) \, dz(s),$$

with $L(t, s)$ continuously differentiable with respect to t.

Then

$$(6.65) \qquad \frac{\partial L}{\partial t} (t, s) = A(t) L(t, s) - L(t, t) C(t) L(t, s)$$

for $t_0 \leq s \leq t$.†

Proof. Since $\hat{x}(t|t)$ is a solution of the filtering problem, we have, by virtue of (6.43),

$$\frac{d}{dt} \operatorname{cov} [x(t), z(\sigma) - z(\tau)] \equiv \frac{d}{dt} \operatorname{cov} \left[\int_{t_0}^{t} L(t, s) \, dz(s), z(\sigma) - z(\tau) \right].$$

It follows, from (6.57), (6.62), and (6.43), that

$$\operatorname{cov} \left[A(t) \int_{t_0}^{t} L(t, s) \, dz(s), z(\sigma) - z(\tau) \right]$$
$$\equiv \operatorname{cov} \left[\int_{t_0}^{t} \frac{\partial L}{\partial t} (t, s) \, dz(s), z(\sigma) - z(\tau) \right]$$
$$+ \operatorname{cov} \left[L(t, t) C(t) \int_{t_0}^{t} L(t, s) \, dz(s), z(\sigma) - z(\tau) \right],$$

or equivalently, that

$$(6.66) \quad \operatorname{cov} \left\{ \int_{t_0}^{t} \left[A(t) L(t, s) - \frac{\partial L}{\partial t} (t, s) \right. \right.$$
$$\left. \left. - L(t, t) C(t) L(t, s) \right] dz(s), z(\sigma) - z(\tau) \right\} = 0.$$

† Note that we are assuming that $R(t) = r(t) W_1 r^*(t)$ is positive definite.

Setting

$$\Delta(t, s) = A(t)L(t, s) - \frac{\partial L}{\partial t}(t, s) - L(t, t)C(t)L(t, s),$$

we observe that (6.66) implies that

$$\int_{t_0}^{t} [L(t, s) + \Delta(t, s)] \, dz(s) = \hat{y}(t|t)$$

also satisfies (6.43) and hence is an optimal filter. As a consequence of the well-known orthogonal projection lemma,

$$E\{\langle x^*, \hat{y}(t|t) - \hat{x}(t|t)\rangle^2\} = 0$$

for all x^*. In other words,

$$x^* \operatorname{cov} \left[\int_{t_0}^{t} \Delta(t, s) \, dz(s), \int_{t_0}^{t} \Delta(t, s) \, dz(s) \right] x = 0$$

for all x. But

$$\operatorname{cov} \left[\int_{t_0}^{t} \Delta(t, s) \, dz(s), \int_{t_0}^{t} \Delta(t, s) \, dz(s) \right]$$
$$= \operatorname{cov} \left[\int_{t_0}^{t} \Delta(t, s) \, C(s)x(s) \, ds, \int_{t_0}^{t} \Delta(t, s) \, C(s)x(s) \, ds \right]$$
$$+ \int_{t_0}^{t} \Delta(t, s) \, R(s) \, \Delta^*(t, s) \, ds.$$

Since $R(s)$ is positive definite for all s, we immediately conclude that $\Delta(t, s) = 0$.

(6.67)   **Corollary.** *Under the hypotheses of the theorem, $\hat{x}(t)$† satisfies the stochastic differential equation*

(6.68) $d\hat{x} = [A(t) - K(t)C(t)]\hat{x} \, dt + K(t)C(t)x(t) \, dt + K(t)r(t) \, dw_1,$

where $K(t) = L(t, t)$.

Proof. We know that

$$\hat{x}(t) = \int_{t_0}^{t} L(t, s) \, dz(s) = \int_{t_0}^{t} L(t, s)C(s)x(s) \, ds + \int_{t_0}^{t} L(t, s)r(s) \, dw_1(s).$$

Now, let us observe that

$$\int_{t_0}^{t} [A(s) - K(s)C(s)]x(s) \, ds$$
$$= \int_{t_0}^{t} [A(s) - K(s)C(s)] \int_{t_0}^{s} L(s, a)C(a)x(a) \, da \, ds$$
$$+ \int_{t_0}^{t} [A(s) - K(s)C(s)] \int_{t_0}^{s} L(s, a)r(a) \, dw_1(a) \, ds.$$

By the various versions of the Fubini theorem [DUNFORD and SCHWARTZ,

---

† For simplicity we drop the $|t$.

1958; Doob, 1953; Falb, 1967], we have

$$\int_{t_0}^t [A(s) - K(s)C(s)] \int_{t_0}^s L(s, a)C(a)x(a) \, da \, ds$$
$$= \int_{t_0}^t \left\{ \int_a^t [A(s) - K(s)C(s)]L(s, a) \, ds \right\} C(a)x(a) \, da,$$

$$\int_{t_0}^t [A(s) - K(s)C(s)] \int_{t_0}^s L(s, a)r(a) \, dw_1(a) \, ds$$
$$= \int_{t_0}^t \left\{ \int_a^t [A(s) - K(s)C(s)]L(s, a) \, ds \right\} r(a) \, dw_1(a),$$

and hence, in view of (6.66),

$$\int_{t_0}^t [A(s) - K(s)C(s)]\hat{x}(s) \, d = \int_{t_0}^t \left[\int_a^t \frac{\partial L}{\partial s} (s, a) \, d \right] dz(a),$$
$$= \int_{t_0}^t [L(t, a) - L(a, a)] \, dz(a),$$
$$= \hat{x}(t) - \int_{t_0}^t K(s) \, dz.$$

The corollary follows immediately. □

(6.69) **Corollary.** *Under the hypotheses of the theorem, $\tilde{x}(t)$ satisfies the stochastic differential equation*

(6.70) $d\tilde{x} = [A(t) - K(t)C(t)]\tilde{x} \, dt + B(t)q(t) \, dw - K(t)r(t) \, dw_1.$

Corollaries (6.67) and (6.69) are at the heart of the development in [Kalman and Bucy, 1961]. Continuing in the same vein, we can derive the remaining results (for full details see [Falb, 1967]). As a typical example, we have the following

(6.71) **Theorem.** *Suppose that the conditions of Theorem (6.63) are satisfied. Then*

(6.72) $K(t) = P(t)C^*(t)R^{-1}(t),$

where $P(t)$ is a solution of the Riccati type equation

(6.73) $\dot{P} = A(t)P + PA^*(t) - PC^*(t)R^{-1}(t)C(t)P + B(t)Q(t)B^*(t),$
with $P(t_0) = P_0 = \text{cov}[x_0, x_0]$.

Proof. Let

$$y(s) = \int_{t_0}^s C(a)x(a) \, da.$$

Then a simple calculation, which is left to the reader, yields for $\sigma < t$

(6.74) $\frac{d}{d\sigma} \text{cov}[\tilde{x}(t), y(\sigma) - y(\tau)]$
$$= [\psi(t, t_0) - \Phi(t, t_0)]P_0\Phi^*(\sigma, t_0)C^*(\sigma) + \int_{t_0}^t [\Phi(t, s)$$
$$- \psi(t, s)]B(s)Q(s)B^*(s)\Phi^*(t_0, s) \, ds \, \Phi^*(\sigma, t_0)C^*(\sigma)$$
$$= L(t, \sigma)R(\sigma),$$

where $\psi(t, s)$ is given by (6.60). By our basic assumptions, the terms in (6.74) are continuous functions of σ, and so, taking limits as $\sigma \to t$, we deduce that (6.72) holds, with

$$(6.75) \quad P(t) = \left\{ [\Phi(t, t_0) - \psi(t, t_0)]P_0 \right.$$
$$\left. + \int_{t_0}^{t} [\Phi(t, s) - \psi(t, s)]B(s)Q(s)B^*(s)\Phi^*(t_0, s)\, ds \right\} \Phi^*(t, t_0).$$

Clearly, $P(t_0) = P_0$. The differentiation of (6.75), which leads to (6.73), is left to the reader (note that (6.65) and (6.72) are used frequently in the calculations). $\qquad\square$

Under the hypotheses of Theorem (6.63) the optimal filter may be viewed as generating an optimal regulator for a dynamical system which is "dual" to the stochastic system (6.36) and (6.37); that is, time is reversed, adjoints are taken, and "sure" functions replace the noise terms. Several examples are worked out in [KALMAN and BUCY, 1961], and a detailed development appears in [FALB, 1967] and [FALB, 1968a]. See also Section 2.6.

4 Necessary conditions for optimality

In this chapter we shall discuss necessary conditions for optimality which are analogous to the familiar Euler equations of the calculus of variations (Section 4.1), study the maximum principle of Pontryagin (Section 4.2), present a simple theorem on the existence of optimal controls (Section 4.3), and make some general comments on necessary conditions for more general problems. In Section 4.2 we use a perturbation approach, and the hamiltonian functional again plays a crucial role. The maximum principle of Pontryagin, which represents a considerable strengthening of the necessary conditions in the case of a finite-dimensional dynamical system, is quite useful in control-system design, as we shall see in Chapter 5. The simple existence theorem given in Section 4.3 is based on the notions of attainability and lower semicontinuity. Section 4.4 gives a brief indication of the various beautiful and very powerful generalizations of the maximum principle due to [HALKIN, 1967; NEUSTADT, 1966–67; CANON, CULLUM, and POLAK, 1967] (see Appendix A also).

4.1   Necessary conditions for optimality

Let us suppose that Σ is our smooth dynamical system and that $f(x, u, t)$ is the generator of Σ. We shall assume that X is itself a Banach space and that Ω is an open subset of the space of all regulated functions from (T_1, T_2) into $\mathscr{B}_U$. Let us for the moment fix a t_0 in (T_1, T_2), an x_0 in X, and a $\hat{u}(\cdot)$ in Ω. We denote by $\hat{x}(\cdot)$ the trajectory of our system generated by $\hat{u}(\cdot)$ and starting at x_0 at t_0; that is, $\hat{x}(t) = \varphi(t; t_0, x_0, \hat{u}(\cdot))$. We let $|\epsilon| > 0$ be such that the sphere $S(\hat{u}(\cdot), |\epsilon|) \subset \Omega$ and we let $h(\cdot) \in S(0(\cdot), 1)$, so that $\hat{u}(\cdot) + \epsilon h(\cdot) \in \Omega$. Letting $x_h(\cdot)$ denote the trajectory of our system generated by $\hat{u}(\cdot) + \epsilon h(\cdot)$, we observe that

(1.1) $$\lim_{\epsilon \to 0} \|x_h(t) - \hat{x}(t)\| = 0$$

uniformly in t, since the transition function of Σ is assumed continuous in all its arguments. We have

$$x_h(t) = x_0 + \int_{t_0}^t f(x_h(\tau), \hat{u}(\tau) + \epsilon h(\tau), \tau) \, d\tau,$$

$$\hat{x}(t) = x_0 + \int_{t_0}^t f(\hat{x}(\tau), \hat{u}(\tau), \tau) \, d\tau,$$

so that

(1.2) $$x_h(t) - \hat{x}(t) = \int_{t_0}^t [f(x_h(\tau), \hat{u}(\tau) + \epsilon h(\tau), \tau) - f(\hat{x}(\tau), \hat{u}(\tau), \tau)] \, d\tau.$$

Now, assuming that f is continuously differentiable† with respect to x and u, we may write‡

$$f(x_h(\tau), \hat{u}(\tau) + \epsilon h(\tau), \tau) - f(\hat{x}(\tau), \hat{u}(\tau), \tau)$$
$$= \frac{\partial f}{\partial x}\Big|_{\wedge} [x_h(\tau) - x(\tau)] + \frac{\partial f}{\partial u}\Big|_{\wedge} \epsilon h(\tau) + o(\epsilon, \tau),$$

where $o(\epsilon, \tau)$ is of order greater than ϵ; that is,

$$\lim_{\epsilon \to 0} \frac{\|o(\epsilon, \tau)\|}{\epsilon} = 0$$

uniformly in τ. However, it follows from (1.2) that

(1.3) $$x_h(t) - \hat{x}(t) = \int_{t_0}^t \left\{ \frac{\partial f}{\partial x}\Big|_{\wedge} [x_h(\tau) - \hat{x}(\tau)] + \frac{\partial f}{\partial u}\Big|_{\wedge} \epsilon h(\tau) \right\} d\tau$$
$$+ \int_{t_0}^t o(\epsilon, \tau) \, d\tau,$$

---

† Having regulated derivatives is sufficient.

‡ The $|_{\wedge}$ indicates that the quantity in question is evaluated along the $\wedge$ trajectory, that is, at $(\hat{x}(t), \hat{u}(t), t)$. Also, recall that $\partial f/\partial x \in \mathscr{L}(X, X)$ and $\partial f/\partial u \in \mathscr{L}(\mathscr{B}_U, X)$.

and hence, by virtue of (1.1), that

$$\lim_{\epsilon \to 0} \frac{\left\| \int_{t_0}^t o(\epsilon, \tau)\, d\tau \right\|}{\epsilon} = 0$$

uniformly in t. Moreover, in view of (1.1) we may write

(1.4) $$x_h(t) - \hat{x}(t) = \epsilon\psi(t) + o(\epsilon, t),$$

where $o(\epsilon, t)$ is of order greater than ϵ uniformly in t. (*Caution:* The notation $o(\epsilon)$, $o(\epsilon, t)$, etc., is used only to indicate the order of terms, and it should *not* be assumed, for example, that the $o(\epsilon, t)$ in (1.4) and the $o(\epsilon, \tau)$ in (1.3) are equal.) It follows that

$$\epsilon\psi(t) + o(\epsilon, t) = \epsilon \int_{t_0}^t \left[\frac{\partial f}{\partial x} \Big|_{\wedge} \psi(\tau) + \frac{\partial f}{\partial u} \Big|_{\wedge} h(\tau) \right] d\tau + \int_{t_0}^t o(\epsilon, \tau)\, d\tau,$$

and hence, by division and passing to the limit as $\epsilon \to 0$, that

(1.5) $$\frac{d\psi(t)}{dt} = \frac{\partial f}{\partial x} \Big|_{\wedge} \psi(t) + \frac{\partial f}{\partial u} \Big|_{\wedge} h(t),$$

$$\psi(t_0) = 0.$$

The *linear* equation (1.5) is called the *perturbation equation* about the trajectory $\hat{x}(\cdot)$. For convenience we shall often write

$$x_h(\cdot) = \hat{x}(\cdot) + \epsilon\psi(\cdot) + o(\epsilon)$$

in place of (1.4), and we shall frequently omit the detailed verification of the manipulations which we make with $o(\epsilon)$ terms.

Let us now turn our attention to the control problem. We shall assume for the moment that our target set is of the form $\{t_1\} \times X$, where t_1 is fixed. Consequently, we shall also suppose that the terminal-cost function depends only on x. We shall further require that K be continuously differentiable and that L be continuously differentiable with respect to x and u.† It will follow that the hamiltonian functional is continuously differentiable in x and u. In particular, since

$$H(x, \lambda, u, t) = L(x, u, t) + \langle \lambda, f(x, u, t) \rangle$$

(recall that, in accordance with the notation previously introduced in Chapter 3, $\langle \lambda, f(x, u, t) \rangle$ denotes the operation of $\lambda \in X^*$ on $f(x, u, t)$ and is not a scalar product) we have

$$\frac{\partial H}{\partial x} = \frac{\partial L}{\partial x} + \left(\frac{\partial f}{\partial x} \right)^* \lambda,$$

$$\frac{\partial H}{\partial u} = \frac{\partial L}{\partial u} + \left(\frac{\partial f}{\partial u} \right)^* \lambda,$$

---

† Again, having regulated derivatives would be sufficient, since all calculations could be done with integrals.

and, viewing X as embedded in X^{**},

$$\frac{\partial H}{\partial \lambda} = f(x, u, t).$$

We observe that since the norm of the adjoint of a bounded linear transformation is the same as the norm of the transformation (see [DUNFORD and SCHWARTZ, 1958]), the mappings

$$(x, u, t) \rightarrow \left(\frac{\partial f}{\partial x}\right)^* (x, u, t),$$

$$(x, u, t) \rightarrow \left(\frac{\partial f}{\partial u}\right)^* (x, u, t)$$

are continuous (or regulated).   We now have

(1.6)   **Lemma.**   *Let $\hat{u}(\cdot)$ be an element of Ω and let $\hat{u}(\cdot) + \epsilon h(\cdot)$ be a perturbation of $\hat{u}(\cdot)$.   Then*

$$J(t_0, x_0, \hat{u} + \epsilon h) - J(t_0, x_0, \hat{u}) = \epsilon \int_{t_0}^{t_1} \left\langle \frac{\partial H}{\partial u}\Big|_{\wedge}, h(\tau) \right\rangle d\tau + o(\epsilon).$$

Proof.   Since K is differentiable and since $\partial H/\partial x$ is continuous, the linear differential equation

(1.7) $$\dot{\lambda}(t) = \left(-\frac{\partial f}{\partial x}\Big|_{\wedge}\right)^* \lambda(t) - \frac{\partial L}{\partial x}\Big|_{\wedge}$$

has a unique solution $\hat{\lambda}(t)$ satisfying the boundary equation

(1.8) $$\hat{\lambda}(t_1) = \frac{\partial K}{\partial x}(\hat{x}(t_1)).$$

We shall call $\hat{\lambda}(\cdot)$ a "costate trajectory"† corresponding to $\hat{x}(\cdot)$ and $\hat{u}(\cdot)$. Since both $x_h(\cdot)$ and $\hat{x}(\cdot)$ are trajectories of our system, we have

$$\int_{t_0}^{t_1} \left[\langle \hat{\lambda}(\tau), f(\hat{x}(\tau), \hat{u}(\tau), \tau) \rangle - \left\langle \hat{\lambda}(\tau), \frac{d\hat{x}(\tau)}{d\tau} \right\rangle \right] d\tau = 0,$$

$$\int_{t_0}^{t_1} \left[\langle \hat{\lambda}(\tau), f(x_h(\tau), \hat{u}(\tau) + \epsilon h(\tau), \tau) \rangle - \left\langle \hat{\lambda}(\tau), \frac{dx_h(\tau)}{d\tau} \right\rangle \right] d\tau = 0,$$

and it follows that

$$J(t_0, x_0, \hat{u} + \epsilon h) = K(x_h(t_1)) + \int_{t_0}^{t_1} H(x_h(t), \hat{\lambda}(t), \hat{u}(t) + \epsilon h(t), t) \, dt$$
$$+ \int_{t_0}^{t_1} \langle \hat{\lambda}(t), \dot{x}_h(t) \rangle \, dt,$$

$$J(t_0, x_0, \hat{u}) = K(\hat{x}(t_1)) + \int_{t_0}^{t_1} H(\hat{x}(t), \hat{\lambda}(t), \hat{u}(t), t) \, dt + \int_{t_0}^{t_1} \langle \hat{\lambda}(t), \dot{\hat{x}}(t) \rangle \, dt.$$

† The motivation for this terminology will be discussed presently.

Now, if we use integration by parts (see [DUNFORD and SCHWARTZ, 1958]),

$$\int_{t_0}^{t_1} \langle \dot{\hat{\lambda}}(t), \, \dot{x}_h(t) \rangle \, dt = \langle \hat{\lambda}(t_1), \, x_h(t_1) \rangle - \langle \hat{\lambda}(t_0), \, x_h(t_0) \rangle - \int_{t_0}^{t_1} \langle \dot{\hat{\lambda}}(t), \, x_h(t) \rangle \, dt,$$

$$\int_{t_0}^{t_1} \langle \hat{\lambda}(t), \, \dot{\hat{x}}(t) \rangle \, dt = \langle \hat{\lambda}(t_1), \, \hat{x}(t_1) \rangle - \langle \hat{\lambda}(t_0), \, \hat{x}(t_0) \rangle - \int_{t_0}^{t_1} \langle \dot{\hat{\lambda}}(t), \, \hat{x}(t) \rangle \, dt.$$

From these equations, the perturbation equation, and the continuous differentiability of K and H, we can readily see that

$$J(t_0, x_0, \hat{u} + \epsilon h) - J(t_0, x_0, \hat{u})$$

$$= \epsilon \left\langle \frac{\partial K}{\partial x}(\hat{x}(t_1)), \, \psi(t_1) \right\rangle + \epsilon \int_{t_0}^{t_1} \left\langle \frac{\partial H}{\partial u} \Big|_{\wedge}, \, h(t) \right\rangle$$

$$- \epsilon \int_{t_0}^{t_1} \left\langle \frac{\partial H}{\partial x} \Big|_{\wedge}, \, \psi(t) \right\rangle dt - \epsilon \langle \dot{\hat{\lambda}}(t_1), \, \psi(t_1) \rangle - \epsilon \int_{t_0}^{t_1} \langle \dot{\hat{\lambda}}(t), \, \psi(t) \rangle \, dt + o(\epsilon).$$

Since $\hat{\lambda}(\cdot)$ satisfies (1.7) and (1.8), the lemma is established. $\qquad \square$

(1.9) **Corollary.** *If $u^0(\cdot)$ is an optimal control, then*

$$\int_{t_0}^{t_1} \left\langle \frac{\partial H}{\partial u} \Big|_{0}, \, h(t) \right\rangle dt = 0$$

for all $h(\cdot) \in S(0(\cdot), 1)$ (a fortiori for all bounded $h(\cdot)$).

Proof. If $u^0(\cdot)$ is optimal, then

$$J(t_0, x_0, u^0 + \epsilon h) - J(t_0, x_0, u^0) = \epsilon \int_{t_0}^{t_1} \left\langle \frac{\partial H}{\partial u} \Big|_{0}, \, h(t) \right\rangle dt + o(\epsilon) \geq 0.$$

As ϵ can be positive or negative and the $o(\epsilon)$ term may be neglected, the corollary follows. $\qquad \square$

During the course of the proof of Lemma (1.6) we introduced the term "costate trajectory." In essence, this described the solution of a differential equation in X^* which was "adjoint" to the perturbation equation (1.5). More formally, we have

(1.10) **Definition.** *If $\hat{u}(\cdot)$ is an element of Ω and $\hat{x}(\cdot)$ is the trajectory of Σ generated by $\hat{u}(\cdot)$, then any solution of the differential equation (in X^*)*

$$\dot{\lambda} = \left(-\frac{\partial f}{\partial x} \Big|_{\wedge} \right)^* \lambda - \frac{\partial L}{\partial x} \Big|_{\wedge}$$

shall be called a **costate** *(or* **adjoint**) **trajectory** *corresponding to $\hat{x}(\cdot)$ and $\hat{u}(\cdot)$.*

In line with this terminology, we shall often speak of elements of X^* as "costates," and we shall call the λ-argument of the hamiltonian $H(x, \lambda, u, t)$ the "costate" (or "adjoint") variable.

We now prove an important lemma:

(1.11) **Lemma.** *Let V be a Banach space and let $v(\cdot)$ be a regulated function from $[t_0, t_1]$ into V^* such that*
(a) *v is continuous from the right (that is, $v(t+) = v(t)$),*
(b) *$\int_{t_0}^{t_1} \langle v(t), k(t) \rangle \, dt = 0$ for all regulated mappings $k(\cdot)$ of $[t_0, t_1]$ into V.*
Then $v(\cdot) = 0$ on $[t_0, t_1]$.

Proof. Let τ be an element of $[t_0, t_1)$ and suppose that $v(\tau) \neq 0$. Then there is an element k of v such that

$$\langle v(\tau), k \rangle > \delta > 0$$

for some δ. The mapping $\varphi(\cdot)$ of $[t_0, t_1)$ into $\mathbf{R}$ defined by

$$\varphi(t) = \langle v(t), k \rangle$$

is regulated and continuous from the right. It follows that there is an $\epsilon > 0$ such that

$$\varphi(t) \geq \delta > 0 \qquad \text{for } t \text{ in } [\tau, \tau + \epsilon).$$

Now let $k(\cdot)$ be given by

$$k(t) = \begin{cases} 0, & t \notin [\tau, \tau + \epsilon), \\ k, & t \in [\tau, \tau + \epsilon). \end{cases}$$

Then $k(\cdot)$ is regulated and

$$\int_{t_0}^{t_1} \langle v(t), k(t) \rangle \, dt = \int_{\tau}^{\tau+\epsilon} \langle v(t), k \rangle \, dt \geq \epsilon\delta > 0.$$

But this contradicts hypothesis (b). Thus $v(\tau) = 0$ for τ in $[t_0, t_1)$, and the lemma is established.   □

This lemma and Corollary (1.9) combine to yield the following theorem:

(1.12) **Theorem.** *Suppose that $\mathcal{B}_U$ is reflexive (in particular, that $\mathcal{B}_U$ is a Hilbert space) and that $u^0(\cdot)$ is an optimal control. Then*
(a) *There is a costate $\lambda^0(\cdot)$ corresponding to $x^0(\cdot)$ and $u^0(\cdot)$.*
(b) *$\dfrac{\partial H}{\partial u} (x^0(\cdot), \lambda^0(\cdot), u^0(\cdot), \cdot) = 0$ almost everywhere on $[t_0, t_1]$.*

Thus for a control to be optimal it is necessary that the hamiltonian have an extremum with respect to u when evaluated along the optimal trajectory and a suitable corresponding costate trajectory.

Now let us suppose that our target set is of the form $\{t_1\} \times \{x_1\}$, where t_1 and x_1 are fixed, and that $K = 0$, that is, there is no terminal cost. We then could attempt to find a continuously differentiable function $\hat{K}(x)$ with the property that the problem with target set $\{t_1\} \times X$ and terminal-cost function $\hat{K}$ was equivalent to our original problem in the sense that a control was optimal for one if and only if it were optimal for the other. If such a $\hat{K}$ existed, then the theorem could be applied. An alternative approach is based upon the idea of showing that in a suitable neighborhood of the optimal control $u^0(\cdot)$ there are "enough" perturbations to ensure that the hamiltonian is actually minimized as a function of u. This is the approach used by [PONTRYAGIN et al., 1961] in their proof of the maximum principle. We shall explicate some of these points in the next section.

We note now that if the target set S is arbitrary and if $u^0(\cdot)$ is an optimal control transferring (t_0, x_0) to S with t_1 as transfer time, then $u^0(\cdot)$ will be optimal for the problem with target set $\{t_1\} \times \{x^0(t_1)\}$ and no terminal cost, so that suitable necessary conditions can also be obtained in the general case. Interesting general results have recently been obtained by [GAMKRELIDZE, 1965; NEUSTADT, 1965, 1966–67; HALKIN, 1967; CANON, CULLUM, and POLAK, 1967] (see Appendix A also).

4.2 The maximum principle of Pontryagin

We now turn our attention to necessary conditions for control problems involving finite-dimensional dynamical systems. If Σ is our dynamical system, then we shall suppose that

1. The domain (T_1, T_2) of Σ is all of $\mathbf{R}$.
2. The state space X of Σ is $\mathbf{R}_n$.
3. The input space U is a subset of $\mathbf{R}_m$, with $m \leq n$.
4. The control space Ω is the set of all regulated functions from $\mathbf{R}$ into U.
5. The equation of Σ is of the form

$$\dot{x}(t) = f(x(t), u(t))$$

(that is, autonomous), where, with the components of f and x denoted by $f_1, \ldots, f_n$ and $x_1, \ldots, x_n$, respectively,

$$f_i(x, u), \frac{\partial f_i}{\partial x_j}(x, u) \in C(\mathbf{R}_n \times \bar{U}),\dagger \quad i, j = 1, \ldots n.$$

In addition, we shall assume that our trajectory-cost function L is of the form $L(x, u)$ (that is, is independent of t) and that

$$L(x, u), \frac{\partial L}{\partial x_j}(x, u) \in C(\mathbf{R}_n \times \bar{U}).$$

$\dagger$ $C(\mathbf{R}_n \times \bar{U})$ is the set of all continuous real-valued functions in $\mathbf{R}_n \times \bar{U}$, where $\bar{U}$ is the closure of U.

Observe that we make no assumptions about the differentiability of f_i and L with respect to u.

We shall consider only the following two particular control problems, which we call Problem 1 and Problem 2:

Problem 1. There is no terminal cost (that is, $K = 0$), and the target set S is of the form $\mathbf{R} \times \{x_1\}$, where x_1 is fixed.

Problem 2. There is no terminal cost ($K = 0$), and the target set S is of the form $\mathbf{R} \times \tilde{S}$, where $\tilde{S}$ is either a smooth k-fold† in $\mathbf{R}_n$ or all of $\mathbf{R}_n$.

Although these various assumptions may appear quite restrictive, it is possible through suitable changes of variable to reduce many problems of interest to one of the cases examined here. For example, if, say, we were originally faced with a time-dependent problem—that is, if the generator of our system were of the form $f(x, u, t)$ and if L were of the form $L(x, u, t)$—then we would introduce a new state variable x_{n+1} with $\dot{x}_{n+1}(t) = 1$, and we would consider the system in $\mathbf{R}_{n+1}$ with generator $g(x, x_{n+1}, u)$ given by $g_i(x, x_{n+1}, u) = f_i(x, u, x_{n+1})$, $i = 1, \ldots, n$, and $g_{n+1}(x, x_{n+1}, u) = 1$, and with trajectory-cost function

$$M(x, x_{n+1}, u) = L(x, u, x_{n+1}).$$

(For more details concerning these changes of variable the reader should consult [ATHANS and FALB, 1966] or [PONTRYAGIN et al., 1961]. Note also that the only difference between the two problems is in the specification of the target set.)

Before giving a precise statement of the Pontryagin principle, let us interpret Problem 1 in an intuitive and geometric fashion. Letting $Z = \mathbf{R} \times \mathbf{R}_n = \mathbf{R}_{n+1}$ and denoting the coordinates of a typical z in Z by $z_0, \ldots, z_n$, we may consider the system

$$(2.1) \qquad \begin{aligned} \dot{z}_0(t) &= L(z_1(t), \ldots, z_n(t), u_1(t), \ldots, u_m(t)), \\ \dot{z}_i(t) &= f_i(z_1(t), \ldots, z_n(t), u_1(t), \ldots, u_m(t)). \end{aligned}$$

If we let $\hat{z} = (0, x_0)$ (that is, $\hat{z}_0 = 0$ and $\hat{z}_i = x_{0,i}$ for $i = 1, \ldots, n$), then Problem 1 may be phrased as follows: determine the control $u(\cdot)$ in Ω which transfers $(t_0, \hat{z})$ to $t_1 \times l$, where l is the line $z_i = x_{1,i}$, $i = 1, \ldots, n$, in such a way that the z_0-coordinate is minimized (see Figure 4.1). Problem 2 may also be given a similar geometric interpretation in that the line l is replaced by the set $\{z: (z_1, \ldots, z_n) \in \tilde{S}\}$. In effect, our problems can be viewed as problems for a system Σ_1 [defined by (2.1)] which involve no trajectory cost and a terminal-cost function of the simple form $K(z) = z_0$. This is, as we shall see, a very fruitful point of view.

---

† A subset G of $\mathbf{R}_n$ is called a *smooth k-fold* if $G = \{x: g_i(x) = 0, i = 1, \ldots, n - k\}$, where $g_i(x)$, $(\partial g_i / \partial x_j)(x)$ are continuous and the gradient vectors $\nabla_x g_i(y)$ are linearly independent if $y \in G$ for $i = 1, \ldots, n - k$ and $j = 1, \ldots, n$.

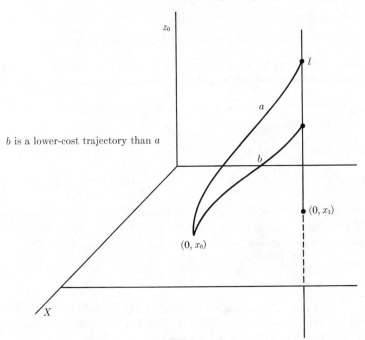

Figure 4.1

To formalize, we have the following lemma:

(2.2)　**Lemma.**　*Let Σ_1 be a dynamical system with domain **R**, state space $\mathbf{R}_{n+1}$, control space Ω, and with equation (2.1).　Let $(t_0, \hat{z})$ be an element of the event space for Σ_1 with $\hat{z} = (0, x_0)$ and let $S_1 \subset \mathbf{R} \times \mathbf{R}_{n+1}$ be of the form $\mathbf{R} \times \mathbf{R} \times \tilde{S}$, where $\hat{S}$ is either $\{x_1\}$ or $\tilde{S}$.　Let $K_1(z) = z_0$ for z in $\mathbf{R}_{n+1}$ and let $L_1 = 0$.　Let P_1 be the corresponding control problem for Σ_1.　Then a control $u^0(\cdot)$ is optimal for P_1 relative to $(t_0, \hat{z})$ if and only if $u^0(\cdot)$ is optimal for P relative to (t_0, x_0) (where P is the corresponding control problem for Σ).*

The proof of this lemma is left to the reader.
The hamiltonian functional for the problem P_1 is given by

$$H_1(z, \lambda, u) = \lambda_0 L(x, u) + \sum_{i=1}^{n} \lambda_i f_i(x, u), \dagger$$

where $z = (z_0, x)$.　We shall call $H_1(z, \lambda, u)$ the hamiltonian of our original problem.　Letting $p = (\lambda_1, \ldots, \lambda_n)$ and noting that $H_1(z, \lambda, u)$

† We observe that since $\mathbf{R}_{n+1}^* = \mathbf{R}_{n+1}$, we can take $\lambda = (\lambda_0, \lambda_1, \ldots, \lambda_n)$ to be an element of $\mathbf{R}_{n+1}$.

does not depend on z_0, we often write

$$H(x, p, u, \lambda_0) = \lambda_0 L(x, u) + \langle p, f(x, u) \rangle$$

in place of $H_1(z, \lambda, u)$ and, by abuse of language, speak of H as the hamiltonian of our original problem. We now have

(2.3) **Definition.** *The $2n^{\text{th}}$-order system of ordinary differential equations given by*

(2.4) $$\dot{x} = \frac{\partial H}{\partial p}(x, p, u, \lambda_0) = f(x, u),$$

(2.5) $$\dot{p} = -\frac{\partial H}{\partial x}(x, p, u, \lambda_0) = -\lambda_0 \frac{\partial L}{\partial x}(x, u) - \left(\frac{\partial f}{\partial x}\right)^*(x, u)p$$

is called the **canonical** *(or* **hamiltonian***) system. If $u(\cdot) \in \Omega$ and $x_u(\cdot)$ is the corresponding solution of (2.5), then any solution of (2.5) is said to correspond to $u(\cdot)$ and λ_0 (cf. Definition (1.10)).*

We observe that (2.4), being our original system equation, is independent of p and λ_0, and that (2.5) is *linear* in p.

(2.6) **Theorem (Pontryagin principle for Problem 1).** *If $u^0(\cdot)$ is an optimal control for Problem 1, then there is a nonnegative constant λ_0^0 (that is, $\lambda_0^0 \geq 0$) and a function $p^0(\cdot)$ such that*
(a) *$p^0(\cdot)$ corresponds to $u^0(\cdot)$ and λ_0^0;*
(b) *$H(x^0(t), p^0(t), u, \lambda_0^0)$ has an absolute minimum as a function of u over U at $u = u^0(t)$ for t in $[t_0, t_1]$,† where t_1 is the transfer time associated with $u^0(\cdot)$;*
(c) *$H(x^0(t), p^0(t), u^0(t), \lambda_0^0) = 0$ for t in $[t_0, t_1]$.†*

(2.7) **Theorem (Pontryagin principle for Problem 2).** *If $u^0(\cdot)$ is an optimal control for Problem 2, then there is a nonnegative constant λ_0^0 (that is, $\lambda_0^0 \geq 0$) and a function $p^0(\cdot)$ such that*
(a) *$p^0(\cdot)$ corresponds to $u^0(\cdot)$ and λ_0^0;*
(b) *$H(x^0(t), p^0(t), u, \lambda_0^0)$ has an absolute minimum as a function of u over U at $u = u^0(t)$ for t in $[t_0, t_1]$, where t_1 is the transfer time associated with $u^0(\cdot)$;*
(c) *$H(x^0(t), p^0(t), u^0(t), \lambda_0^0) = 0$ for t in $[t_0, t_1]$;*
(d) *$p^0(t_1) = 0$ if $\tilde{S} = \mathbf{R}_n$ and $p^0(t_1)$ is normal to $\tilde{S}$‡ at $x^0(t_1)$ if $\tilde{S}$ is a smooth k-fold in $\mathbf{R}_n$.*

We now give a plausibility argument for Theorem (2.6). For a complete proof the reader should consult [HALKIN, 1963a; PONTRYAGIN et al., 1961].

---

† This is to be construed in an almost-everywhere sense.
‡ The term "transversal to $\tilde{S}$" is also used quite frequently here.

"Proof" of *Theorem* (2.6). For ease of exposition, we shall deal with the problem P_1 in $\mathbf{R}_{n+1}$. Thus we consider the system

$$(2.8) \quad \begin{aligned} \dot{z}_0(t) &= L(z_1(t), \ldots, z_n(t), u_1(t), \ldots, u_m(t)) = g_0(z(t), u(t)), \\ \dot{z}_i(t) &= f_i(z_1(t), \ldots, z_n(t), u_1(t), \ldots, u_m(t)) = g_i(z(t), u(t)), \end{aligned}$$

which we often write in the form

$$\dot{z}(t) = g(z(t), u(t)).$$

We assume that $u^0(\cdot)$ is an optimal control relative to $(t_0, \hat{z})$, where $\hat{z} = (0, x_0)$, and that t_1 is the transfer time associated with $u^0(\cdot)$. We also suppose that $u^0(\cdot)$ is left continuous; [that is, $u^0(t-) = u^0(t)$].†

We are now going to construct a suitable set π of perturbations of $u^0(\cdot)$. Let $\epsilon > 0$ and let a and b be elements of $\mathbf{R}$, with $b \geq 0$. If $\tau \in (t_0, t_1]$ and $\omega \in U$, then the control $u(\cdot\,; a, b, \tau, \omega)$ defined by

$$u(t; a, b, \tau, \omega) = \begin{cases} \omega, & t \in (\tau - \epsilon b, \tau], \\ u^0(t), & t \notin (\tau - \epsilon b, \tau], t \in [t, t_1 + \epsilon a], a \leq 0, \\ u^0(t), & t \notin (\tau - \epsilon b, \tau], t \in [t, t_1], a > 0, \\ u^0(t_1), & t \in [t_1, t_1 + \epsilon a], a > 0, \end{cases}$$

is called a *basic perturbation* of $u^0(\cdot)$. If $a = 0$ the perturbation is said to be *spatial*, and if $b = 0$ the perturbation is said to be *temporal* (see [ATHANS and FALB, 1966]). We observe that a basic perturbation has the property that the corresponding solution $z(\cdot\,; a, b, \tau, \omega)$ of (2.8), when evaluated at $t_1 + \epsilon a$, is given by

$$z(t_1 + \epsilon z; a, b, \tau, \omega) = z^0(t_1) + \epsilon \xi[a; b, \tau, \omega] + o(\epsilon),$$

where $\xi[a; b, \tau, \omega]$ does not depend on ϵ.

Now, if Ψ is a set of perturbations $u(\cdot\,; \alpha)$ of $u^0(\cdot)$ depending on a parameter α, then we shall say that Ψ is *complete* if

$$z(t_1 + \epsilon \alpha; \alpha) = z^0(t_1) + \epsilon \xi[\alpha] + o(\epsilon),$$

where $\xi[\alpha]$ does not depend on ϵ, and if there is a perturbation in Ψ which corresponds to $\xi[c_1\alpha_1 + c_2\alpha_2]$ for $c_1, c_2 \geq 0$. We let π be a complete set of perturbations (with a as parameter) containing all the basic perturbations. The set of vectors ξ corresponding to elements of π forms a convex cone C, which may be viewed as attached to $z^0(t_1)$. Moreover, every element of C is generated by an element of π. (These last two statements, which here are taken on faith, are vital points of the actual proof.)

If $\zeta = (-1, 0, \ldots, 0)$, then we claim that ζ is not an element of the interior of C. For if ζ were in the interior of C, then, letting $\bar{u}(\cdot)$ be the

---

† This causes no loss of generality as every admissible control is equivalent in an almost-everywhere sense to a left-continuous control.

element of π corresponding to ζ, we would have

$$(2.9) \qquad \bar{z}_0 = z_0^0(t_1) - \epsilon + o(\epsilon), \ \bar{z}_i = x_{1,i} + o(\epsilon) \qquad i = 1, \ldots, n,$$

where $\bar{z}$ is the point in $\mathbf{R}_{n+1}$ to which $\hat{z}$ is transferred by $\bar{u}(\cdot)$. Making (2.9) "sharp," we could find a perturbation $u^\#(\cdot)$ such that $z_0^\# < z_0^0(t_1)$, $z_i^\# = x_{1,i}$, $i = 1, \ldots, n$, which would contradict the optimality of $u^0(\cdot)$ (this making (2.9) "sharp" is also a crucial point of the actual proof).

Since ζ is not in the interior of C, there is a hyperplane Λ with normal $\lambda^0 = (\lambda_0^0, \lambda_1^0(t_1), \ldots, \lambda_n^0(t_1))$ which separates ζ from C. In other words,

$$\langle \lambda^0, \zeta \rangle \leq 0,$$
$$\langle \lambda^0, \xi \rangle \geq 0 \qquad \text{for all } \xi \text{ in } C.$$

Clearly, $\lambda_0^0 \geq 0$. We let $p^0(\cdot)$ be the solution of (2.5) corresponding to $u^0(\cdot)$ and λ_0^0 with the property that $p^0(t_1) = (\lambda_1^0(t_1), \ldots, \lambda_n^0(t_1))$. We note that $p^0(\cdot)$ is well-defined, since (2.5) is linear.

Now suppose that $u(\cdot; 0, b, \tau, \omega)$ is a spatial perturbation of $u^0(\cdot)$. Then

$$z(\tau; 0, b, \tau, \omega) = z^0(\tau) + \epsilon[g(z^0(\tau), \omega) - g(z^0(\tau), u^0(\tau))] + o(\epsilon).$$

If $\xi(t)$ denotes the solution of the perturbation equation† about the optimal trajectory satisfying the condition

$$\xi(\tau) = g(z^0(\tau), \omega) - g(z^0(\tau), u^0(\tau)),$$

then

$$z(t_1; 0, b, \tau, \omega) = z^0(t_1) + \epsilon\xi(t_1) + o(\epsilon).$$

Since $\xi(t_1)$, which equals $\xi[0; b, \tau, \omega]$, is an element of C, we have

$$\langle \lambda_0^0, \xi(t_1) \rangle \geq 0.$$

However, the equation for $\xi(t)$ is the adjoint of the equation for $(\lambda_0^0, p^0(t))$,‡ and therefore

$$\langle (\lambda_0^0, p^0(\tau)), \xi(\tau) \rangle \geq 0,$$

† In other words,

$$\dot{\xi} = \frac{\partial g}{\partial z}\bigg|_0 \xi.$$

‡ This means that, letting $\lambda^0(t) = (\lambda_0^0, p^0(t))$,

$$\dot{\lambda}^0 = -\left(\frac{\partial g}{\partial z}\bigg|_0\right)^* \lambda^0,$$

where

$$\dot{\xi} = \frac{\partial g}{\partial z}\bigg|_0 \xi.$$

or

$$\lambda_0^0 g_0(z^0(\tau),\, \omega) + \sum_1^n p_i^0(\tau) g_i(z^0(\tau),\, \omega)$$

$$\geq \lambda_0^0 g_0(z^0(\tau),\, u^0(\tau)) + \sum_1^n p_i^0(\tau) g_i(z^0(\tau),\, u^0(\tau)).$$

Thus we have established (a) and (b) of Theorem (2.6).

As for (c), if $u(\cdot\,; a, 0, \tau, \omega)$ is a temporal perturbation of $u^0(\cdot)$, then

$$\xi[a; 0, \tau, \omega] = ag(z^0(t_1),\, u^0(t_1)) \in C,$$

so that

$$a\langle \lambda^0,\, g(z^0(t_1),\, u^0(t_1))\rangle \geq 0.$$

But a can be any element of $\mathbf{R}$, and so (c) holds at t_1.   Now we let $H^0(t)$ be given by

$$H^0(t) = \langle(\lambda_0^0,\, p^0(t)),\, g(z^0(t),\, u^0(t))\rangle.$$

If $(t_2, t_3) \subset [t_0, t_1]$ and $u^0(t)$ is continuous on $(t_2, t_3]$, then, by condition (b), we have

$$\langle(\lambda_0^0,\, p^0(\tau)),\, g(z^0(\tau),\, u^0(\sigma))\rangle - H^0(\sigma) \geq H^0(\tau) - H^0(\sigma)$$
$$\geq H^0(\tau) - \langle(\lambda_0^0 p^0(\sigma)),\, g(z^0(\sigma),\, u^0(\tau))\rangle, \qquad \sigma, \tau \in (t_2, t_3].$$

However, the function

$$\langle(\lambda_0^0,\, p^0(t)),\, g(z^0(t),\, u^0(\sigma))\rangle = \varphi(t)$$

is differentiable on $(t_2, t_3]$ with $\dot{\varphi}(t) = 0$.   It follows that

$$H^0(\tau) = H^0(\sigma),$$

and hence that $H^0(\cdot)$ is constant on $(t_2, t_3]$.   On the other hand, if t_4 is a discontinuity of $u^0(\cdot)$ and if $\tau = t_4 + \delta$ with δ small, then $H^0(\tau) > H^0(t_4)$ would imply that

$$\langle(\lambda_0^0, p^0(t_4)),\, g(z^0(t_4),\, u^0(\tau))\rangle < H^0(t_4)$$

(a contradiction of (b)), and $H^0(\tau) < H^0(t_4)$ would imply that

$$\langle(\lambda_0^0,\, p^0(\tau)),\, g(z^0(\tau),\, u^0(t_4))\rangle < H^0(\tau)$$

(again a contradiction of (b)).   Thus

$$H^0(t_4) = H^0(t_4+).$$

It follows that $H^0(t) = H^0(t_1) = 0$ for t in $[t_0, t_1]$, and (c) is established. This completes our plausibility argument for Theorem (2.6) (we reiterate, this is not a (rigorous) proof). $\square$

We conclude this section with some comments on the Pontryagin principle.   In the next section we shall discuss some generalizations of this principle, and in the next chapter, its use in control-system design.

(2.10)   **Comment.**   We observe that Theorem (2.6) actually represents necessary conditions for local optimality which are similar to the vanishing of the derivative in ordinary minimization problems.

(2.11)   **Comment.**   The canonical system has solutions along *any* trajectory of the system, not just along optimal or extremal trajectories.

(2.12)   **Comment.**   We can readily see that the minimization of the hamiltonian (Theorem (2.6*b*)) may be viewed as a reduction of our functional minimization problem to an ordinary minimization problem. This is of considerable practical value.

(2.13)   **Comment.**   The conclusions of Theorem (2.6) are independent of the target set S, and the only difference between the theorems is the transversality condition (*d*) of Theorem (2.7). This condition may be viewed as providing "enough" boundary conditions to integrate the canonical system. More will be said about this point in the next chapter.

(2.14)   **Comment.**   Supposing that our problem is nondegenerate in the sense that $\lambda_0^0 \neq 0$, we immediately conclude that there is no loss of generality in taking $\lambda_0^0 = 1$, since the so-called *costate equation* (2.5) is linear in $p^0(\cdot)$. The design problems, which we consider later on, are nondegenerate, and so we shall assume without further ado that $\lambda_0^0 = 1$ in these problems.

(2.15)   **Comment.**   If we examined the function $\tilde{H}(x, p, u, \lambda_0)$ given by

$$\tilde{H}(x, p, u, \lambda_0) = \lambda_0 L(x, u) - \langle p, f(x, u) \rangle$$

in place of H, then the only change in our results would be the replacement of "minimum" by "maximum" in Theorem (2.6*b*). As Pontryagin dealt with $\tilde{H}$, he obtained a maximum principle, while we derived a minimum principle.

4.3   An existence theorem

Let us now examine the question of the existence of an optimal control. We shall begin with a discussion of the notion of attainability and then show how this notion is related to existence. Suppose that Σ is our smooth dynamical system and that Ω is the set of control functions for Σ. We then have

(3.1)   **Definition.**   *Let (t_0, x_0) be a given element of the event space $(T_1, T_2) \times X$. Then an event (t_1, x_1) with $t_1 \geq t_0$ is* **attainable** *(or* *reachable) from* (t_0, x_0) *relative to* Ω *if there is a* $u(\cdot) \in \Omega$ *such that*

$$x_1 = \varphi(t_1; t_0, x_0, u(\cdot)).$$

The set $A(t_0, x_0, \Omega) = \{(t_1, x_1): (t_1, x_1)$ *is attainable from* (t_0, x_0) *relative to* $\Omega\}$ *is called the* **attainable set** *relative to* t_0, x_0, *and* Ω. *The* t_1-*section*† *of* $A(t_0, x_0, \Omega)$, *denoted by* $A(t_1; t_0, x_0, \Omega)$, *is called the attainable set at* t_1 *relative to* t_0, x_0, *and* Ω.

We immediately observe that if we are considering a control problem with target set S and admissible control set Γ, then $S \cap A(t_0, x_0, \Gamma) \neq \varnothing$ is a necessary condition for the existence of an optimal control relative to (t_0, x_0) and Γ. Supposing without loss of generality that there is only a terminal-cost term $K(t, x)$,‡ we easily deduce the following:

(3.2) **Theorem.** *Suppose that* $S \cap A(t_0, x_0, \Gamma) \neq \varnothing$ *and that there is a topology (not necessarily inherited from the product topology on* $(T_1, T_2) \times X$) *such that* $S \cap A(t_0, x_0, \Gamma)$ *is compact and* $K(t, x)$ *is lower semicontinuous.*§ *Then there is an optimal control relative to* (t_0, x_0) *and* Γ.

The theorem is an immediate consequence of the fact that a lower-semicontinuous real-valued function on a compact set has a minimum [ROTHE, 1948].

Since the cost functionals in most control problems are suitably smooth (that is, are at least lower semicontinuous), our interest naturally focuses on the properties of $S \cap A(t_0, x_0, \Omega)$ and $A(t_0, x_0, \Omega)$. Various useful results, particularly those relating to the closure of $A(t_0, x_0, \Omega)$ and $A(t; t_0, x_0, \Omega)$, can be found in [FALB, 1964; HALKIN, 1963b; ROXIN, 1962]. Two of these results may be used to establish the existence of an optimal control for the example of Section 5.1.

Let us now suppose that our state space is of the form $X = \mathbf{R} \times \hat{X}$, and let us write $x = (\psi, \hat{x})$. Assuming that our target set S is of the form $\{t_1\} \times \mathbf{R} \times \hat{S}$ (that is, a "cylinder" which is "parallel" to the ψ-axis) and the cost is given simply by $K(x) = \psi$, we have the following rather obvious lemma:

(3.3) **Lemma.** *Let* $A(\psi; t_0, x_0, \Omega)$ *denote the* ψ-*section of* $A(t_1; t_0, x_0, \Omega)$ *viewed as a subset of* X. *Suppose that if* $\hat{S}(\psi) = \hat{S} \cap A(\psi; t_0, x_0, \Omega)$,

† If $A \cap (T_1, T_2) \times X$ and π is the projection of $(T_1, T_2) \times X$ on (T_1, T_2), then $\pi^{-1}(t_1) \cap A = \{(t_1, x): (t_1, x) \in A\}$ is the t_1-*section* of A.

‡ If there were a trajectory-cost term, then we would adjoin a new variable $\bar{x}$, say, with $\dot{\bar{x}}(t) = L(x(t), u(t), t)$.

§ Recall that a function K is *lower semicontinuous* at $(\hat{t}, \hat{x})$ if $\epsilon > 0$ implies there is a $\delta > 0$ such that $\|(t, x) - (\hat{t}, \hat{x})\| < \delta \Rightarrow K(t, x) \leq K(\hat{t}, \hat{x}) + \epsilon$.

there is a ψ^0 such that $\hat{S}(\psi) = \varnothing$ for $\psi < \psi^0$ and $\hat{S}(\psi^0) \neq \varnothing$. Then there is an optimal control $u^0(\cdot)$, and the cost of $u^0(\cdot)$ is ψ^0.

Proof. Since $\hat{S}(\psi^0) \neq \varnothing$, there is a $u^0(\cdot)$ which transfers (t_0, x_0) to S, and the cost associated with $u^0(\cdot)$ is ψ^0. On the other hand, if $u^1(\cdot)$ transfers (t_0, x_0) to S with cost ψ^1, then $\hat{S}(\psi^1) \neq \varnothing$, and so $\psi^1 \geq \psi^0$. □

An immediate consequence of this lemma is the following

(3.4) Theorem. *Suppose that*
(a) $\hat{S}$ *is compact.*
(b) $A(\psi; t_0, x_0, \Omega)$ *is closed for all ψ.*
(c) $A(\psi_1; t_0, x_0, \Omega) = \bigcap\limits_{\psi > \psi_1} A(\psi; t_0, x_0, \Omega)$ *for all ψ_1.*
(d) *There is a $\bar{\psi}$ such that $\hat{S}(\psi) = \varnothing$ for all $\psi < \bar{\psi}$ (that is, the cost is bounded below).*
(e) *There is a control which transfers (t_0, x_0) to S (hence the cost is bounded above).*
Then there is an optimal control.

Proof. Let $\psi^0 = glb \{\psi : \hat{S}(\psi) \neq \varnothing\}$. ψ^0 exists, in view of conditions (d) and (e). We now claim that $\hat{S}(\psi^0) \neq \varnothing$. To verify this claim we let ψ_i, $i = 1, t, \ldots$, be a monotone-decreasing sequence which converges to ψ^0. Then $\hat{S}(\psi_i) \neq \varnothing$ for $i = 1, 2, \ldots$, and so we let $\hat{x}_i$, $i = 1, 2, \ldots$, be an element of $\hat{S}(\psi_i)$. Since $\hat{x}_i \in \hat{S}$ for all i, the sequence $\{\hat{x}_i\}$ has a subsequence which converges to an element $\hat{x}^0$ of $\hat{S}$. Now, (c) implies that $A(\psi; t_0, x_0, \Omega) \subset A(\psi'; t_0, x_0, \Omega)$ whenever $\psi \leq \psi'$ and so $\hat{x}_0 \in A(\psi, t_0, x_0, \Omega)$ for all $\psi > \psi^0$. It follows that

$$\hat{x}^0 \in A(\psi^0; t_0, x_0, \Omega) = \bigcap\limits_{\psi > \psi^0} A(\psi, t_0, x_0, \Omega)$$

and, in view of the lemma, that the theorem is established. □

The following typical result for finite-dimensional dynamical systems is due to [ROXIN, 1962]:

(3.5) Theorem. *Consider the finite-dimensional dynamical system with generator $f(x, u, t)$. Suppose that*
(a) U *is compact.*
(b) f *is integrable in t for each $(x, u) \in X \times U$.*
(c) f *is lipschitzian in x; that is, there is a constant $C > 0$ such that for each (t, u) in $\mathbf{R} \times U$*

$$\|f(x_1, u, t) - f(x_2, u, t)\| \leq C\|x_1 - x_2\|.$$

(d) $f(x, U, t)$ *is convex for each (t, x) in $\mathbf{R} \times X$.*

(e) $\|f(x, u, t)\| \leq m(t)h(\|x\|)$, *where* $m(t)$ *is integrable and* $h(\cdot)$ *is bounded and of order* $\|x\|$ *as* $\|x\| \to \infty$.

Then $A(t_0, x_0, \Omega)$ *is closed*.

The proof of this theorem is based on properties of the weak topology on L_1 and can be found in [ROXIN, 1962].

We shall use this theorem to show that an optimal control exists for the problem to be considered in Section 5.1. In general, the important question of the existence of optimal controls is quite difficult, and we have here given only a very small indication of what is involved.

4.4 Remarks on necessary conditions in control problems

We have already observed (Section 3.3) that control problems can be viewed as special cases of the problem of minimizing a real-valued functional defined on a subset of a normed linear space. The necessary conditions for optimality that we have considered, when viewed from this general standpoint, are similar in that the existence of a linear functional generating a suitable inequality is crucial. For example, the maximum principle involved the demonstration that there was a costate $(\lambda_0^0, p^0(\cdot))$ with the property

$$H(x^0, p^0, u^0, \lambda_0^0) \leq H(x^0, p^0, u, \lambda_0^0)$$

for all u in U; moreover, as we saw in Section 4.2, this demonstration was based on the idea of finding a linear functional on $\mathbf{R}_{n+1}$ (that is, an $(n + 1)$-vector) which separated a ray in the direction of decreasing cost from a convex cone generated by special variations that represented an approximation to the set of changes due to all control variations. These ideas can be generalized quite a bit (see [CANON, CULLUM, and POLAK, 1967; HALKIN, 1967; NEUSTADT, 1966–67]). Here we shall only briefly comment on some aspects of these generalizations.

We consider the following basic optimization problem:†

(4.1) **Problem.** Let P and Q be normed linear spaces, let Ω be a subset of P, let π be a convex subset of Q, let f map P into $\mathbf{R}$, and let g map P into Q. Then determine $\hat{p}$ in P such that $\hat{p} \in \Omega$, and $g(\hat{p}) \in \pi$ and such that $f(\hat{p}) \leq f(p)$ if $p \in \Omega$ and $g(p) \in \pi$.

In other words, we seek the minimum of f on Ω subject to the "constraint" that g be in π (by abuse of language). An element $\hat{p}$ of Ω which achieves this minimum shall be called *optimal*. The necessary conditions for opti-

---

† Greater generality would involve concepts beyond the scope of this book.

mality of [CANON, CULLUM, and POLAK, 1967; HALKIN, 1967; NEUSTADT, 1966–67] are all, loosely speaking, of the following type: If $\hat{p}$ is optimal, if there is a suitable approximation M to Ω at $\hat{p}$, and if there is a suitable approximation h to the map $(f(\cdot), g(\cdot))$ of P into $\mathbf{R} \times Q$; then there is a nonzero λ in $(\mathbf{R} \times Y)^*$ which separates $h(M)$, and

$$\rho = \{(b, \pi): b < 0, \pi \in \Pi\}.$$

Of course, the difficulties lie in defining the notion of "suitable approximation."   In the case where $P = \mathbf{R}_n$, $Q = \mathbf{R}_m$, $\Pi = \{0\}$, and f and g are continuously differentiable [CANON, CULLUM, and POLAK, 1967], h is simply the jacobian matrix of the map (f, g) of $\mathbf{R}_n$ into $\mathbf{R}_{m+1}$, so that

$$(f(p + \delta p), g(p + \delta p)) = (f(p), g(p)) + h(p)\delta p + o(\|\delta p\|)$$

and M satisfies the following condition: if $\delta p^1, \ldots, \delta p^k$ are linearly independent elements of M, then there is an $\epsilon > 0$ and a continuous map ξ of the convex hull of $\{\hat{p}, \hat{p} + \epsilon\, \delta p^1, \ldots, \hat{p} + \epsilon\, \delta p^k\}$† into Ω such that $\xi(\hat{p} + \delta p) = \hat{p} + \delta p + o(\|\delta p\|)$.

The proofs of the various results relating to necessary conditions all depend on properties of convex sets and on the Brouwer fixed-point theorem.   For more details the reader should consult the indicated references (or Appendix A).

APPENDIX

4.A   Necessary conditions for optimality

We discuss the general results on necessary conditions for optimality of [CANON, CULLUM, and POLAK, 1967; HALKIN, 1967; NEUSTADT, 1966–67] in this appendix.   Our treatment is based on [FALB and POLAK, 1968] and assumes a reasonable degree of mathematical sophistication on the part of the reader.

Let us suppose that X is a locally convex real topological vector space.   We shall consider the problem of minimizing a real valued function f on a subset Ω of X subject to a finite dimensional side constraint $g(x) = 0$.   In order to derive necessary conditions for such a problem, we

---

† The *convex hull* of $\{\hat{p}, \hat{p} + \epsilon\, \delta p^1, \ldots, \hat{p} + \epsilon\, \delta p^k\}$ is the set of all p such that $p = r_0\hat{p} + r_1(\hat{p} + \epsilon\, \delta p^1) + \cdots + r_k(\hat{p} + \epsilon\, \delta p^k)$, where $\displaystyle\sum_{i=0}^{k} r_i = 1$ and $r_i \geq 0$, $i = 0, \ldots, k$ (see [ATHANS and FALB, 1966] or [DUNFORD and SCHWARTZ, 1958]).

need the notion of an approximation.   So we have

(A.1)   **Definition.**   *Let f map X into R and let g map X into* $\mathbf{R}_m$. *A convex cone* $A(\hat{x}, \Omega) \subset X$ *is an* **approximation of** Ω *at* $\hat{x}$ **relative to** *f* **and** *g if there are*
  (i) *a continuous linear functional* $f'(\hat{x})$ *on X, and*
  (ii) *a continuous linear transformation* $g'(\hat{x})$ *from X into* $\mathbf{R}_m$ *which satisfy the following conditions:*
    Given any finite set $\{\delta x_1, \ldots, \delta x_k\}$ *of linearly independent elements of* $A(\hat{x}, \Omega)$, *there are*
  (a) *an* $\epsilon_1 > 0$,
  (b) *a continuous map* ψ *of* co $\{\epsilon_1 \delta x_1, \ldots, \epsilon_1 \delta x_k\}$ *into* $\Omega - \{\hat{x}\}$ *(where* ψ *may depend on* ϵ_1 *and* $\{\delta x_1, \ldots, \delta x_k\}$*), and*
  (c) *continuous maps* o_f *and* o_g *of X into* **R** *and* $\mathbf{R}_m$, *respectively, such that*

(A.2) $$\lim_{\epsilon \to 0} |o_f(\epsilon \delta x)|/\epsilon = 0, \qquad \lim_{\epsilon \to 0} \|o_g(\epsilon \delta x)\|/\epsilon = 0$$

uniformly for δx *in* co $\{\delta x_1, \ldots, \delta x_k\}$;

(A.3)
$$f(\hat{x} + \psi(\epsilon \delta x)) = f(\hat{x}) + \epsilon f'(\hat{x}) \delta x + o_f(\epsilon \delta x),$$

$$g(\hat{x} + \psi(\epsilon \delta x)) = g(\hat{x}) + \epsilon g'(\hat{x}) \delta x + o_g(\epsilon \delta x)$$

for all δx *in* co $\{\delta x_1, \ldots, \delta x_k\}$ *and* $0 < \epsilon \leq \epsilon_1$.

We observe that if f and g are differentiable at $\hat{x}$ and if $C(\hat{x}, \Omega)$ is a "linearization of Ω at $\hat{x}$," (i.e., $C(\hat{x}, \Omega)$ is a convex cone contained in X and for which conditions (analogous to) (a) and (b) of definition (A.1) are satisfied), then $C(\hat{x}, \Omega)$ is also an approximation of Ω at $\hat{x}$ relative to f and g.   We also note that if g is written in the form $g(x) = (g_1(x), \ldots, g_m(x))$, then some of the components $g_i'(\hat{x})$ can be replaced by convex functionals (rather than linear functionals) and the definition will still make sense (see [HALKIN, 1967; NEUSTADT, 1966–67]).   This generalization would lead to a somewhat stronger version of the basic necessary condition lemma.   Another generalization would involve the replacement of $\mathbf{R}_m$ by any normed linear space E; however, in that case, we would also require that the cone $C(\hat{x}) = \{(f'(\hat{x})\delta x, g'(\hat{x})\delta x): \delta x \in A(\hat{x}, \Omega)\}$ in $\mathbf{R} \oplus E$ (the direct sum of $\mathbf{R}$ and E) have (relative) interior points in the smallest closed subspace of $\mathbf{R} \oplus E$ containing the cone $C(\hat{x})$.   No essential change is required by this generalization.

Having a satisfactory notion of approximation at hand, we are now ready to examine the question of necessary conditions.   We consider the following

Problem.   Let X be a locally convex topological vector space.   Let f and g be mappings of X into **R** *and* $\mathbf{R}_m$, *respectively, and let* Ω *be a subset*

of X. *Let* X_g *be the subset of* X *given by* $X_g = \{x\colon g(x) = 0\}$. *Then determine* x^0 *in* $X_g \cap \Omega$ *such that* $f(x^0) \leq f(x)$ *for all* x *in* $X_g \cap \Omega$.

A solution x^0 of this problem shall be referred to as an *optimal element*. We then have:

(A.4)   **Lemma.**   *If* x^0 *is an optimal element and if* $A(x^0, \Omega)$ *is an approximation of* Ω *at* x^0 *relative to* f *and* g, *then there is a nonzero continuous linear functional* p *on* $\mathbf{R} \oplus \mathbf{R}_m$ *such that* $p[(f'(x^0)\delta x,\, g'(x^0)\delta x)] \leq 0$ *for all* δx *in the closure* $\overline{A(x^0,\, \Omega)}$ *of* $A(x^0,\, \Omega)$, *i.e., there is a nonzero vector* $p = (p_0,\, p_1,\, \ldots\,,\, p_m)$ *in* $\mathbf{R} \oplus \mathbf{R}_m$ *such that* $p_0 \leq 0$ *and*

(A.5) $$\langle p,\, (f'(x^0)\delta x,\, g'(x^0)\delta x) \rangle \leq 0$$

for all δx *in* $\overline{A(x^0,\, \Omega)}$.

Proof [Dacunha and Polak, 1967].   The proof, which we sketch here, is based on the separation theorem for convex sets ([Dunford and Schwartz, 1958]) and the Brouwer fixed point theorem ([Dieudonne, 1960]).

Let x^0 be our optimal element and $C(x^0)$ be the cone $\{(f'(x^0)\delta x,\, g'(x^0)\delta x)\colon \delta x \in A(x^0,\, \Omega)\}$ in $\mathbf{R} \oplus \mathbf{R}_m$. Since $A(x^0,\, \Omega)$ is a convex cone and the maps $f'(x^0)$, $g'(x^0)$ are linear, $C(x^0)$ is also a convex cone. We let ρ be the open half-ray in $\mathbf{R} \oplus \mathbf{R}_m$ given by

(A.6) $$\rho = \{(y_0,\, 0,\, \ldots\,,\, 0)\colon y_0 < 0\}.$$

We note that ρ is a convex cone. We claim that $C(x^0)$ and ρ are separated in $\mathbf{R} \oplus \mathbf{R}_m$.

If our claim is valid, then there is a nonzero vector

$$p = (p_0,\, p_1,\, \ldots\,,\, p_m)$$

in $\mathbf{R} \oplus \mathbf{R}_m$ such that

(A.7) $\langle p,\, (f'(x^0)\delta x,\, g'(x^0)\delta x) \rangle \leq 0$ for all δx in $A(x^0,\, \Omega)$,

and

(A.8) $\langle p,\, y \rangle \geq 0$ for all y in ρ.

In view of the continuity of $f'(x^0)$ and $g'(x^0)$, (A.7) and (A.8) establish the lemma.

So let us assume that $C(x^0)$ and ρ are not separated. Then $C(x^0) \cap \rho$ is not empty and there is a δx_1 in $A(x^0,\, \Omega)$ such that

(A.9) $f'(x^0)\delta x_1 < 0$ and $g'(x^0)\delta x_1 = 0$.

Moreover, $g'(x^0)A(x^0,\, \Omega) = \mathbf{R}_m$, i.e., contains 0 as an interior point (otherwise there would be a vector of the form $(0,\, q)$, $q \neq 0$, separating

$C(x^0)$ and ρ). It follows that there is a simplex $[z_1, z_2, \ldots, z_{m+1}]$ in $\mathbf{R}_m$ containing 0 as an interior point such that

(A.10) $[z_1, \ldots, z_{m+1}] \subset g'(x^0) A(x^0, \Omega)$, i.e., $z_i = g'(x^0)\delta x_i$, where

$\delta x_i \in A(x^0, \Omega)$ for $i = 1, 2, \ldots, m + 1$;

(A.11) $\psi(\delta x) \in (\Omega - \{x^0\})$ for all δx in co $\{\delta x_1, \ldots, \delta x_{m+1}\}$, where ψ is the map occurring in the definition of $A(x^0, \Omega)$,

(A.12) $f'(x^0)\delta x_i < 0$ for $i = 1, 2, \ldots, m + 1$,

and

(A.13) $\{z_2 - z_1, \ldots, z_{m+1} - z_1\}$ is a basis of $\mathbf{R}_m$.

The existence of a simplex $[z_1, \ldots, z_{m+1}]$ satisfying (A.10), (A.11), and (A.12) is not difficult to establish (see [Dacunha and Polak, 1967]). Condition (A.13) holds for any simplex. Since the $z_j - z_1, j = 2, \ldots, m + 1$ form a basis of $\mathbf{R}_m$, we can define a linear mapping L of $\mathbf{R}_m$ into X by setting

(A.14) $L(z_j - z_1) = \delta x_j - \delta x_1$ for $j = 2, \ldots, m + 1$.

If $z = \Sigma r_j z_j + (1 - \Sigma r_j)z_1$ is an element of $[z_1, \ldots, z_{m+1}]$, then we let

(A.15) $\varphi(z) = L(z - z_1) + \delta x_j = \Sigma r_j \delta x_1 + (1 - \Sigma r_j)\delta x_j$.

Clearly φ is a continuous map of the simplex into co $\{\delta x_1, \ldots, \delta x_{m+1}\}$.

Now, for $0 < \alpha \leq 1$, we define a continuous map h_α from $\alpha[z_1, \ldots, z_{m+1}]$ into $\mathbf{R}_m$ by

(A.16) $h_\alpha(\alpha z) = -g(x^0 + \psi(\alpha[L(z - z_1) + \delta x_1])) + \alpha z$.

In view of the properties of an approximation, we have

(A.17) $h_\alpha(\alpha z) = -o_g(\alpha L(z - z_1) + \alpha \delta x_1)$

and

(A.18) $o_g(\alpha L(z - z_1) + \alpha \delta x_1) \in \alpha[z_1, \ldots, z_{m+1}]$

for all α with $0 < \alpha \leq \alpha_0$ and some α_0. Moreover, since $f'(x^0)\delta x_i < 0$, $i = 1, \ldots, m + 1$ and since $A(x^0, \Omega)$ is an approximation, we also have

(A.19) $f(x^0 + \psi(\alpha L(z - z_1) + \alpha \delta x_1)) < f(x^0)$

for all α with $0 < \alpha \leq \alpha_1$ and some α_1.

Letting $\beta = \min\{\alpha_0, \alpha_1\}$, we deduce from the Brouwer fixed point theorem ([Dieudonne, 1960]) that there is an element z_β in $\beta[z_1, \ldots,$

z_{m+1}] such that $h_\beta(\beta z_\beta) = \beta z_\beta$. It now follows from (A.16) that if

$$x = x^0 + \psi(L(z_\beta - \beta z_1) + \beta \delta x_1),$$

then $g(x) = 0$, $x \in \Omega$ (since $x - x^0 \in \psi(\text{co } \{\delta x_1, \ldots, \delta x_{m+1}\}) \subset \Omega - \{x^0\}$), and $f(x) < f(x^0)$ by (A.19). This contradicts the optimality of x^0 and thus the lemma is established. $\square$

We shall use this basic lemma to obtain the maximum principle. So we now turn our attention to the optimal control problem and indicate how this problem can be recast in such a way that the Lemma (A.4) will apply. Thus, we consider a system described by the differential equation

(A.20) $\dot{x} = f(x, u), \qquad t \in [t_0, t_1],$

where $x(t) \in \mathbf{R}_n$ is the state, $u(t) \in \mathbf{R}_m$ is the control vector, and f maps $\mathbf{R}_n \times \mathbf{R}_m$ into $\mathbf{R}_n$. We assume that

(A.21) $u(t) \in U \subset \mathbf{R}_n$ for (almost) all t,

(A.22) $u(\cdot)$ is measurable and essentially bounded,

(A.23) $f(x, u)$ is continuous on $\mathbf{R}_n \times \bar{U}$, and

(A.24) $f(x, u)$ is continuously differentiable in x.

We suppose that the initial manifold is a single point x_0 (i.e., $x(t_0) = x_0$) and that the terminal manifold is defined by a C^1 function h from $\mathbf{R}_n$ into $\mathbf{R}_p$ such that

(A.25) $\partial h/\partial x$ has maximum rank.

In other words, the terminal manifold S_1 is given by

(A.26) $S_1 = \{x : h(x) = 0\},$

and S_1 is a smooth manifold. We further require that the cost functional be defined by a trajectory cost term $L(x, u)$ where L maps $\mathbf{R}_n \times \mathbf{R}_m$ into $\mathbf{R}$ and is continuous in x and u and continuously differentiable in x. The cost of a control $u(\cdot)$ is then given by

(A.27) $J(u(\cdot)) = \int_{t_0}^{t_1} L(x_u(t), u(t)) dt,$

where $x_u(t)$ is the solution of (A.20) starting from x_0 at t_0 generated by $u(\cdot)$ and where it is assumed that $u(\cdot)$ transfers x_0 to S_1 (i.e., $x_u(t_1) \in S_1$). With these assumptions, our problem becomes: *determine a control $u(t)$ (with $u(t) \in u$ for all t) and a corresponding trajectory $x_u(t)$ of (A.20) such that $x_u(t_0) = x_0$, $x_u(t_1) \in S_1$ and $J(u(\cdot))$ is minimized.*

Now let us see how this problem can be transcribed in such a way that the Lemma (A.4) will apply to it. We first imbed the problem in an augmented state space $\mathbf{R}_{n+1}$. In other words, denoting a typical element y of $\mathbf{R}_{n+1}$ by

(A.28) $$y = (x_0, x), \qquad x \in \mathbf{R}_n,$$

and letting F be the map of $\mathbf{R}_{n+1} \times \mathbf{R}_m$ into $\mathbf{R}_{n+1}$ given by

(A.29) $$F(y, u) = (L(x, u), f(x, u)),$$

we consider the differential equation

(A.30) $$\dot{y} = F(y, u)$$

with initial condition

(A.31) $$y(t_0) = y_0 = (0, x_0)$$

and with terminal manifold $\mathbf{R} \times S_1$. We then seek a control $u(t)$ (with $u(t) \in U$ for all t) and a corresponding trajectory $y_u(t)$ of (A.30) such that $y_u(t_0) = y_0$, $y_u(t_1) \in \mathbf{R} \times S_1$ and $y_0(t_1)(= J(u(\cdot)))$ is minimized. Clearly this problem is equivalent to our original problem.

We next observe that the solutions $y_u(\cdot)$ of (A.30) are absolutely continuous functions on $[t_0, t_1]$ and so we let Ω be the set of all absolutely continuous functions $y_u(\cdot)$ which satisfy (A.30) and (A.31) for some $u(\cdot)$ satisfying (A.22). Before defining the maps f and g, we shall introduce a locally convex linear topological space X containing Ω. To do this, we let $C[t_0, t_1]$ denote the normed linear space of continuous $\mathbf{R}_{n+1}$-valued functions $y(\cdot)$ on $[t_0, t_1]$ with

$$\|y(\cdot)\| = \sup_{t \in [t_0, t_1]} \{\|y(t)\|\}$$

as norm. We let $\mathfrak{u}$ denote the set of all $\mathbf{R}_{n+1}$ valued functions $z(\cdot)$ on $[t_0, t_1]$ for which there is a sequence $x_j(\cdot)$ in $C[t_0, t_1]$ such that

(A.32) $\quad x_{j,i}(t) \le x_{j+1,i}(t)$ for all t, $i = 0, \ldots, n$ and $j = 1, \ldots,$

where $x_{j,i}$ is the i-th component of x_j, and

(A.33) $$\lim_{j \to \infty} x_j(t) = z(t)$$

for all t in $[t_0, t_1]$. Note that $\mathfrak{u} = \Pi \, \mathcal{S}_i$ where $\mathcal{S}_i$ is the set of upper semi-continuous real-valued functions of t on $[t_0, t_1]$. Letting $X = \mathfrak{u} - \mathfrak{u}$, we observe that X is a linear space. We define a topology on X by choosing as sub-base the family of sets

(A.34) $$\{z(\cdot): z(t) \in \mathcal{O}\}$$

where t is an element of $[t_0, t_1]$ and $\mathcal{O}$ is an open set in $\mathbf{R}_{n+1}$.† X is a

---

† This is the so-called *pointwise topology*.

locally convex topological vector space with respect to this topology and $\Omega \subset X$. We now define the mappings f and g by setting

(A.35) $f(z(\cdot)) = z_0(t_1),$

(A.36) $g(z(\cdot)) = h(z_1(t_1), \ldots, z_n(t_1)),$

where $z_i(\cdot)$, $i = 0, \ldots, n$ is the ith component of $z(\cdot)$. The mappings f and g are clearly continuous.

Now let $\hat{y}_{\hat{u}}(\cdot)$ be an element of Ω and let us construct an approximation $A(\hat{y}_{\hat{u}}, \Omega)$ of Ω at $\hat{y}_{\hat{u}}$ relative to f and g. We let $I \subset [t_0, t_1]$ be the set of regular points of $\hat{u}$, i.e., $t \in I$ if and only if $t_0 < t < t_1$ and

(A.37) $\displaystyle \lim_{\mu(J) \to 0} \frac{\mu(\hat{u}^{-1}(N) \cap J)}{\mu(J)} = 1$

holds for every neighborhood N of $\hat{u}^{-1}(t)$ and every subinterval J of I with $t \in J$ (where μ is the Lebesgue measure on I).† We note that $\mu(I) = t_1 - t_0 = \mu([t_0, t_1])$. We denote the transition matrix of the linear differential equation

(A.38) $\delta \dot{y} = \dfrac{\partial F}{\partial y}(\hat{y}_{\hat{u}}, \hat{u}) \delta y$

by $\Phi(t, \tau)$. In other words, $\Phi(t, \tau)$ is the solution of the matrix differential equation

(A.39) $\dfrac{d}{dt} \Phi(t, \tau) = \dfrac{\partial F(\hat{y}_{\hat{u}}(t), \hat{u}(t))}{\partial y} \Phi(t, \tau)$

satisfying the condition

(A.40) $\Phi(t, t) = \mathbf{I},$

where $\mathbf{I}$ is the $(n + 1) \times (n + 1)$ identity matrix. If $s \in I$ and $u \in U$ then we let $\delta y_{s,u}$ be the solution of (A.38) satisfying the condition

(A.41) $\delta y_{s,u}(s) = F(\hat{y}_{\hat{u}}(s), u) - F(\hat{y}_{\hat{u}}(s), \hat{u}(s)),$

and we let $\delta z_{s,u}$ be the element of X given by

(A.42) $\delta z_{s,u}(t) = \begin{cases} 0, & t_0 \le t < s; \\ \delta y_{s,u}(t), & s \le t \le t_1. \end{cases}$

Finally, we let

(A.43) $A(\hat{y}_{\hat{u}}, \Omega) = \{ \delta z : \delta z = \Sigma \alpha_i \delta z_{s_i, u_i}, \; s_i \in I, \; u_i \in U, \; \alpha_i \ge 0 \},$

---

† The significance of a regular point lies in the following observation: if τ is a regular point of $u(\cdot)$ and if $g(t, u)$ is continuous, then $\displaystyle \int_{\tau + a\epsilon}^{\tau + b\epsilon} g(t, u(t)) dt = \epsilon(b - a) g(\tau, u(\tau)) + o(\epsilon)$ where a, b are elements of $\mathbf{R}$ and ϵ is small.

and we define the linear maps $f'(\hat{y}_{\hat{u}})$ and $g'(\hat{y}_{\hat{u}})$ by setting

(A.44) $$f'(\hat{y}_{\hat{u}})\delta z = \delta z_0(t_1),$$

(A.45) $$g'(\hat{y}_{\hat{u}})\delta z = \frac{\partial h}{\partial x}(\delta z_1(t_1), \ldots, \delta z_n(t_1))(\delta z_1(t_1), \ldots, \delta z_n(t_1)),$$

where $\delta z_0, \delta z_1, \ldots, \delta z_n$ are the components of δz. The basic work of [PONTRYAGIN, 1961] includes a proof of the fact that the convex cone $A(\hat{y}_{\hat{u}}, \Omega)$ is indeed an approximation of Ω at $\hat{y}_{\hat{u}}$ relative to f and g. To illustrate what is involved in the proof, we shall exhibit the map ψ for the case of two linearly independent vectors δz_1 and δz_2 of $A(\hat{y}_{\hat{u}}, \Omega)$. We may assume (reordering and inserting zeros if necessary) that

(A.46) $$\delta z_i = \sum_{j=1}^{k} \alpha_j^i \delta z_{s_j, u_j}$$

for $i = 1, 2$ and that $s_1 \leq s_2 \leq \cdots \leq s_k$. We then note that $\delta z \in$ co $\{\delta z_1, \delta z_2\}$ implies that

(A.47) $$\delta z = \lambda_1 \delta z_1 + \lambda_2 \delta z_2 = \sum_{j=1}^{k} \delta t_j(\lambda_1, \lambda_2) \delta z_{s_j, u_j},$$

where $\lambda_1 + \lambda_2 = 1$, $\lambda_i \geq 0$ for $i = 1, 2$, and

(A.48) $$\delta t_j(\lambda_1, \lambda_2) = \lambda_1 \alpha_j^1 + \lambda_2 \alpha_j^2$$

for $j = 1, \ldots, k$. Clearly co $\{\delta z_1, \delta z_2\}$ is homeomorphic to the set $\Lambda = \{(\lambda_1, \lambda_2): \lambda_1 + \lambda_2 = 1, \lambda_i \geq 0\}$. Let $\epsilon > 0$ be a small positive number and let

(A.49) $$v_j = \begin{cases} -(\delta t_j + \cdots + \delta t_k) & \text{if } s_j = s_k, \\ -(\delta t_j + \cdots + \delta t_r) & \text{if } s_j = \cdots = s_r < s_{r+1}(j < k). \end{cases}$$

We consider the half-open intervals I_j given by

(A.50) $$I_j = \{t \in I: s_j + \epsilon v_j < t \leq s_j + \epsilon(v_j + \delta t_j)\},$$

and we suppose that ϵ is small enough to ensure that the I_j are disjoint. We define a perturbation $u(t)$ of $\hat{u}(t)$ by setting

(A.51) $$u(t) = \begin{cases} \hat{u}(t), & t \notin \cup I_j; \\ u_j, & t \in I_j. \end{cases}$$

We let $y(t; \epsilon, \lambda_1, \lambda_2)$ be the solution of (A.40) satisfying the initial conditions (A.41) for which the control is $u(t)$. We note that $u(t)$ is an admissible control and that $y(t; \epsilon, \lambda_1, \lambda_2)$ depends on λ_1, λ_2 since the δt_j depend on λ_1, λ_2. Using the standard result on dependence of solutions of a differential equation on parameters (see, for example, [DIEUDONNE, 1960]), we can deduce that $y(t; \epsilon, \lambda_1, \lambda_2)$ is a continuous function of ϵ, λ_1

and λ_2. It then follows that the mapping ψ of co $\{\epsilon\delta z_1, \epsilon\delta z_2\}$ into $\Omega - \{\hat{y}_{\hat{u}}\}$ defined by

(A.52) $$\psi(\epsilon\delta z) = y(t; \epsilon, \lambda_1, \lambda_2) - \hat{y}_{\hat{u}}(t)$$

is continuous. The proof that ψ has the other requisite properties is based upon the properties of regular points and some standard results on differential equations. The details are given in [PONTRYAGIN, 1961].

Applying the lemma (A.4), we deduce the following theorem of Pontryagin:

(A.53)   **Theorem (Maximum Principle).**   *If $u^0(\cdot)$ is an optimal control with $x^0(\cdot)$ as corresponding optimal trajectory, then there are a vector valued function $p(\cdot)$, a vector $\lambda(t_1)$ in $\mathbf{R}_p$ and a number $p_0 \leq 0$ such that the following conditions are satisfied:*
   (a) p_0, $p(t)$ is not identically zero.
   (b) $x^0(\cdot)$ and p_0, $p(\cdot)$ satisfy the canonical system of differential equations, i.e.,

(A.54) $$\dot{x}^0(t) = f(x^0(t), u^0(t)),$$

(A.55) $$\dot{p}(t) = - \frac{\partial L}{\partial x}(x^0(t), u^0(t))p_0 - \frac{\partial f}{\partial x}(x^0(t), u^0(t))'p(t).$$

   (c) $u^0(t)$ maximizes the Hamiltonian over U for almost all t, i.e.,

(A.56) $$p_0 L(x^0(t), u^0(t)) + \langle p(t), f(x^0(t), u^0(t)) \rangle \geq p_0 L(x^0(t), u) + \langle p(t), f(x^0(t), u) \rangle$$
   for all u in U;
   (d) $p(t_1)$ is transversal to S_1 at $x^0(t_1)$, i.e.,

(A.57) $$p(t_1) = \frac{\partial h}{\partial x}(x^0(t_1))\lambda(t_1).$$

As we have remarked before, a more general version of this theorem involving (for example) state space constraints has been proven by [NEUSTADT, 1966–67] and [HALKIN, 1967].

5 Control system design

We are now going to discuss the ways in which the theoretical considerations of the previous chapter are applied in control-system design. We begin with the examination of the simple problem of the time optimal control of a double-integral plant. Then in Section 5.2 we consider the Pontryagin principle (or necessary-condition) approach to deriving optimal controls in a general control problem. As the "practical" solution of most control problems requires the use of numerical techniques, we devote the remainder of the chapter to these techniques, considering both indirect and direct methods.

5.1 A simple example

Let us consider the dynamical system

(1.1) $$\dot{x}_1(t) = x_2(t), \qquad \dot{x}_2(t) = u(t)$$

with the control functions constrained to satisfy the relation

$$|u(t)| \leq 1 \quad \text{for all } t.$$

The target set S will be $\mathbf{R} \times \{(0, 0)\}$, so that we deal with a free-time fixed-end-point problem. We suppose that there is no terminal cost $(K = 0)$ and that the trajectory-cost term is exactly

$$L(x, u) = 1.$$

Thus our control problem is to determine the admissible control that transfers any given initial state (ξ_1, ξ_2) to the origin $(0, 0)$ in minimum time.†

We now assert that a solution of this problem exists. Theorem (3.5) of Chapter 4 can be applied to verify the correctness of our assertion, for from this theorem we deduce that $A(0, (\xi_1, \xi_2), \Omega)$, where Ω is our set of admissible controls, is closed. We then note that the system is completely controllable relative to Ω, so that $\bar{t} \times (0, 0)$ is an element of $A(0, (\xi_1, \xi_2), \Omega)$ for some $\bar{t} > 0$. Now, if we consider the problem with target set $\tilde{S} = [0, \bar{t}] \times \{(0, 0)\}$ and cost functional t, then this problem has a solution, as $\tilde{S} \cap A(0, (\xi_1, \xi_2), \Omega)$ is compact. *A fortiori*, our original time optimal control problem has a solution. Having established the existence of an optimal control, we develop a step-by-step procedure for finding it.

Step 1   Determination of the H-minimal control. The hamiltonian for this problem is given by

$$H(x, p, u) = 1 + x_2 p_1 + u p_2.$$

If x_2, p_1, and p_2 are given, and if $p_2 \neq 0$, then the minimum of H as a function of u for $u \in U = \{v: |v| \leq 1\}$ occurs when $u = - \text{sgn} \{p_2\}$. Thus the H-minimal control $u^0(x, p)$ is given by

$$u^0(x, p) = - \text{sgn} \{p_2\}$$

(provided that $p_2 \neq 0$). We note that H is not quite regular.

Step 2   Integration of the canonical equations. Consider the system of equations

$$\dot{x}_1(t) = x_2(t), \qquad \dot{x}_2(t) = - \text{sgn} \{p_2(t)\},$$
$$\dot{p}_1(t) = 0, \qquad \dot{p}_2(t) = -p_1(t)$$

with the boundary conditions

$$x_1(0) = \xi_1, \qquad x_2(0) = \xi_2,$$
$$x_1(t^*) = 0, \qquad x_2(t^*) = 0,$$

---

† This is the so-called *time-optimal control problem* for the *double-integral plant*. Since the system is linear and autonomous, we may suppose that the initial time t_0 is zero.

where t^* is the unknown terminal time. Can we determine the solution of these equations? Let us denote the unknown values of $p_1(0)$ and $p_2(0)$ by π_1 and π_2, respectively. Then

$$p_1(t) = \pi_1, \qquad p_2(t) = \pi_2 - \pi_1 t.$$

We observe that unless $p_1(t) = p_2(t) = 0$, $p_2(t)$ has at most one zero. As an optimal control exists, $p_1(t) = p_2(t) = 0$ is impossible. Thus, letting $\Delta(t) = - \operatorname{sgn} \{p_2(t)\}$, we deduce that

(1.2)
$$x_1(t) = \xi_1 + \xi_2 t + \frac{t^2}{2} \Delta(t),$$
$$x_2(t) = \xi_2 + t \, \Delta(t),$$

and moreover, that $\Delta(t)$ is a piecewise constant function which can take on one of the following four sequences of values: $\{1\}$, $\{-1\}$, $\{1, -1\}$, $\{-1, 1\}$. We call these sequences the candidates for optimal control (see [ATHANS and FALB, 1966]). If we eliminate t in (1.2), we obtain trajectories in the state plane; these trajectories are parabolas and are

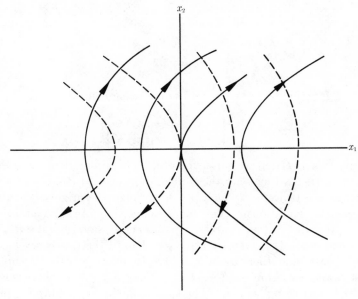

Figure 5.1

illustrated in Figure 5.1. The arrows indicate increasing t, the solid trajectories are for $\Delta \equiv 1$, and the dashed trajectories are for $\Delta \equiv -1$.

Step 3 Computation of the control law. We observe from Figure 5.1 that there is a unique trajectory through the origin for $\Delta \equiv 1$ and for

$\Delta \equiv -1$.   We let $\alpha = \alpha_- \cup \alpha_+$ be the curve in the state plane defined by

$$\alpha_- = \{(x_1, x_2): x_1 = -\tfrac{1}{2}x_2^2, \ x_2 \geq 0\},$$
$$\alpha_+ = \{(x_1, x_2): x_1 = \tfrac{1}{2}x_2^2, \ x_2 \leq 0\},$$

and we let R_- and R_+ be the regions above and below α, respectively, as illustrated in Figure 5.2. We now assert that the optimal control is

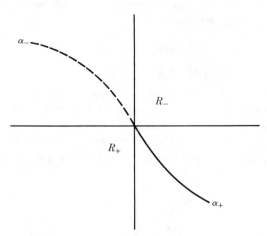

Figure 5.2

given as a function of the state by the following control law:

$$(1.3) \qquad u^0(x_1, x_2) = \begin{cases} 1, & (x_1, x_2) \in \alpha_+ \cup R_+, \\ -1, & (x_1, x_2) \in \alpha_- \cup R_-. \end{cases}$$

To verify this assertion we can use a process of elimination. For example, if $(x_1, x_2) \in \alpha_+$, then only the control sequence $\{1\}$ (that is, $\Delta(t) \equiv 1$) will generate a trajectory which passes through the origin and satisfies the necessary conditions. (For more details see [ATHANS and FALB, 1966]). Now, if (ξ_1, ξ_2) is our given state and if, say, $(\xi_1, \xi_2) \in R_-$, then the optimal control is given by $u^0(t) = -1$ until the trajectory meets α (actually, α_+), and then by $u^0(t) = 1$. In other words, the optimal control "switches" from -1 to $+1$ at the curve α. For this reason α is called the "switching" curve. The time required to get to the origin by using the control law (1.3) is given as a function of the initial state by

$$t^*(\xi_1, \xi_2) = \begin{cases} \xi_2 + \sqrt{4\xi_1 + 2\xi_2^2}, & \xi_1 > -\tfrac{1}{2}\xi_2|\xi_2|, \\ -\xi_2 + \sqrt{-4\xi_1 + 2\xi_2^2}, & \xi_1 \leq -\tfrac{1}{2}\xi_2|\xi_2|. \end{cases}$$

We note that $\partial t^*/\partial \xi_1$ has a discontinuity along the curve α, so that the Hamilton-Jacobi sufficiency condition cannot be applied globally.

Now let us review what has been done. We examined the hamiltonian, and we determined the H-minimal control. We then integrated the canonical equations and in effect constructed all the extremals (the trajectories satisfying all the necessary conditions) of our problem. Next we used a process of elimination to show that the extremals were unique; coupling this uniqueness with the demonstrated existence of an optimal control (which, of course, must be extremal), we computed the desired optimal control law. Thus we can isolate three important stages in the process of deriving the optimal control: (1) the determination of the H-minimal control, (2) the construction of all the extremals through solution of the two-point boundary-value problem for the canonical system, and (3) computation of the optimal control law. The entire process was carried out under the condition that an optimal control existed.

5.2 Control-system design via the Pontryagin principle

The Pontryagin principle, or necessary-condition approach, to deriving optimal controls in a general control problem is similar to the procedure of Section 5.1. However, this simple example does not indicate or illustrate many of the difficulties which may be encountered at various stages of the process. Let us therefore discuss each step in turn, indicating some of the attendant problems which may arise.

Step 1. Determination of the H-minimal control. We begin by forming the hamiltonian, and then we attempt to determine the H-minimal control. If H is regular, then this is, at least in theory, always possible. On the other hand, if H is not regular, then difficulties arise. For example, there may be ranges of values for x and p for which the relation $H(x, p, \omega) = $ constant for all ω in U is satisfied. This means, in effect, that the hamiltonian minimization necessary condition provides no information. Such problems are called *singular problems*.† Alternatively, for various values of x and p there may be several points in U at which H has its absolute minimum. In such a case there may be several extremals through a given initial point, and the cost for each must be evaluated (examples illustrating these points may be found in [ATHANS and FALB, 1966]). As a trivial illustration, we have

(2.1) **Example.** Consider the system $\dot{x}(t) = u(t)$ with $|u(t)| \leq 1$. Let $S = \mathbf{R} \times \{0\}$, $K = 1$, and $L(x, u) = |u|$. Then $H = |u| + pu$ and

---

† For a detailed discussion of singular problems see [JOHNSON, 1965; JOHNSON and GIBSON, 1963].

$\dot{p}(t) = 0$, so that p is a constant. If, for example, $p = 1$, then any u in $[-1, 0]$ will minimize H. It is easy to see that if $\xi_1 > 0$ is our initial state, then any control of the form $u(t) = -k$, $k \in [0, 1]$, will be optimal.

Step 2. Integration of the canonical system. In the second stage we attempt to find the extremals through our initial point by solving (integrating) the two-point boundary-value problem for the canonical system. In general this must be done numerically. Various iterative techniques such as the gradient method [KELLEY, 1962], convexity methods (particularly in time optimal problems [EATON, 1962; NEUSTADT, 1960], and the general Newton or Newton-Raphson procedure [KANTOROVICH and AKILOV, 1964; VAN DINE, 1965] have been used successfully in a number of problems. A frequently used approach is to "guess" an initial value of the costate $p(t_0)$, integrate the canonical equations forward in time, check to see if the terminal conditions on x and p are met, and if not, to alter the initial value of p and repeat the procedure. Naturally, questions of convergence and accuracy immediately come to mind in this context. Considerable current research is devoted to these problems, and we shall have more to say about numerical techniques in Sections, 5.3 to 5.5.

Step 3. Computation of the optimal control law. When all the extremals have been computed, we evaluate the cost along each extremal, compare these costs, and thus determine the optimal control. If the optimal control and/or the extremal controls are not unique, this can be a rather difficult matter. Thus there is considerable utility in uniqueness results for extremals. Some interesting work along these lines appears in [ATHANS, 1966] and in [LEE and MARKUS, 1961].

We again observe that use of necessary conditions presupposes the existence of an optimal control. In general the question of existence is quite difficult, and a certain wariness of the "blind" application of the necessary conditions is called for.

For numerous examples of the application of the Pontryagin principle to particular control-system design problems see [ATHANS and FALB, 1966].

5.3 Numerical techniques in control theory: general remarks

We have already noted that the actual solution of most control problems requires the use of numerical techniques, particularly iterative methods.

These iterative methods can be loosely divided into two categories: indirect methods and direct methods. Indirect methods are generally based on a reduction of the control problem to a problem involving a differential equation (such as the Hamilton-Jacobi partial differential equation) or a system of differential equations (such as the canonical system of the Pontryagin principle). Direct methods generally revolve around the construction of suitable minimizing families. *No matter what type of method or algorithm is proposed, the basic questions of convergence, error estimation, and computer effects must be answered.* In other words, if we are given an iterative (numerical) procedure for solving a control problem, then we want to know the following:

1. Does the procedure converge, and are there reasonable conditions ensuring convergence?

2. Is it possible to determine suitable bounds (or estimates) on the error at each stage of the iteration?

3. How do the truncation (roundoff) errors of a digital computation procedure and the inherent inaccuracies (noise) of an analog-computation procedure affect the results obtained?

As might be expected, there are no generally applicable answers to these questions. We shall therefore examine, in a rather general way, some of the approaches which have been applied to obtain extremals in control problems (that is, to solve the two-point boundary-value problem for the canonical system). In the next section we shall turn our attention to some more specific aspects of the application of these basic iterative methods in control problems.

Now, in considering the iterative methods which have been used to obtain extremals for control problems, we shall discuss three features of the methods: (1) the particular iterative or adjustment technique, (2) the type of quantity adjusted, and (3) the particular extremal condition (or conditions) at which the iteration is aimed. The three major iterative techniques that we shall briefly examine are the gradient, Newton (or Newton-Raphson), and successive-approximation methods. We shall also mention in passing the very important penalty-function method for the handling of constraints. As a matter of fact, the approximation theorem of Moser [COURANT, 1962], which we give here in Section 5.5, often provides a sound theoretical justification for using penalty functions. The two types of quantities, adjusted in accordance with one of the iterative techniques, are vectors, such as the initial conditions for p, and functions, such as the control. In view of the Pontryagin principle, an extremal control and trajectory must satisfy the following conditions:

1. The canonical system of differential equations,

2. The hamiltonian minimization relation,

3. The boundary conditions determined by the initial point, the target set, and the transversality condition.[†]

The most frequently used procedures involve the generation of trajectories satisfying two of these conditions which are iteratively changed until the third condition is satisfied. In the so-called "neighboring-optimum" method [BRYSON and DENHAM, 1962; DENHAM, 1963; BREAKWELL, SPEYER, and BRYSON, 1963], trajectories satisfying conditions (1) and (2) are modified until (3) is satisfied. The popular "gradient" technique (see, for example, [BLUM, 1964; BUSHAW, 1963; DENHAM, 1963; KANTOROVICH and AKILOV, 1964; KELLEY, 1960, 1962]) is a procedure in which trajectories satisfying conditions (1) and (3) are changed until (2) is satisfied. The remaining possibility, iteration on solutions of (2) and (3) until the canonical system is solved [that is, until (1) holds], is often referred to as the quasilinearization method [KALABA, 1959; McGILL and KENNETH, 1963]. In any of these methods the iteration may be performed in accordance with a suitable choice of technique operating on either a vector or a function type of quantity. See also [KALMAN, 1966a].

Now let us turn our attention to the basic iteration techniques viewed in an abstract way. We suppose that Π is a subset of a Banach space $\mathcal{P}$, and we call the elements of Π (or $\mathcal{P}$) *parameters*. We assume that E is a mapping of Π into a Banach space $\mathcal{E}$, and we shall call E the *error mapping*. The goal of the iteration methods is to make the "error" $E(\pi)$ zero, or at least very near zero. The key idea behind each of the three techniques is linear approximation.[‡] In other words, if π_0 is our old guess of the parameter value, then our new guess π_n is given by

$$\pi_n = \pi_0 - \Phi\{E(\pi_0)\},$$

where Φ is a suitable element of $\mathcal{L}(\mathcal{E}, \mathcal{P})$, (that is, Φ is a continuous linear transformation from $\mathcal{E}$ into $\mathcal{P}$). Thus the choice of the linear transformation Φ is what distinguishes the various methods.

The gradient method. Here let us assume that $\mathcal{P}$ and $\mathcal{E}$ are Hilbert spaces. Then, if E is differentiable on Π, we have

$$E(\pi + h) = E(\pi) + \frac{\partial E}{\partial \pi}\Big|_\pi h + o(\|h\|),$$

where $\partial E/\partial \pi$ is an element of $\mathcal{L}(\mathcal{P}, \mathcal{E})$. Since $\mathcal{P}$ and $\mathcal{E}$ are Hilbert spaces,

[†] The additional extremal condition $H = 0$ is rarely applied in iterative procedures, as these are used directly in nonautonomous problems.

[‡] Frequently, going to "higher-order" terms is extremely useful [COLLATZ, 1964] However, we shall not consider such ramifications here.

the adjoint $(\partial E/\partial \pi)^*$ of $\partial E/\partial \pi$ is an element of $\mathcal{L}(\mathcal{E}, \mathcal{P})$, so we can consider the iteration scheme

$$(3.1) \qquad \pi_{k+1} = \pi_k - r_k \left(\frac{\partial E(\pi_k)}{\partial \pi} \right)^* E(\pi_k),$$

where r_k is a real number. If $\mathcal{E} = \mathbf{R}$, then $(\partial E/\partial \pi)^*$ can be identified with the gradient ∇E (which is an element of $\mathcal{P}$), and choosing $r_k = s_k/E(\pi_k)$, we may write (3.1) in the form

$$(3.2) \qquad \pi_{k+1} = \pi_k - s_k \, \nabla E(\pi_k).$$

This is the familiar "steepest-descent" or "gradient" formula, and s_k is called the *step size* at the kth step. Results relating to the convergence of (3.2) can be found in [KANTOROVICH and AKILOV, 1964]. In a little while we shall see one way in which (3.2) has been applied in control problems.

Newton's method. Let us suppose that E is differentiable at a point $\hat{\pi}$. Then

$$(3.3) \qquad E(\hat{\pi} + h) = E(\hat{\pi}) + \left(\frac{\partial E(\hat{\pi})}{\partial F} \right) h + o(\|h\|),$$

where $\partial E/\partial \pi$ is an element of $\mathcal{L}(\mathcal{P}, \mathcal{E})$. Assuming for the moment that (3.3) is true without the correction term $o(\|h\|)$, we have

$$E(\hat{\pi} + h) = E(\hat{\pi}) + \frac{\partial E(\hat{\pi})}{\partial \pi f} \, h,$$

and letting $\hat{h} = (-\partial E(\hat{\pi})/\partial \pi)^{-1} E(\hat{\pi})$,

$$E(\hat{\pi} + \hat{h}) = 0.$$

This observation leads us to the following iteration scheme:

$$(3.4) \qquad \pi_{k+1} = \pi_k - \left(\frac{\partial E(\pi_k)}{\partial \pi} \right)^{-1} E(\pi_k)$$

(assuming, of course, that the indicated inverse exists). The procedure of (3.4) is called Newton's method. Frequently (3.4) is replaced by

$$(3.5) \qquad \pi_{k+1} = \pi_k - \left(\frac{\partial E(\pi_0)}{\partial \pi} \right)^{-1} E(\pi_k),$$

where π_0 is the initial guess. This is called the modified Newton's method. Convergence theorems for (3.4) and (3.5) can be found in [KANTOROVICH and AKILOV, 1964]. We shall later examine a use of this method in the "neighboring-optimum" approach to control problems.

The successive-approximation method.   The familiar procedure of successive approximation, also called Picard's method, is quite useful. We let Φ be an element of $\mathcal{L}(\mathcal{E}, \mathcal{P})$, and we simply let

$$(3.6) \qquad \pi_{k+1} = \pi_k - \Phi\{E(\pi_k)\}.$$

Naturally, an effort is made to choose Φ in a judicious fashion.   For example, the modified Newton's method, (3.5), may be viewed as a Picard iteration with $\Phi = [\partial E(\pi_0)/\partial \pi]^{-1}$.   Often Φ is chosen to be contractive (norm-reducing), although this is not necessary for convergence. Properties of the successive-approximation method are discussed in [COLLATZ, 1964; KANTOROVICH and AKILOV, 1964].   Unfortunately, little use has been made of this method in control problems, and we shall indicate in the sequel a possible use of this technique in a quasi-linearization type of approach to control problems (however, see [FALB and DEJONG, 1968]).

5.4   Numerical techniques in control theory: indirect methods in control problems

We now turn our attention to the application of these general iterative methods to control problems.   Consider a dynamical system with equation

$$(4.1) \qquad \dot{x} = f(x, u)$$

and suppose that (t_0, x_0) is our initial event, that $S = \{t_1\} \times S_X$ is our target set, and that the cost involves only a terminal-cost term of the form $K(x)$.   Recall (see Section 4.1) that if $\hat{u}(\cdot)$ is a given control and if $\hat{x}(\cdot)$ is the corresponding trajectory of equation (4.1), then the equation

$$(4.2) \qquad \dot{\psi} = \frac{\partial f}{\partial x}\Big|_\wedge \psi + \frac{\partial f}{\partial u}\Big|_\wedge h$$

is called the *perturbation equation*.   We write (4.2) in the more familiar form

$$(4.3) \qquad \delta\dot{x} = \frac{\partial f}{\partial x}\Big|_\wedge \delta x + \frac{\partial f}{\partial u}\Big|_\wedge \delta u,$$

and since our initial event (t_0, x_0) is fixed, the initial condition for (4.3) is

$$\delta x(t_0) = 0.$$

Now let us examine three uses of the iterative methods.

The gradient method. Here the quantity over which we shall iterate is the control function. Thus we shall suppose that $\mathcal{B}_U$ is a Hilbert space and that $\Omega \subset L_2([t_0, t_1], \mathcal{B}_U)$ is a Hilbert space with inner product given by

$$\langle u(\cdot), v(\cdot) \rangle = \int_{t_0}^{t_1} \langle u(t), v(t) \rangle_U \, dt.$$

In other words, the parameter space $\mathcal{P}$ is $L_2([t_0, t_1], \mathcal{B}_U)$, the space $\mathcal{E}$ is simply $\mathbf{R}$, the real numbers, and the error mapping E is given by

$$E(u(\cdot)) = K(x_u(t_1)) - K^0,$$

where $K^0 = \inf \{ K(x_u(t_1)) : u(\cdot) \in \Omega \}$. We observe that if $\hat{u}(\cdot) \in \Omega$, then $\nabla E(\hat{u}(\cdot))$ is an element of $L_2([t_0, t_1], \mathcal{B}_U)$ given by

$$\nabla E(\hat{u}(\cdot))(\cdot) = \left(\frac{\partial f}{\partial x} \Big|_{\wedge} \right)^* p(\cdot),$$

where $p(\cdot)$ is a solution of the adjoint equation

(4.4) $$\frac{dp}{dt} = \left(\frac{\partial f}{\partial x} \Big|_{\wedge} \right)^* p$$

[note that $p(t) \in X^*$, the dual space of the state space.] The adjoint (or costate) variable $p(\cdot)$ is sometimes called the *influence function*. Assuming that the target set S_X is given by

$$S_X = \{ x : g_i(x) = 0, \, i = 1, \ldots, k \},$$

where the g_i have continuous derivatives which are linearly independent on S_X, we take the boundary condition for $p(\cdot)$ to be

$$p(t_1) = \frac{\partial K}{\partial x} (\hat{x}(t_1)) - \sum_{i=1}^{k} \alpha_i \frac{\partial g_i}{\partial x} (\hat{x}(t_1)),$$

where the α_i are suitable constants. We observe then that

$$E(\hat{u}(\cdot) + \delta u(\cdot)) = E(\hat{u}(\cdot)) + \int_{t_0}^{t_1} \langle \nabla E(\hat{u}(\cdot))(t), \delta u(t) \rangle \, dt + o(\|\delta u\|).$$

The greatest change in E for a given value of $\|\delta u(\cdot)\|^2$ is obtained when

$$\delta u(\cdot) = s \, \nabla E(\hat{u}(\cdot))(\cdot),$$

where s is a constant. We note also that since (4.4) is linear and homogeneous, we may write $\nabla E(\hat{u}(\cdot))$ in the form

$$\nabla E(\hat{u}(\cdot)) = \nabla_K E(\hat{u}(\cdot)) + \sum_{i=1}^{k} \nabla_i E(\hat{u}(\cdot)),$$

where

$$\nabla_K E(\hat{u}(\cdot)) = \left(\frac{\partial f}{\partial u}\Big|_{\wedge}\right)^* p_K(\cdot),$$

$$\nabla_i E(\hat{u}(\cdot)) = \left(\frac{\partial f}{\partial u}\Big|_{\wedge}\right)^* p_i(\cdot),$$

and $p_K(\cdot)$, $p_i(\cdot)$ are solutions of (4.4) satisfying the boundary conditions

$$p_K(t_1) = \frac{\partial K}{\partial x}(\hat{x}(t_1)), \qquad p_i(t_1) = -\alpha_i \frac{\partial g_i}{\partial x}(\hat{x}(t_1)).$$

This exhibits the change in E as generated by changes in K and the terminal point. We can now indicate the steps in the "gradient" method:

1. Choose a nominal control function $u_0(\cdot)$ which generates a trajectory of (4.1) that satisfies the boundary conditions (in practice, only approximately).

2. Integrate the adjoint equation backward in time to obtain the functions $p_K(\cdot)$ and $p_i(\cdot)$, $i = 1, \ldots, k$.

3. Adjust the control function according to the gradient procedure. In other words, choose

$$u_1(\cdot) = u_0(\cdot) - s_{0K} \nabla_K E(u_0(\cdot)) - \sum_{i=1}^{k} s_{0i} \nabla_i E(u_0(\cdot)),$$

where the s_{0K}, and s_{0i} are suitable step-size constants. Recompute the trajectory, using the control $u_1(\cdot)$ to obtain a new nominal trajectory $x_1(\cdot)$.

4. Iterate until the value of E is sufficiently small (or until $\nabla E = 0$).

The choice of the step sizes is frequently a matter of judgment and intuition. Moreover, the method is not very useful in a small neighborhood of the optimal trajectory. Finally, in practice, satisfaction of the terminal conditions often presents difficulties; the penalty-function method is frequently used to overcome these difficulties. For example, instead of attempting to satisfy the conditions $g_i(x) = 0$, we might adjoin a term of the form

$$\sum_{i=1}^{k} a_i g_i^2(x), \ a_i > 0, \qquad i = 1, \ldots, k$$

to our cost $K(x)$ (this term would then be the penalty function). For more details and examples of the use of the gradient method the reader should consult [BLUM, 1964; DENHAM, 1963; BRYSON and DENHAM, 1962; KANTOROVICH and AKILOV, 1964; KELLEY, 1960].

The Newton-Raphson method. Here the quantity over which we shall iterate is the initial condition for the adjoint equation. In order to set up the procedure, we shall suppose that the hamiltonion

$$H = \langle p, f(x, u) \rangle$$

is regular, so that an H-minimal control $u^0(x, p)$ exists. We let

$$Y = X \times X^*,$$

with a typical element y of Y written in the form $y = (x, p)$. If we substitute the H-minimal control $u^0(x, p)$ into the canonical equations, then we obtain the system

(4.5) $$\frac{dy(t)}{dt} = \mathcal{Y}(y(t))$$

in the space Y. We note that $\mathcal{Y}$ is given by

$$\mathcal{Y}(x, p) = \left(f(x, u^0(x, p)), -\frac{\partial f}{\partial x}(x, u^0(x, p))^* p \right).$$

Now the boundary conditions for an *optimal* trajectory are "split" between the initial time and the terminal time. For example, if $y^0(t)$ is an optimal trajectory, then we know that $y^0(t_0)$ must be of the form $(x_0, \cdot)$, where x_0 is our given initial state, and that $y^0(t_1) = (x_1, p_1)$, where x_1 is an element of the target set S_X and p_1 is transversal to S_X. Now we observe that if we knew the "right" value for the initial costate, then we could integrate (4.5) forward in time and obtain the optimal solution (since solutions of the initial value problem are unique). If π is any element of X^*, then (4.5) has a unique solution $y(t; \pi)$ satisfying the initial condition

$$y(t_0; \pi) = (x_0, \pi),$$

and the "error" in the terminal conditions will be

$$y(t_1; \pi)_f - y^0(t_1)_f,$$

where the subscript f indicates that only the "proper" portion of the element is considered. For example, if $S_X = \{x_1\}$ is a fixed point, then $y^0(t_1)_f = x_1$ and $y(t_1; \pi)_f = x(t_1; \pi)$. So we let X^* be our parameter space, Y (or, better, Y_f) be the space $\mathcal{E}$, and the mapping E given by

$$E(\pi) = y(t_1; \pi)_f - y^0(t_1)_f$$

be our error mapping. We observe that

$$\frac{\partial E}{\partial \pi} = \frac{\partial y(t_1; \pi)_f}{\partial \pi},$$

so that the Newton-Raphson iterative procedure becomes

$$\pi_{k+1} = \pi_k - \left(\frac{\partial y(t_1; \pi)_f}{\partial \pi} \right)^{-1} [y(t_1; \pi_k)_f - y^0(t_1)_f].$$

Thus the determination of $\partial y(t_1; \pi)_f / \partial \pi$ is crucial. The perturbation equation

$$(4.6) \qquad\qquad \dot{\delta y} = \frac{\partial \dot{y}}{\partial y} \delta y$$

is frequently used here. For example, if the system is finite dimensional, then, letting $\epsilon_1, \ldots, \epsilon_n$ denote the natural basis of $\mathbf{R}_n$ (which we identify with X^*), we readily deduce that the solution of (4.6) satisfying the initial condition

$$\delta y(t_0) = (0, \epsilon_i), \qquad i = 1, \ldots, n$$

is the vector

$$\frac{\partial y(t; \pi)}{\partial \pi_i}, \qquad i = 1, \ldots, n.$$

Thus in this case the matrix $\partial y(t_1; \pi)/\partial \pi$ (*and a fortiori* $\partial y(t_1; \pi)_f/\partial \pi$) can be found by no more than n integrations of the perturbation equation (4.6). We can now indicate, just as before, the steps required:

1. Guess an initial value for the costate.

2. Integrate the canonical equations forward in time, using the given initial value for the state, the guessed initial value of the costate, and the H-minimal control.

3. Adjust the initial value of p in accordance with Newton's method. In particular, this involves determination of $\partial y(t_1; \pi_k)_f/\partial \pi$, which usually requires suitable integrations of the perturbation equation. The adjustment takes the form

$$\pi_{k+1} = \pi_k - \left(\frac{\partial y(t_1; \pi_k)_f}{\partial \pi} \right)^{-1} [y(t_1; \pi_k)_f - y^0(t_1)_f].$$

4. Iterate until the error is sufficiently small.

Here no judgment of step size is necessary; however, each step requires the computation of the inverse of a linear transformation (a matrix in the finite-dimensional case), which often involves a number of integrations of the linear-perturbation equation. Since convergence of this method is usually quadratic, it is a good procedure to apply in a small neighborhood of the optimum. Also, in many cases the modified Newton's method can be used to alleviate the difficulties involved in inverting the linear transformation $\partial E/\partial \pi$. For additional information and examples consult [BALAKRISHNAN and NEUSTADT, 1964; BRYSON and DENHAM, 1962; KANTOROVICH and AKILOV, 1964].

The successive-approximation method. Here the quantities to be iterated on are the trajectory and the adjoint trajectory. In effect, we replace the original boundary-value problem by a sequence of linear boundary-value problems in which the linearization is fixed. Again we shall suppose that the hamiltonian H is regular, so that an H-minimal control $u^0(x, p)$ exists. Letting $Y = X \times X^*$ and substituting $u^0(x, p)$ into the canonical equations, we obtain the system

$$\frac{dy(t)}{dt} = \mathcal{Y}(y(t))$$

in the space Y. We suppose that the boundary conditions for an optimal trajectory $y^0(t)$ are given by

$$y^0(t_0) = (x_0, \cdot), \qquad y^0(t_1) = (\cdot, p_1),$$

or, in other words, that the initial state and the terminal costate are known.† If we let $A(t)$ be a continuous map of $[t_0, t_1]$ into $\mathcal{L}(Y, Y)$, then we can consider the linearization based on $A(t)$. In other words, if $y_1(t)$ is a differentiable curve in Y such that

$$y_1(t_0) = y^0(t_0) = (x_0, \cdot), \qquad y_1(t_1) = y^0(t_1) = (\cdot, p_1),$$

then we can consider the *linear* two-point boundary-value problem for the equation

(4.7) $$\frac{d \, \delta y(t)}{dt} = A(t) \, \delta y(t) + \mathcal{Y}(y_1(t)) - \frac{dy_1(t)}{dt}$$

with boundary conditions

(4.8) $$\delta x(t_0) = 0, \qquad \delta p(t_1) = 0.$$

If we let $\delta y_1(t)$ be the solution of this linear two-point boundary-value problem we can see that $y_2(t) = y_1(t) + \delta y_1(t)$ will also satisfy the requisite boundary conditions. We can now indicate the steps in the procedure:

1. Determine a differentiable curve satisfying the boundary conditions.

2. Solve the linear two-point boundary-value problem represented by (4.7) and (4.8).

3. Adjust the trajectory in accordance with the iteration

$$y_{k+1}(t) = y_k(t) + \delta y_k(t)$$

(note that this is *not* a linear iteration procedure because of the forcing term in (4.7)).

4. Iterate until $\delta y = 0$ or becomes very small.‡

Very little of a practical nature is known about the procedure out-

---

† Other combinations are possible of course, and we leave them to the reader.
‡ A simple calculation shows that if, say, $\delta y_k = 0$, then $\dot{y}_k = \mathcal{Y}(y_k)$.

lined here.   The more familiar method of quasi-linearization involves the replacement of (4.7) by the equation

$$\frac{d\,\delta y}{dt} = \frac{\partial \mathcal{Y}}{\partial y}\,(y)\,\delta y + \left[\mathcal{Y}(y) - \frac{dy}{dt}\right].$$

Results pertaining to the quasi-linearization approach can be found in [McGILL and KENNETH, 1963].   Additional work along these lines would probably prove fruitful and some results are given in [FALB and DEJONG, 1968].

The various approaches to the determination of optimal controls that we have examined so far all involve solution of the two-point boundary-value problem for the canonical system.   Thus these approaches are indirect and, moreover, lead only to extremals.   Independent proofs of existence and a consideration of the uniqueness of the extremals is also required if these methods are to furnish us with the optimal control (or a reasonable approximation to it).   Frequently such demonstrations are not given in the literature.   Furthermore, in many cases it is difficult to estimate the error (or value) of the cost functional at each stage of the iteration.   So we now turn our attention to direct methods.

5.5   Numerical techniques in control theory: direct methods

The key idea in so-called direct methods is the notion of a minimizing family.   Several results, including the approximation theorem of Moser [COURANT, 1962], will be presented during the course of our discussion.

We observed in Chapter 3 that the control problem may be viewed as a special case of the problem of minimizing a functional on a subset of a normed linear space; let us adopt this point of view now.   We consider a normed linear space $\mathfrak{B}$, a nonempty subset Φ of $\mathfrak{B}$, a real-valued functional J on $\mathfrak{B}$, and the problem of minimizing J on Φ.   Suppose that the problem makes sense, that is, that

$$(5.1) \qquad\qquad -\infty < \inf_{\Phi} J(\cdot) = J^0 < \infty.$$

We now have

(5.2)   **Definition.**   *Let* $D = \{\delta : \delta \in [0,\ \infty)\}$ *be an unbounded upwardly directed† subset of the nonnegative half line.   Then a family of elements*

---

† An *upwardly directed set* D is defined as follows: If δ_1 and δ_2 are elements of D, then there is an element δ_3 of D such that $\delta_1 < \delta_3$ and $\delta_2 < \delta_3$.   The set of positive integers is a typical unbounded upwardly directed set.

v_δ of $\mathcal{B}$ *is called a* **minimizing family** *for J over Φ if*

$$\lim_{\delta \to \infty} J(v_\delta) = J^0 = \inf_{\Phi} J(\cdot).$$

If, in addition,

$$\lim_{\delta \to \infty} v_\delta = v^0,$$

where $v^0 \in \Phi$, then $\{v_\delta\}$ is called a **convergent minimizing family.**

In view of (5.1), minimizing families always exist. However, there are not always convergent minimizing families. If we knew that there were a convergent minimizing family and that J were sufficiently "smooth" to ensure that

$$J(v^0) = J(\lim_{\delta \to \infty} v_\delta) = \lim_{\delta \to \infty} J(v_\delta) = J^0,$$

then we could conclude that v^0 was a solution of our minimization problem. These observations are at the heart of the direct-method approach which, in essence, consists of (1) the construction of a suitable convergent minimizing family and (2) the demonstration that J is smooth. The following proposition is indicative of the type of smoothness required of J for the applicability of the direct method technique:

(5.3) **Proposition.** *Suppose that $\{v_\delta\}$ is a convergent minimizing family with limit v^0 and that J is lower semicontinuous at v^0.[†] Then v^0 minimizes J on Φ. In other words, $J(v^0) = J^0 = \inf_{\Phi} J(\cdot)$.*

Proof. Since $\{v_\delta\}$ is a minimizing family and $v^0 \in \Phi$, we have

(5.4) $$J(v^0) \geq J^0 = \inf_{\Phi} J(\cdot) = \lim_{\delta \to \infty} J(v_\delta).$$

On the other hand, if $\epsilon > 0$, then there is a $\delta(\epsilon)$ such that $\delta > \delta(\epsilon)$ implies that

$$J(v^0) \leq J(v_\delta) + \epsilon,$$

as J is lower semicontinuous at v^0. It follows that

$$J(v^0) \leq \lim_{\delta \to \infty} J(v_\delta) + \epsilon = J^0 + \epsilon$$

and hence that

$$J(v^0) \leq J^0,$$

which, together with (5.4), establishes the proposition. $\qquad \Box$

It is often easy to verify that the cost functional in a control problem is lower semicontinuous (or even continuous). Thus we can see that the

[†] Recall that J *lower semicontinuous* at v^0 is defined as follows: if $\epsilon > 0$ is given, then there is a $\delta > 0$ such that $\|v - v^0\| < \delta \Rightarrow J(v^0) \leq J(v) + \epsilon$.

construction of convergent minimizing families is crucial. There are several standard techniques for the systematic construction of minimizing families. One class of these techniques entails the approximation of our original problem by a sequence (or family) of simpler problems. We shall examine two such methods here. The first is aimed at replacing a problem with side conditions or constraints by a problem without side conditions, and the second consists of the replacement of the general problem with a suitable sequence of finite-dimensional minimization problems. For simplicity, we shall speak of these methods as Method 1 and Method 2.

Method 1. Let us suppose for the moment that our subset Φ of $\mathfrak{B}$ is given by

$$\Phi = \{v \colon \varphi(v) = 0\},$$

where φ is a nonnegative (that is, $\varphi(v) \geq 0$ for all v in $\mathfrak{B}$) real-valued function which is lower semicontinuous with respect to some topology $\mathfrak{F}$ on $\mathfrak{B}$. For example, if ψ is a weakly-continuous mapping of $\mathfrak{B}$ into another normed linear space $\mathfrak{B}_1$, then $\varphi(v) = \|\psi(v)\|$ is a function of the required type with respect to the weak topology on $\mathfrak{B}$ [Rothe, 1948].

(5.5) **Theorem.** *Suppose that J is lower semicontinuous with respect to the topology $\mathfrak{F}$. Suppose there is an unbounded upwardly directed subset $D = \{\delta\}$ of $[0, \infty)$ with the properties:*

(a) *$\delta \in D$ implies that there is a $v_\delta \in \mathfrak{B}$ such that*

$$J(v_\delta) + \delta\varphi(v_\delta) \leq J(v) + \delta\varphi(v) \quad \text{for all } v \in \mathfrak{B}.\dagger$$

(b) *There is a $v^0 \in \mathfrak{B}$ such that*

$$\lim_{\delta \to \infty} v_\delta = v^0,$$

where the limit is taken with respect to the topology $\mathfrak{F}$.
Then v^0 is an element of Φ and $J(v^0) = J^0 = \inf_{v \in \Phi} \{J(v)\};$

hence v^0 is a solution of our problem.

A proof of this theorem can be found in [Courant, 1962].

The major role of this theorem in control problems lies in its use in the replacement of problems involving specified conditions such as boundary conditions by problems which are "free." For example, we saw in Sections 4.1 and 4.2 that it was easy to show that an optimal control led to an extremum of the hamiltonian when the terminal state was free but that difficulties arose when the terminal state was fixed. We

---

† Actually, we need only assume that this condition is satisfied on some subset Ψ of $\mathfrak{B}$ which contains Φ (that is, $\Psi \supset \Phi$).

could try to use Theorem (5.5) in this situation in the following way: Supposing that the target set is of the form $\{t_1\} \times \{x_1\}$, where t_1 and x_1 are fixed, we let $\varphi(u(\cdot)) = \frac{1}{2}\|x_u(t_1) - x_1\|^2$, say, and we find necessary conditions for the free-end-point problem with cost functional $J(u(\cdot)) + \delta\varphi(u(\cdot))$. Assuming the existence of a suitably convergent family of solutions for these free-end-point problems, we can deduce necessary conditions for the original fixed-end-point problem. To formalize, let us suppose that we consider a control problem with target set $\{t_1\} \times \{x_1\}$, where x_1 is fixed, with no terminal cost (that is, $K = 0$) and with $\mathfrak{B}_U$, Ω, and X Hilbert spaces.

(5.6)   **Corollary.**   *Suppose that J is lower semicontinuous and that there is an unbounded upwardly directed subset $D = \{\delta\}$ of $[0, \infty)$ with the following properties:*

   (a) $\delta \in D$ implies that there is a $u_\delta(\cdot) \in \Omega$ such that

$$J(u_\delta(\cdot)) + \frac{\delta}{2}\|x_{u_\delta}(t_1) - x_1\|^2 \le J(u(\cdot)) + \frac{\delta}{2}\|x_u(t_1) - x_1\|^2.$$

   for all $u(\cdot) \in \Omega$.
   (b)   There is an m in $[0, \infty)$ such that

(5.7) $$\lim_{\delta \to \infty} \delta\|x_{u_\delta}(t_1) - x_1\| = m.$$

   (c)   There is a $u^0(\cdot) \in \Omega$ such that

(5.8) $$\lim_{\delta \to \infty} u_\delta(\cdot) = u^0(\cdot)$$

   (in the norm topology on Ω).
   Then $u^0(\cdot)$ is an optimal control and there is a costate $\lambda^0(\cdot)$ corresponding to $u^0(\cdot)$ such that

$$\frac{\partial H}{\partial u}(x^0(\cdot), \lambda^0(\cdot), u^0(\cdot)) = 0$$

   almost everywhere on $[t_0, t_1]$.

Proof. The optimality of $u^0(\cdot)$ follows from Theorem (5.5). Since $\partial H/\partial u$ is continuous and since (5.8) implies that

$$\lim_{\delta \to \infty} x_{u_\delta}(\cdot) = x^0(\cdot),$$

we need only show that

(5.9) $$\lim_{\delta \to \infty} \lambda_\delta(\cdot) = \lambda^0(\cdot)$$

exists and is a solution of the differential equation

$$(5.10) \qquad \dot{\lambda} = -\left(\frac{\partial f}{\partial x}\Big|_0\right)^* \lambda - \frac{\partial L}{\partial x}\Big|_0, ^\dagger$$

where $\lambda_\delta(\cdot)$ is the costate corresponding to $u_\delta(\cdot)$ such that

$$\frac{\partial H}{\partial u}\,(x_{u_\delta}(\cdot),\, \lambda_\delta(\cdot),\, u_\delta(\cdot)) = 0$$

almost everywhere on $[t_0, t_1]$. $\lambda_\delta(\cdot)$ exists by virtue of Theorem 4.4, since $u_\delta(\cdot)$ is optimal for a free-end-point problem.   Moreover (see Section 3.5),

$$\lambda_\delta(t_1) = \delta(x_{u_\delta}(t_1) - x_1),$$

and so

$$\lim_{\delta \to \infty} \lambda_\delta(t_1) = \hat{\lambda}$$

exists, by (5.7).   If we let $\lambda^0(\cdot)$ be the solution of (5.10) satisfying the terminal condition

$$\lambda^0(t_1) = \hat{\lambda},$$

we may conclude, from the continuity of $\partial f/\partial x$ and $\partial L/\partial x$, that

$$\lim_{\delta \to \infty} \frac{\partial f}{\partial x}\Big|_\delta = \frac{\partial f}{\partial x}\Big|_0$$

and

$$\lim_{\delta \to \infty} \frac{\partial L}{\partial x}\Big|_\delta = \frac{\partial L}{\partial x}\Big|_0$$

together imply that (5.9) and (5.10) are satisfied.   Thus the corollary is established. $\square$

We observe that the $u_\delta(\cdot)$ form a convergent minimizing family for J. Therein lies the chief difficulty in using the corollary, since it is often quite hard to show that (5.7) and (5.8) are satisfied.   The following example illustrates the method; further work in connection with this method would probably prove interesting and useful.

(5.11)   **Example.**   Consider the (scalar) system $\dot{x} = -x + u$, with $x(0) = \xi \ne 0$.   Suppose that our target set $S = \{T\} \times \{0\}$ with T fixed and that our cost functional J is given by

$$J(u) = \tfrac{1}{2} \int_0^T (x^2 + u^2)\, dt.$$

If we consider the problem with target set $\{T\} \times \mathbf{R}$ and cost functional

$$\frac{\delta}{2}\, x(T)^2 + \tfrac{1}{2} \int_0^T (x^2 + u^2)\, dt$$

---

† $L(x, u)$ is the trajectory-cost term.   In other words, $J(u) = \int_{t_0}^{t_1} L(x_u(t),$ $u(t))\, dt.$

for $\delta \in (0, \infty)$, then (see [ATHANS and FALB, 1966]) this problem has the solution

$$(5.12) \qquad u_\delta(t) = -k_\delta(t)x_\delta(t),$$

where $k_\delta(t)$, the solution of a suitable Riccati equation, is given by

$$k_\delta(t) = \frac{(\sqrt{2} + 1) + (\sqrt{2} - 1)\dfrac{(\delta + 1 - \sqrt{2})}{(\delta + 1 + \sqrt{2})} e^{2\sqrt{2}\,(t - T)}}{1 - \dfrac{(\delta + 1 - \sqrt{2})}{(\delta + 1 + \sqrt{2})} e^{2\sqrt{2}\,(t - T)}}$$

and $x_\delta(t)$ is given by

$$(5.13) \qquad x_\delta(t) = \xi \exp\left\{ \int_0^t [-1 - k_\delta(\tau)]\, d\tau \right\}$$

(see [ATHANS and FALB, 1966]). Since $k_\delta(\tau)$ is positive, property (c) of Corollary 5.6 will follow from (5.12) and (5.13) once we have shown that property (b) is valid (that is, $\lim\limits_{\delta \to \infty} x_\delta(T)$ exists and is finite). To do this it will be enough to establish the following claim:

$$(5.14) \qquad \lim_{\delta \to \infty} \delta \exp\left[- \int_0^T k_\delta(\tau)\, d\tau \right] \text{ exists.}$$

Now, a somewhat complicated calculation (which is left to the reader) shows that

$$\exp\left\{ - \int_0^T k_\delta(\tau)\, d\tau \right\}$$

$$= \exp\left\{ -\left(\frac{\sqrt{2} + 1}{2\sqrt{2}}\right) \log\left[\frac{\delta(e^{2\sqrt{2}\,T} - 1) + (\sqrt{2} + 1)e^{2\sqrt{2}\,T} + (\sqrt{2} - 1)}{2\sqrt{2}} \right] \right\}$$

$$\times \exp\left\{ -\left(\frac{\sqrt{2} - 1}{2\sqrt{2}}\right) \log\left[\frac{\{\delta(e^{2\sqrt{2}\,T} - 1) + (\sqrt{2} + 1)e^{2\sqrt{2}\,T} + (\sqrt{2} - 1)\}\{(\sqrt{2} + 1) + \delta\}}{2\sqrt{2}\,\{\delta + (\sqrt{2} + 1)e^{2\sqrt{2}\,T}\}} \right] \right\}$$

It follows that

$$\lim_{\delta \to \infty} \delta \exp\left[- \int_0^T k_\delta(\tau)\, d\tau \right] = \frac{2\sqrt{2}}{\exp(2\sqrt{2}\,T) - 1},$$

and our claim is established. Thus an optimal control for our original problem exists and can be calculated by taking the limit of (5.12) as $\delta \to \infty$.

Finally, let us note that if Φ is given by

$$\Phi = \{v \colon \varphi_i(C) = 0\},$$

where the φ_i are continuous, then it may still be possible to apply Theorem (5.5). For example, if there are positive constants m_i such that

$$|\varphi_i(v)| \leq m_i\|v\|, \qquad \sum_i m_i^2 < \infty,$$

then

$$\Phi = \{v: \varphi(v) = 0\},$$

where

$$\varphi(v) = \sum_i m_i^2 \varphi_i^2(v),$$

and the theorem can be used. Thus the method is considerably more general than might be expected from a hasty perusal of the constraint.

Method 2. We now turn our attention to the Ritz method, which consists of the replacement of our general problem with a nested sequence of finite-dimensional minimization problems whose solutions form the desired minimizing sequence.

Let us suppose that $w_0, w_1, \ldots$, is a sequence of elements of $\mathcal{B}$, and let us denote the span of $w_0, w_1, \ldots, w_k$ by $\mathcal{B}_k$. In other words,

$$\mathcal{B}_k = \Big\{v: v = \sum_0^k a_j w_j, a_j \in \mathbf{R}\Big\}.$$

Then $\mathcal{B}_k$ is a finite-dimensional subspace of $\mathcal{B}$ and $\mathcal{B}_k \cap \Phi$ is a subset of Φ, which we shall denote by Φ_k. It is clear that the problem of minimizing J on Φ_k is equivalent to minimizing a function of $k+1$ variables subject to constraints. Since $\Phi_0 \subseteq \Phi_1 \subseteq \cdots$, we note that

$$\inf_{\Phi_0} J(\cdot) \geq \inf_{\Phi_1} J(\cdot) \geq \cdots.$$

Now, if we knew that

$$\lim_{k \to \infty} \inf_{\Phi_k} J(\cdot) = J_0$$

and that for each k there was an element v_k of Φ_k such that

$$J(v_k) = \inf_{\Phi_k} J(\cdot),$$

then the v_k would form a minimizing family. This is the basic idea behind the Ritz method. We now give conditions which ensure the viability of the method. To begin with, we have

(5.15) **Definition.** *If $\mathfrak{F}$ is a topology on $\mathcal{B}$, then we say that $\{w_0, w_1, \ldots,\}$ is* **complete** *in Φ (with respect to $\mathfrak{F}$) if Φ is a subset of the $\mathfrak{F}$-closure of $\bigcup_k \mathcal{B}_k$. In other words, if v is an element of Φ and $N(v)$ is an $\mathfrak{F}$-neighborhood of v, then for some k (which may well depend on $N(v)$), there is a v_k in $\Phi_k \cap N(v)$.*

(5.16) **Theorem.** *Suppose that the following conditions are satisfied:*
  (a) *For each $k = 0, 1, \ldots$ there is a v_k^0 in Φ_k such that*

$$J(v_k^0) = \inf_{\Phi_k} J(\cdot).$$

  (b) $\{w_0, w_1, \ldots,\}$ *is complete in Φ with respect to a topology $\mathfrak{F}$.*
  (c) $J(\cdot)$ *is continuous with respect to $\mathfrak{F}$.*
Then $\{v_k \in k = 0, 1, \ldots\}$ is a minimizing family for J over Φ.

Proof. Let $\epsilon > 0$ be given and let $\hat{v}$ be an element of Φ such that

$$J(\hat{v}) < J^0 + \frac{\epsilon}{2}$$

(recall that $J^0 = \inf_{\Phi} J(\cdot)$). Since J is continuous, there is an $\mathfrak{F}$-neighborhood $N(\hat{v})$ such that $v \in \Phi \cap N(\hat{v})$ implies that

$$|J(v) - J(\bar{v})| < \frac{\epsilon}{2}.$$

As $\{w_0, w_1, \ldots\}$ is complete in Φ, there is a $\hat{v}_k$ in $\Phi_k \cap N(\hat{v})$. It follows that

$$J(\hat{v}_k) < J^0 + \epsilon.$$

Now, by condition (a), $J(v_k^0) \leq J(\hat{v}_k)$, and so we have

$$J^0 \leq J(v_k^0) < J^0 + \epsilon.$$

We immediately deduce that

$$\lim_{k \to \infty} J(v_k^0) = J^0,$$

and the theorem is established. □

Although the Ritz method is quite well known, it has not been adequately exploited in control theory, and further work would undoubtedly prove fruitful. With this in mind, we conclude with an example of the way in which the Ritz method may be used in control problems.

(5.17) **Example.** Let us consider a linear dynamical system with state space $X = L_2(\Delta, \mathbf{R})$, where Δ is a suitable domain in $\mathbf{R}_n$, and with Ω given by

$$\Omega = \{u(\cdot) : u(\cdot) \in L_2([t_1, t_2]x\Delta, \mathbf{R}), \|u(\cdot)\|_\infty \leq 1\}$$

(where $\|u(\cdot)\|_\infty = \text{ess sup } |u(\cdot)|$). Let us suppose that x_0 is a given element of X and that $x_d(\cdot)$ is a given curve in X. We assume that our cost functional $J(u(\cdot))$ is given by

$$J(u(\cdot)) = \tfrac{1}{2} \int_{t_1}^{t_2} \|x_u(t) - x_d(t)\|^2 \, dt,$$

where $x_u(\cdot)$ is the trajectory of our linear system starting from x_0 at t_1 and generated by $u(\cdot)$.   We observe that Ω is weakly compact and convex and that $J(\cdot)$, being continuous and convex, is weakly lower semicontinuous.   It follows that $J(\cdot)$ has a minimum on Ω.   Now if we divide $[t_1,\, t_2] \times \Delta$ by a $1/k$-mesh (see Fig. 5.3) and if we let $\chi_1^k,\, \ldots$ be the

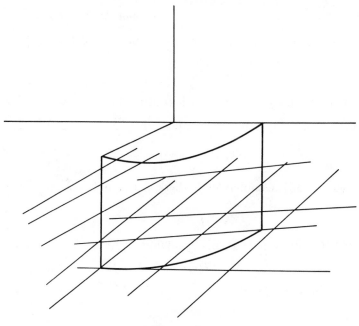

Figure 5.3

characteristic functions of the (disjoint) subsets of $[t_1,\, t_2] \times \Delta$ generated by this mesh, then the set

$$\Omega_k = \{u(\cdot)\colon u(\cdot) = \Sigma a_j \chi_j^k,\, |a_j| \leq 1\}$$

is a weakly-compact convex subset of Ω.

Thus $J(\cdot)$ has a minimum $u_k^0(\cdot)$ on Ω_k.   Since $\cup \Omega_k$ is (weakly) dense in Ω, we can apply the theorem to show that

$$\lim_{k \to \infty} J(u_k^0) = \min_{\Omega} J(\cdot).$$

The method here amounts to a discretization procedure, and some actual numerical results of its use can be found in [FALB, 1968b].   Additional material relating to the use of the Ritz method in control problems is also given in [FALB, 1968b].

PART THREE
Automata theory

M. A. Arbib

6 Automata theory: the rapprochement with control theory

In Chapters 7, 8, and 9, we shall show how algebraic methods may be used to explore the structure of finite automata. Before we turn to structure theory, we shall show in this chapter that automata theory and control theory are not as disparate as may appear from casual inspection.[†] We shall find that concepts presented earlier in the book for control systems receive an intuitively attractive formulation in the context of finite systems.

6.1 Semigroups

In Part Three, we shall consider a system to be stationary (that is, time-invariant) and to operate in discrete time $T = \{0, 1, \ldots\}$—except where we explicitly mention the use of the real half line $[0, \infty)$—and to be specified by the quintuple (note the change of notation[‡])

$$S = (\Omega, Y, Q, \lambda, \delta),$$

[†] The basic ideas of this chapter were developed in [ARBIB, 1965, 1966].

[‡] Since the writing of this part of the book, a different notation convention has become entrenched: $S = (X, Y, Q, \delta, \lambda)$.

where   Ω = the set of admissible input segments,
            Y = the set of outputs,
            Q = the set of states,
$\lambda: Q \times \Omega \to Q$ = the "next-state function,"
$\delta: Q \times \Omega \to Y$ = the "next-output function."

We interpret this formal quintuple as being a mathematical description of a machine which if at time t is in state q and receives input segment ω from time t to time t_0† will at time t_0 be in state $\lambda(q, \omega)$ and will emit output $\delta(q, \omega)$.

We now list a few basic definitions, and some simple assertions whose easy proofs are left to the reader:

(1.1) **Definition.** *Two states q and q' belonging to systems S and S', where S and S' may or may not be identical but have common Ω and Y, are said to be* **equivalent** *if and only if for all input segments $\omega[t_0, t)$ from Ω the response segment of S starting in state q is identical with the response segment of S' starting in state q'; that is,*

$$q \cong q' \Leftrightarrow \delta(q; \omega_{[t_o,t)}) = \delta'(q'; \omega_{[t_o,t)})$$

for all times t and t_0, $t_0 \leq t$, and all input segments $\omega_{[t_o,t)}$ of S and S'.

(1.2) **Definition.** *A system S is in* **reduced form** *if there are no distinct states in its state space which are equivalent to each other.*

(1.3) **Assertion.** *If q and q' are* **equivalent,** *so are the states into which they are taken by any input segment of S and S'.*

(1.4) **Definition.** *A state q' of S is* **reachable** *from a state q of S if and only if there exists an input segment $\omega_{[t_o,t)}$ in Ω such that*

$$q' = \lambda(q, \omega_{[t_o,t)}).$$

(1.5) **Definition.** *S is said to be* **strongly connected** *if every state is reachable from every other state.*

(1.6) **Definition.** *Systems S and S' are* **equivalent,** *$S \equiv S'$, if and only if to every state in the state space of S there corresponds an equivalent state in the state space of S', and vice versa.*

Our systems become the automata of automata theory if we quantize time and study their behavior at successive moments, $t = 0, 1, \ldots$, on some appropriate discrete time scale, and further require that the input and output sets be finite.

For a discrete time scale the set of admissible input segments becomes just the set of finite sequences on a finite input set X. This set X thus

---

† We may then find it convenient to denote the input segment by $\omega_{[t,t_o)}$.

determines Ω, and so we usually denote a finite automaton by the quintuple $(X, Y, Q, \lambda, \delta)$ rather than by $(\Omega, Y, Q, \lambda, \delta)$.

We shall not necessarily demand that there be only finitely many states. If Q does have only finitely many members, we shall say that M is a *finite automaton*, or *finite-state machine*. It will be an interesting question whether, given an automaton, there exists an equivalent *finite* automaton.

We devote the remainder of this section to developing for systems in general a number of ideas usually encountered only in automata theory, as prolegomenon to the semigroup† theory of machines.

We define $\hat{T}$ to be the set of finite initial segments of T, that is,

$$\{0, 1, \ldots\}^\wedge = \{\{0, 1, \ldots, n\} : n = 0, 1, \ldots\},$$
$$[0, \infty)^\wedge = \{[0, a) : a \geq 0\}.$$

Adopting the notation $[a, b) = \{t \in T : a \leq t < b\}$, we have that $[0, n) = \{0, 1, \ldots, n - 1\}$ if $T = \{0, 1, \ldots,\}$, whereas $[0, b)$ is the usual half-open interval if T is the real half line. Then

$$\hat{T} = \{[0, t) : t \in T\}.$$

Given T and a set A, we define $A^{\hat{T}}$ to be simply the set of all functions from $\hat{T}$ to A. If $\alpha : [0, a) \to A$, and $\beta : [0, b) \to A$, then we define

$$\alpha\beta : [0, a + b) \to A$$

by

$$\alpha\beta(t) = \begin{cases} \alpha(t), & 0 \leq t < a, \\ \beta(t - a), & a \leq t < a + b. \end{cases}$$

$A^{\hat{T}}$ is clearly a semigroup under this operation and has for identity the null function $\Lambda : \varnothing \to A$.‡ If α is defined on $[a, b)$ we set $l(\alpha) = b - a$, the "length" of α.

In the remainder of this section we assume that our set Ω of admissible initial input segments is a *subsemigroup* X^* of $X^{\hat{T}}$.

In case $T = \{0, 1, \ldots\}$, X^* is the familiar "free semigroup on X," consisting of finite sequences of elements of X, composed under concatenation.

---

† Or, as M. P. Schützenberger would insist, the *monoid theory*, since all our semigroups have identities, but no topological structure. A semigroup for us is just a set on which is defined a binary associative operation, $(xy)z = x(yz)$. Basic concepts of semigroups and the further development of our present ideas for finite automata will be presented in the next chapter.

‡ This definition is appropriate for *time-invariant* systems. In the general case we would have to consider $\hat{T} = \{[a, b) : a < b; a, b \in T\}$ and define $\alpha\beta$ only in case α is defined on $[a, c)$ and β on $[c, b)$ for some a and b and a mutual c.

A time-invariant system is really defined by two functions,

(1.7)
$$\lambda: Q \times X^* \to Q,$$
$$\delta: Q \times X^* \to Y.$$

For an automaton we are usually given the state-transition and output functions as $\lambda: Q \times X \to Q$ and $\delta: Q \times X \to Y$, but these extend immediately to form (1.7) in which we shall feel free to use them.

Returning now to our general time-invariant systems, the consistency condition for input segments is

$$\lambda(\lambda(q, x), x') = \lambda(q, xx') \qquad \text{for all } q \in Q; x, x' \in X^*.$$

The input/output function of the time-invariant system S started in state q is the function

$$S_q: X^* \to Y$$

defined by $S_q(x) = \delta(q, x)$ for $x \in X^*$. Defining $L_{x'}(x) = x'x$ for all $x', x \in X^*$ and noting that $\delta(q, x'x) = \delta(\lambda(q, x'), x)$, we see that

$$S_{\lambda(q,x')}(x) = \delta(\lambda(q, x'), x) = \delta(q, x'x) = S_q(x'x) = S_q L_{x'}(x).$$

Thus

$$S_{\lambda(q,x')} = S_q L_{x'}.$$

When our interest in a system is in how it transforms input sequences into output sequences, all that we wish to know about q is contained in the function S_q. Returning to our notion of system equivalence, we then have the following:

(1.8) **Assertion.** *Two time-invariant systems S and S' with state sets Q and Q', respectively, are equivalent if and only if*

$$\{S_q: q \in Q\} = \{S_{r'}: r' \in Q'\}.$$

Clearly, we also have the following:

(1.9) **Assertion.** *S is a system in reduced form if and only if $q \mapsto S_q$ is a one-to-one mapping.*

We say that S is a *state-output system* if there is a function $i: Q \to Y$ such that $\delta(q, x) = i(\lambda(q, x))$; that is, if the output depends only on the state at a given time.

(1.10) **Assertion.** *Let $S = (X^*, Y, Q, \lambda, \delta)$. Then there exists a reduced state-output system equivalent to S. One such system is given by S^0, termed the state-output reduction of S, where*

$$S^0 = (X^*, Y, Q^0, \lambda^0, i^0\lambda^0),$$

where

$$Q^0 = \{S_q : q \in Q\},$$
$$\lambda^0(q^0, x) = q^0 L_x \quad for\ q^0 \in Q^0,\ x \in X^*,$$
$$i^0(q^0) = q^0(\Lambda);$$

for example,

$$i^0(S_q L_x) = \delta(q, x).$$

In the case of a finite automaton, the reduction is again a finite automaton. The reader *au fait* with linear systems will recognize that the reduction of a linear system with state space $Q = E^n$ will be a linear system with state space $E^m (m \leq n)$ (cf. Section 6.2).

However, for a nonlinear system with state space E^n, the reduced system may not be of interest, since the reduction may well destroy the euclidean topology. This statement is a pessimistic one; more optimistically, the topology of the reduced state space may yield valuable insight into the stability properties of the nonlinear system.

We close this section with the definition of the semigroup of the system S. First we recall that an equivalence $\equiv$ on a semigroup A is called a congruence if

$$x \equiv y \Rightarrow xz \equiv yz, \qquad x \equiv y \Rightarrow zx \equiv zy \qquad for\ all\ x,\ y,\ z \in A.$$

In this case we may define the factor semigroup $A/\equiv$ to have elements $[x]_\equiv$, the equivalence classes under $\equiv$, with multiplication defined by $[x]_\equiv[y]_\equiv = [xy]_\equiv$. Given the system $S = (X^*, Y, Q, \lambda, \delta)$, we define an equivalence $\equiv_S$ on the semigroup X^* by $x \equiv_S x'$ if and only if

$$S_q(uxv) = S_q(ux'v)$$

for all $q \in Q$ and $u, v \in X^*$. This is clearly a congruence, and so we may define the semigroup of the system S to be the factor semigroup $X^*/\equiv_S$.

In the following chapters we shall develop the semigroup theory of finite automata in detail. The reader may find it an interesting exercise to extend much of it to general systems, in the fashion of our present treatment.

6.2   Additivity and duality

So much of modern mathematics, from the humble matrix to the most abstract Banach space, is concerned with linearity that it is not surprising that linear systems have been much studied by control theorists. In control theory the spaces X, Y, and Q are usually euclidean, and the theory of linear systems is erected on this basis. Our contribution in this section is to develop some of the basic theory, using only the group property of the relevant spaces.

Thus, except in our treatment of duality for automata, we shall assume in this section that X, Y, and Q are abelian groups and use $+$ for the group operations. Since we make no use of scalar multiplication, we shall use "additivity" to refer to our various analogs of the classical linearity.

(2.1) Definition. *A state θ of the system S is a* **zero state** *if*

$$\delta(\theta, 0^t) = 0 \qquad \text{for all } t,$$

where 0^t is the zero input, $0^t : [0, t) \to Y$, defined by $0^t(\tau) \equiv 0$.

We now have the crucial definition:

(2.2) Definition. *The system S is* **additive** *if and only if it has the following three properties:*
(a) The decomposition property: $\delta(q, u) = \delta(q, 0^{l(u)}) + \delta(\theta, u)$.
(b) Zero-state additivity: $\delta(\theta, x - x') = \delta(\theta, x) - \delta(\theta, x'), l(x) = l(x')$.
(c) Zero-input additivity: $\delta(q' - q'', 0^t) = \delta(q', 0^t) - \delta(q'', 0^t)$.

(2.3) Proposition. *If $\lambda(\theta, 0^t) \equiv \theta$ for all t and all states are reachable from θ, then property (b) of Definition (2.2) implies property (a).*

Proof. Suppose $q = \lambda(\theta, x)$. Then
$$\begin{aligned}
\delta(q, u) &= \delta(\lambda(\theta, x), u), \\
&= \delta(\theta, x \cdot u), \\
&= \delta(\theta, x \cdot 0 + 0 \cdot u), \\
&= \delta(\theta, x \cdot 0) + \delta(\theta, 0 \cdot u) \qquad \text{(by Definition (2.2}b\text{)),} \\
&= \delta(\lambda(\theta, x), 0) + \delta(\lambda(\theta, 0), u), \\
&= \delta(q, 0) + \delta(\theta, u).
\end{aligned}$$
$\square$

(2.4) Assertion. *If S has the decomposition property, state q' is equivalent to state q'' if and only if*

$$\delta(q', 0^t) = \delta(q'', 0^t) \qquad \text{for all } t \geq 0.$$

Since
$$\delta(q' - q'', x) = \delta(q' - q'', 0) + \delta(\theta, x) = \delta(q', 0) - \delta(q'', 0) + \delta(\theta, x),$$

we see further that two states are equivalent if and only if their difference is equivalent to the zero state.

(2.5) Lemma. *The states equivalent to the zero state form a normal subgroup N of the group Q of states.*

Proof. If $r, s \in N$, then $r - s \in N$, since

$$\delta(r - s, x) = \delta(\theta, x) \qquad \text{for all } x \in X^*;$$

that is, N is a subgroup.   But N is also normal, since Q is abelian. $\square$

(2.6) **Assertion.**   *If q is equivalent to q' and $x \in X^*$, then*

$$\lambda(q, x) - \lambda(q', x) \in N.$$

Thus we may set up the factor group Q/N.   The elements of Q/N are the equivalence classes of states of S.   Hence the reduced system of S is simply

$$S^0 = (X^*, Y, Q/N, \lambda^0, \delta^0),$$

where

$$\lambda^0([q], x) = [\lambda(q, x)], \qquad \delta^0([q], x) = \delta(q, x).$$

(2.7) **Corollary.**   *An additive automaton is equivalent to a finite-state machine if and only if Q/N is a finite group.*

(2.8) **Definition.**   *An additive system is said to be* **completely additive** *if each of the three properties of Definition (2.2) still holds on replacing δ by λ.*

A routine proof yields the following:

(2.9) **Theorem.**   *A system $M = (X^*, Y, Q, \lambda, \delta)$ operating on*

$$T = \{0, 1, 2, \ldots\}$$

for abelian groups is completely additive if and only if there exist homomorphisms

$$A : Q \to Q, \qquad B : X \to Q,$$
$$C : Q \to Y, \qquad D : X \to Y,$$

such that for all $q \in Q$ and $x \in X$ we have

(2.10)
$$\lambda(q, x) = Aq + Bx,$$
$$\delta(q, x) = Cq + Dx.$$

We then have

$$\lambda(q, x_1 \cdots x_n) = A^n q + \sum_{m=1}^{n} A^{m-1} B x_{n-m+1},$$

(2.11)

$$\delta(q, x_1 \cdots x_n) = CA^{n-1} q + \sum_{m=1}^{n-1} CA^{m-1} B x_{n-m} + Dx_n.$$

Clearly, Definition (2.2) for λ and δ is satisfied.   Let us set

$$\Phi_0(n) = CA^{n-1},$$
$$h(m) = \begin{cases} D, & m = 0, \\ CA^{m-1}B, & m > 0. \end{cases}$$

and let us set

$$\Phi(n) = A^n,$$
$$h_s(m) = A^m B.$$

We then have

$$\delta(q, x_0 x_1 \cdots x_{n-1}) = \Phi_0(n)q + \sum_{m=0}^{n-1} h(m)x_{n-m}.$$

and

$$\lambda(q, x_0 x_1 \cdots x_n) = \Phi(n)q + \sum_{m=0}^{n-1} h_s(m)x_{n-m}.$$

Now the passage from additive systems to the usual linear systems consists in (1) replacing group homomorphisms by vector-space homomorphisms and (2) replacing discrete time by continuous time. Thus we get a linear system S with state space $Q = E^n$ and time $T = [0, \infty)$ described by

$$y(t) = \Phi_0(t - t_0)x(t_0) + \int_{t_0}^t h(t - \xi)u(\xi)\, d\xi, \qquad t \geq t_0,$$

where the p-vector $y(t)$ is the output at time t, the n-vector $x(t)$ is the state at time t, the r-vector $u(t)$ is the input at time t, $\Phi_0(t)$ is the output-transition matrix whose ith column is the response of S at time t to zero-input when started in state $(0, \ldots, 1, \ldots, 0)$ (1 in ith place), and $h(t)$ is the impulse response of S (that is, the response of S when started in the zero state) to the impulse input

$$\delta(t) = \begin{cases} \text{Dirac delta function for } T = [0, \infty), \\ \text{Kronecker delta } \delta_{0t} \text{ for } T = \{0, 1 \ldots\}. \end{cases}$$

If Φ_0 is differentiable and h is unsullied by delta functions or their derivatives, we may refer to S as a linear differential system. Then the state equations of S in differential form read

$$\dot{x}(t) = \dot{\Phi}(0)x(t) + h_s(0 +)\, u(t),$$
$$y(t) = Cx(t),$$

and the state at time t is given by

$$x(t) = \Phi(t - t_0)x(t_0) + \int_{t_0}^t h_s(t - \xi)u(\xi)\, d\xi, \qquad t \geq t_0,$$

where $h(t)$ is the state impulse response and $\Phi(t)$ is the state-transition matrix and satisfies

$$\dot{\Phi}(t) = \dot{\Phi}(0)\Phi(t), \qquad \Phi(0) = I.$$

The state equations

$$(2.12) \qquad \begin{aligned} x(t) &= \Phi(t - t_0)x(t_0) + \int_{t_0}^t h_s(t - \xi)u(\xi)\,d\xi, \\ y(t) &= Cx(t) \end{aligned}$$

hold for all t and t_0, provided the state-transition matrix $\Phi(t)$ and the state-impulse response $h_s(t)$ are understood to be extended rather than one-sided.

We have the well-known

(2.13) **Corollary.** *If S is characterized by an input/output-state relation of the form (2.12), in which S is in reduced form, then S is initial-state determinable in the following sense: Given $u_{(t_0,t]}$ and $y_{(t_0,t]}$, one can uniquely determine the initial state $x(t_0)$.*

For a proof see [ZADEH and DESOER, 1963, 3.7].
Roughly speaking then, *linear differential systems are reversible.*

(2.14) **Definition.** *The two systems of equations*

$$\begin{aligned} \dot{x} &= A(t)x, \\ \dot{y} &= -A^*(t)y, \end{aligned}$$

where $A^*(t)$ is the conjugate transpose of $A(t)$, are said to be **adjoint** *to one another.*

(2.15) **Theorem.** *Let $\Phi(t, t_0)$ be the state-transition matrix of the system*

$$\dot{x} = A(t)x$$

and let $\Psi(t, t_0)$ be the state-transition matrix of the adjoint system; that is,

$$\frac{d}{dt}\Psi(t, t_0) = -A^*(t)\Psi(t, t_0), \qquad \Psi(t, t_0) = I.$$

Then $\Psi^*(t, t_0)\Phi(t, t_0) = I$ for all t and t_0. Thus

$$(2.16) \qquad \Psi^*(t, t_0) = \Phi(t, t_0)^{-1} = \Phi(t_0, t).$$

Conversely, if this holds, then the corresponding systems are adjoint.

For a proof see [ZADEH and DESOER, 1963, 6.2].

(2.17) **Definition.** *Let the linear system S have a real-valued impulse response $h(t, \xi)$. The linear system $S^{(a)}$ is said to be the* **adjoint** *of S if its impulse response $h^{(a)}(t, \xi)$ satisfies the relation*

$$(2.18) \qquad h^{(a)}(t, \xi) = h(\xi, t) \qquad \text{for all } t, \xi,$$

where $h^{(a)}(t, \xi)$ is the response of $S^{(a)}$ at time t to a unit impulse applied at time ξ.

Thus (2.18) implies that $S^{(a)}$ has a response at time t to a unit impulse applied at time ξ equal to the response of S at time ξ to a unit impulse applied at time t.

Note that in general if S is nonanticipatory, then $S^{(a)}$ has to be anticipatory; that is, the present output of $S^{(a)}$ depends on future inputs.

Consider now the equations

$$\text{(2.19)} \qquad \begin{aligned} \dot{x} &= Ax + Bu, \\ y &= Cx + Du. \end{aligned}$$

They have the solution

$$x(t) = \Phi(t - t_0)x(t_0) + \int_{t_0}^{t} \Phi(t - \tau)B(\tau)u(\tau) \, d\tau,$$

and thus the impulse-response matrix for input $\delta(t - t_0)$ is simply

$$h(t, t_0) = C(t)\Phi(t - t_0)B(t_0).$$

Similarly, the system

$$\text{(2.20)} \qquad \begin{aligned} \dot{\xi} &= A^*\xi + C^*v, \\ \xi &= B^*\xi - D^*v, \end{aligned}$$

has impulse-response matrix to input $\delta(t - t_0)$ as

$$h^{(a)}(t, t_0) = B^*(t)\Psi(t - t_0)C^*(t_0).$$

Thus

$$\text{(2.21)} \qquad h^{(a)}(t, t_0) = h(t_0, t)^*.$$

We accept (2.21) as the appropriate generalization of (2.18) to a system with multidimensional output, and say that system (2.20) is the adjoint of system (2.19) (an alternative has the signs of B^* and C^* changed).

In automata theory, duality has been little studied. The only interesting example is given by [Rabin and Scott, 1959]. We shall slightly modify their definition. For the correct modification, see [Arbib and Zeiger, 1968].

(2.22) **Definition.** *Given $M = (X, Y, Q, \lambda, \delta)$ with $Q = \{q_1, \ldots, q_n\}$, let $\bar{M} = (X, \bar{Y}, \bar{Q}, \lambda^*, \delta^*)$, where $\bar{Q}$ is the set of subsets of Q and $\bar{Y}$ is the set of subsets of Y, and let*

$$\lambda^*(q^*, x) = \bigcup_{q \in q^*} \{t \in Q : \lambda(t, x) = q\},$$
$$\delta^*(q^*, x) = \bigcup_{q \in q^*} \{\delta(t, x) : \lambda(t, x) = q\}.$$

*M^*, the **dual** of M, is defined to be $\bar{M}$ restricted to those states reachable from at least one of the states $\{q_1\}, \ldots, \{q_n\}$.*

In general, if M has n states, then M^* has of the order of 2^n states; that is, the state space "blows up" under taking of duals. It is of some interest to know when we can preserve the state space, as is possible with linear systems. Now

M^* has states $\{q_1\}, \ldots, \{q_n\}$
$\Leftrightarrow \lambda(q, x) = \lambda(q', x) \Rightarrow q = q'$ for all $q, q' \in Q$; $x \in X^*$
$\Leftrightarrow \lambda(q, x) = \lambda(q', x) \Rightarrow q = q'$ for all $q, q' \in Q$; $x \in X$
$\Leftrightarrow$ each $\lambda(\cdot, x) \colon Q \to Q$ is a permutation.

If this is the case, we may identify M^* with

$$M^R = (X, Y, Q, \lambda^R, \delta^R),$$

where $\lambda^R(q, x)$ is the unique q' for which $\lambda(q', x) = q$, and then

$$\delta^R(q, x) = \delta(q', x).$$

We shall say that M is *reversible* and that M^R is the *reverse* of M. Clearly, $(M^R)^R = M$.

(2.23) **Assertion.** *M^* has the same number of states as M if and only if M is reversible. Then $M^* = M^R$.*

(2.24) **Assertion.** *If M is reversible and there is a state q from which all states of M are reachable, then M is strongly connected.*

We should contrast our notion of reversible automata with the general idea of a converse system.

(2.25) **Definition.** *α and β are* **converse systems** *if every input/output pair (u, y) for α† has the property that (y, u), with y as input and u as output, is an input/output pair for β, and vice versa.*

(2.26) **Theorem.** *The converse of a finite automaton is not necessarily finite state.*

Proof. Let M be a finite automaton with m inputs and n outputs, but let $n < m$. Then M has m^k input sequences of length k and at most n^k output sequences of length k.

$(x_1 \cdots x_k, y_1 \cdots y_k)$ is an input output pair for M if there exists a state q of M such that

$$(2.27) \qquad y_n = \delta(q, x_1 \cdots x_n) \qquad \text{for } 1 \leq n \leq k.$$

Let M_c be the converse of M (it may not be completely specified). If M_c has a finite number of states, then (2.27) implies $m^k \leq rn^k$, so that $r \geq (m/n)^k$. Letting k increase, we get a contradiction. $\qquad \square$

In mathematics, if M^{**} is the dual of the dual of some system M, then usually M is isomorphic either to M^{**} or to some subsystem of M^{**}.

† That is, there is a $q \in Q_\alpha$ such that $y(t) = \delta_\alpha(q, u_{[0,t)})$.

Thus in automata theory we might be tempted to claim that M is equivalent to the smallest submachine of M^{**} whose states include $\{\{q_1\}\}$, $\ldots$, $\{\{q_n\}\}$; call it M. However, a state $\{q_1\}$ is not in general reachable from other states of M^*; the reversing action of a nonreversible machine in general increases the cardinality of a state set at each transition. This means that in general $\lambda^{**}(\{\{q_1\}\}, x) = \varnothing$, which is not interesting. To see this, let us follow a computation:

$$\lambda^{**}(R, x) = \bigcup_{T \in R} \{Q' \subseteq Q : \lambda^*(Q', x) = T\},$$

where

$$\lambda^*(Q', x) = \bigcup_{q' \in Q'} \{q \in Q : \lambda(q, x) = q'\}.$$

If $R = \{\{q_i\}\}$, then

$$\lambda^{**}(R, x) = \{Q' \subseteq Q : \lambda^*(Q', x) = \{q_i\}\}.$$

But

$$\{q_i\} = \bigcup_{q' \in Q'} \{q \in Q : \lambda(q, x) = q'\}$$

if and only if $Q' = \{\lambda(q_i, x)\}$ and $\lambda(q, x) = \lambda(q_i, x) \Rightarrow q = q_i$. We thus arrive at the following result:

(2.28) **Theorem.** *M is isomorphic to the submachine $\hat{M}$ of M^{**} if and only if M is reversible.*

6.3   Controllability and observability

In Chapter 2 we saw a development of controllability and observability in terms of linear differential systems S described by such equations as

(3.1)
$$\dot{x} = Ax + Bu,$$
$$y = Cx + Du,$$

where A, B, C, and D are, respectively, $n \times n$, $n \times r$, $p \times n$, and $p \times r$ matrices. The n-vector x is the state of the system, the r-vector u is the input, and the p-vector y is the output of S.

We shall examine as much of the theory as possible in a form applicable to general systems (and so, in particular, to automata). Many results will turn out to require only our additivity conditions, rather than the linearity conditions used elsewhere.

We say that a state q is controllable if we may so choose the input as to bring our system from q to the zero state. More formally, for any system S with a designated state θ, we have the following definition:

(3.2) **Definition.** *State q of system S is* **controllable** *if and only if* *there exists $u \in X^*$ such that*

$$\lambda(q; u) = \theta.$$

The system S is said to be **controllable** *if and only if every state of S is controllable.*

(3.3) **Assertion.** *If S is a system in which every state is reachable from θ, then the fact that S is controllable implies that S is strongly connected, and vice versa.*

Let us recall the additive system M of (2.10). It is reversible (cf. (2.21)) if for each x, $\lambda(\cdot, x)$ is invertible; that is, if for each $r \in Q$ there is a unique solution of

$$Aq = r - Bx,$$

that is, if and only if A is an automorphism of Q with inverse A^{-1}, say. If we define

$$\lambda^R(r, x) = A^{-1}r - A^{-1}Bx,$$
$$\delta^R(r, x) = \delta(\lambda^R(r, x), x),$$
$$= C[A^{-1}(r - Bx)] + Dx,$$
$$= CA^{-1}r + [D - CA^{-1}B]x,$$

then $M^R = (X, Y, Q, \lambda^R, \delta^R)$ is also completely linear. Note that we do indeed have $M^{RR} = M$, since

$$(A^{-1})^{-1} = A, \qquad CA^{-1}AA^{-1}B - CA^{-1}B + D = D, \qquad \text{etc.}$$

Now, our last assertion tells us that the reversible machine M is controllable if and only if all states can be reached from the zero state. Consulting (2.11), we immediately have the following result:

(3.4) **Theorem.** *A reversible, completely additive system is controllable if and only if each state q can be represented as a linear combination*

$$A^{m-1}Bx, \qquad x \in I.$$

Turning now to system (3.1) and recalling that the power A^n of an $n \times n$ matrix may be represented as a linear combination of I, A, A^2, . . . , A^{n-1}, we have "essentially" proved Theorem (3.4) of Chapter 2:

(3.5) **Theorem.** *The system S of (3.1) is controllable if and only if the column vectors of the matrix*

$$[B, AB, \ldots, A^{n-1}B]$$

span the state space of S.

The other crucial concept here is that of observability. For linear systems we shall see that the concept is dual to that of controllability.

In fact, the simplicity of this concept for linear systems has blinded most system theorists to the more complex behavior of the concept for non-linear systems in the form studied by the automata theorist.

We say that a state is *observable* if we can determine it by observing just the input/output behavior of the system. However, there are many notions of what we mean by "observing the input/output behavior." Several are listed in the next definition:

(3.6) **Definition.**

(a) *An* **(ordinary) simple experiment** *is an input/output pair* $(x[t_0, t), y[t_0, t))$; *that is, given the system in the unknown state, we input* $x[t_0, t)$ *and observe the ouput* $y[t_0, t)$.

(b) *A* **branching simple experiment** *is an input/output pair* $(x[t_0, t), y[t_0, t))$, *where for* $t_0 < t' < t$ *we have* $x(t') = f(x[t_0, t'), y[t_0, t'))$.

In (c) and (d) each of the N realizations starts in the same state, although this state may be unknown.

(c) *An* **(ordinary) multiple experiment of size** N *consists of* N *input output pairs* $(x_i[t_0, t), y_i[t_0, t))_{1 \le i \le N}$; *that is, we feed* $x_i[t_0, t)$ *into the ith of the* N *realizations and observe the output* $y_i [t_0, t)$.

(d) *A* **branching multiple experiment of size** N *consists of* N *input output pairs* $(x_i(t_0, t], y_i(t_0, t])_{1 \le i \le N}$, *where for* $t_0 < t' < t$ *we have*

$$\begin{bmatrix} x_1(t') \\ \cdots \\ x_N(t') \end{bmatrix} = f \begin{bmatrix} x_1[t_0, t') & y_1[t_0, t') \\ \cdots\cdots\cdots\cdots \\ x_N[t_0, t') & y_N[t_0, t') \end{bmatrix}.$$

Similarly, one may define *infinite* experiments. In all cases $t - t_0$ is the *length* of the experiment.

(3.7) **Definition.** *Two systems* S_1 *and* S_2 *are* **simply equivalent** *if it is impossible to distinguish them by any simple experiment: for all* q_1 *and for all* $x(t_0, t]$ *there exists a* q_2 *such that*

$$\delta_1(q_1, x[t_0, t)) = \delta_2(q_2, x[t_0, t))$$

and vice versa.

(3.8) **Definition.** *Two systems* S_1 *and* S_2 *are* **multiply equivalent** *if it is impossible to distinguish them by any multiple experiment: that is, if and only if each state of* S_1 *is equivalent to a state of* S_2, *and vice versa.*

(3.9) **Theorem.** *If* S_1 *and* S_2 *are simply equivalent and strongly connected, then they are multiply equivalent.*

For a proof of this theorem see [GINSBURG, 1962, pp. 28–30], which contains a thoroughgoing discussion of this notion of machine experiment.

(3.10) **Definition.** *S is* **initial-state determinable in the strong (wide) sense** *if, no matter what q_0 is, the experimenter can find what q_0 was from any (some) experiment on S started in q_0.*

The reader may test his understanding of these concepts by providing the (straightforward) proof of

(3.11) **Theorem.** *A system is in reduced form if and only if it is initial-state determinable by an infinite multiple experiment.*

It is clear that a system can be initial-state determinable by any type of experiment only if it is in reduced form. However, a nonlinear system in reduced form may not be initial-state determinable by an ordinary simple experiment. We have from [GINSBURG, 1962] that if an (n, m, p) system is one with n states, m inputs, and p outputs, then

(3.12) **Theorem.** *It is possible to determine the initial state of an (n, m, p) system in reduced form by performing a multiple experiment of length $\leq n - 1$ and of size $\leq n - 1$.*

We may add to this two new theorems and a corollary due to [FAURRE, 1965].

(3.13) **Theorem.** *Every conclusion that can be drawn about a reversible system by a multiple experiment can be drawn by a simple experiment.*

Proof. We carry out the experiments sequentially. After executing the jth trial, we feed into the machine the "reverse" of the jth input sequence, thus returning S to the initial state ready for the $(j + 1)$st trial. □
From Theorem (3.12) and the proof of Theorem (3.13) we conclude

(3.14) **Corollary.** *Every reversible (n, m, p) system is initial-state determinable by a simple experiment of length less than $(n - 1)(2n - 3)$.*

(3.15) **Theorem.** *A discrete-time additive system is initial state determinable iff it is in reduced form. If the state space Q is an n-dimensional vector space, it is then possible to deduce the initial state from any simple experiment of length n.*

Proof. The proof is by straightforward computation using (2.10) and (2.11). The second statement follows on recalling that any power of a linear transformation A of a space of dimension n is a linear combination of $A^0, A^1, \ldots, A^{n-1}$. □

Thus we shall use the

(3.16)   **Definition.**   *S is* **observable** *if and only if it is in reduced form.*

The point is that for our additive (linear) systems this is equivalent to *any* form of initial-state determinability.   However, for nonlinear systems there are many nonequivalent notions of initial-state determinability, so the theorist of nonlinear systems must be prepared to deal with several types of controllability.   [VERBEEK, personal communication] has provided the following example of a finite automaton which is reduced but *not* initial-state determinable by a simple experiment (see Figure 6.1).

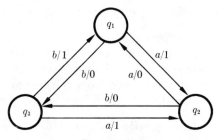

Figure 6.1

Input alphabet $X = \{a, b\}$,
Output alphabet $Y = \{0, 1\}$,
State set $Q = \{q_1, q_2, q_3\}$.

An input sequence starting with a confuses states q_1 and q_3, whereas one starting with b confuses q_1 and q_2.   Hence no simple experiment could assure us that the automaton started in state q_1 (if it did!), whereas a multiple experiment of size 2 could.

Now we know that our completely additive system (2.10) is in reduced form only if $N = \{0\}$; that is, the zero state is equivalent only to itself. But inspection of (2.11) provides the next result:

(3.17)   **Theorem.**   *The completely additive system S is observable if and only if*

$$CA^{k-1}q = 0 \text{ for all } k \quad \Rightarrow \quad q = 0.$$

This immediately yields the familiar consequence of Chapter 2:

(3.18)   **Corollary.**   *The system S of (3.1) is observable if and only if the column vectors of the matrix*

$$[C^*, A^*C^*, \ldots , A^{*(n-1)}C^*]$$

span the state space of S.

Theorem (3.5) and Corollary (3.18) combine to give the following duality result of [KALMAN, 1962]:

(3.19) **Theorem.** *Let Σ be the system dual to S of (3.1); it is defined by*

$$\xi = -A^*\xi + C^*\nu,$$
$$\eta = B^*\xi - D^*\nu,$$

where the state ξ is an n-vector, the input ν is a p-vector, and the output η is an r-vector. Then S is controllable if and only if Σ is observable, and vice versa.

Recalling our discussion of the reduced form of an additive system, we see the following:

(3.20) **Theorem.** *An additive system with state group Q and subgroup N of states equivalent to the zero state can be partitioned into two subsystems, S_1, which is observable, and S_2, which is unobservable, if and only if*

$$Q = N \times Q/N,$$

that is, the decomposition of Q by N splits.

Note that we have only used conditions (a) and (b) of the definition of additive systems, (2.2). In the case of linear systems Q becomes a linear vector space and N becomes a subspace—and such a decomposition always splits.

6.4 Tolerance automata

One thing an automata theorist must often envy a control theorist is the use of continuity. One may state the problem thus: "How can we put a relevant topology on a discrete set which is not the discrete topology?" A person who uses automata in the form of neural nets to provide crude models of the brain might ask the question, "How can we define continuity in a form which can enter the 'life' of a finite automaton?" [ARBIB, 1966] introduced the idea of tolerance automaton in response to this question. The basic idea is to have a form of "continuity" at our disposal for dealing with the state set. This notion was developed in [ARBIB, 1967].

Tolerance and continuity. The basic concept is that of a tolerance, introduced by [ZEEMAN, 1962].

(4.1) **Definition.** *A **tolerance** ξ on a set X is a relation on X that is* *reflexive and symmetric.   A* **tolerance space** *(X, ξ) is a set together with a tolerance on it.*

(4.2)   **Example.**   Let X be the euclidean plane, and let $ξ$ be all pairs of points less than $ε$ apart.

(4.3)   **Example.**   Let X be the visual field and let $ξ$ be the visual acuity tolerance, that is, all pairs of points that are indistinguishable.

Let us now conjure with these notions.   In the spirit of automata theory, let us assume that time is quantized, $T = \{0, 1, \ldots\}$.   Let $(X, ξ)$ be a tolerance space.

(4.4)   **Definition.**   *We shall say a* **motion** *in X is a function m: $T \to X$, and we shall call m(t) the* **position** *of the point (undergoing the motion m) at time t.   We shall say that the motion is ξ-***continuous** *if $[m(t), m(t + 1)] \in ξ$ for all t; that is, if there are "no detectable jumps in the motion."*

We thus see that a discrete automaton M is given an intuitively acceptable idea of continuity if the input set of M is the collection L_X of subsets of a tolerance space $(X, ξ)$.   We think of M as having a "retina" with one receptor for each point of X, the input at any time being the stimulation of a number of receptors.   It will help to consider T as bearing the "*adjacency*" tolerance: $(x, y) \in ξ$ if and only if $|x - y| \leq 1$.

Let $M = (X, Y, Q, λ, δ)$ be an automaton.   With each input sequence $u = (u_1, u_2, \ldots) \in X^ω$ we associate the motion $m_u: T \to Q$, where

$$m_u(0) = q_0 \quad \text{(an initial state in } Q),$$
$$m_u(t + 1) = λ(m_u(t), u_t).$$

Let us fix some tolerance $ξ$ on Q.   Then we say that M is a *tolerance automaton* if each motion m_u is continuous for each $u \in X^ω$.

If we take $ξ$ to be the "smudge" tolerance $e = Q \times Q$, then every automaton is, of course, an e-tolerance automaton.   Thus our theorems below apply to all automata but are often vacuous in case the only applicable tolerance is the smudge tolerance.

In the following discussion $ξ$ will always denote a tolerance, with distinguishing subscripts added where necessary.

(4.5)   **Definition.**   *Let $M = (X, Y, Q, λ, δ)$ be an automaton for which Q is a tolerance space.   We say that M is a* **tolerance automaton** *if for each $x \in X$ and $q \in Q$ we have $(q, λ(q, x)) \in ξ_Q$.*

Thus a tolerance automaton has *inertia;* a sudden change of input cannot cause a sudden change of state.

(4.6)   **Example.**   Consider an ordinary digital computer M in which the state of the machine is given by the contents of the registers.   If two states are within tolerance only if they differ in the contents of a limited number of registers, then M is a tolerance automaton. (Note that here the clock pulse must serve as an input when there is no input from the "environment.")

(4.7) **Definition.** *Let X and Y be tolerance spaces. A function $f: X \to Y$ is said to be ξ-**continuous** if $(x_1, x_2) \in \xi_X \Rightarrow (f(x_1), f(x_2)) \in \xi_Y$.*

Recall that $(x_1, x_2) \in \xi^2$ if there is an x such that $(x_1, x) \in \xi$ and $(x, x_2) \in \xi$; similarly for ξ^n.

(4.8) **Definition.** *$f: X \to Y$ is said to be n-**continuous** if $(x_1, x_2) \in \xi_X \Rightarrow (f(x_1), f(x_2)) \in \xi_Y^n$.*

(4.9) **Definition.** *Let $M = (X, Y, Q, \lambda, \delta)$ be a tolerance automaton. Then we say that M is an n-**tolerance automaton** if $\lambda(\cdot, x): Q \to Q$ is n-continuous for each $x \in X$.*

(4.10) **Theorem.** *Every tolerance automaton is a 3-tolerance automaton.*

Proof. $(q, \lambda(q, x)) \in \xi_Q$, $(q', \lambda(q', x')) \in \xi_Q$, and thus, if $(q, q') \in \xi_Q$, $(\lambda(q, x), \lambda(q', x')) \in \xi_Q^3$. $\qquad\square$

(4.11) **Theorem.** *Let M be a 1-tolerance automaton. Then*

$$\lambda(\cdot, x): Q \to Q$$

is ξ_Q-continuous for every $x \in X^*$.

Proof. By definition of a 1-tolerance automaton, $\lambda(\cdot, x): Q \to Q$ is 1-continuous for all $x \in X$. Now, $\lambda(\cdot, xy) = \lambda(\lambda(\cdot, x), y)$. But then the result is immediate by induction: if f and g are ξ-continuous, then so is their composition. $\qquad\square$

Thus 1-tolerance automata have a *stability* property unshared by general tolerance automata; small differences in initial state cannot give rise to large differences in state at any later time. Each mapping $\lambda(\cdot, x): Q \to Q$ is thus something in the nature of a contraction, as we can see by defining a metric

$$d(x, x') = \min \{n: (x, x') \in \xi^n\}.$$

Output tolerances. In the "black-box" approach to automata, in which our attention focuses on input/output behavior rather than that of the states, it makes sense to focus on new notions of tolerance which relate "small changes" in input to "small changes" in output. We thus make the new definitions:

(4.12) **Definition.** *Let $M = (X, Y, Q, \lambda, \delta)$ be an automaton for which the set of input strings is a tolerance space (X^*, ξ_{X^*}) and the set of* *output strings is a tolerance space* (Y^*, ξ_{Y^*}). *We say that M is an* **input/output tolerance** *automaton if for all states $q \in Q$ and* $x, x' \in X^*$

$$(x, x') \in \xi_X^* \Rightarrow (\delta(q, x), \delta(q, x')) \in \xi_{Y^*}.$$

We are often interested in the case where ξ_{X^*} is induced directly by a tolerance ξ_X on X.

(4.13) **Definition.** *Let (A, ξ) be a tolerance space. Then (A^*, ξ^*) is A^** **made into a tolerance space** *by defining $a, a' \in A^*$ to be within the tolerance ξ^* if and only if*

$$a = (a_1, \ldots, a_n), \qquad a' = (a_1', \ldots, a_{n'}')$$

satisfy $n = n'$ and $(a_1, a_1') \in \xi, \ldots, (a_n, a_n') \in \xi$.

Let us use $LP(X, Y)$ to denote the length-preserving functions from X^* to Y^*. Then ξ_X and ξ_Y induce a natural subset of $LP(X, Y)$, the set of functions continuous with respect to ξ_X^* and ξ_Y^*:

$$f \in LP_\xi(X, Y) \quad \Leftrightarrow \quad (x, x') \in \xi_{X^*} \Rightarrow (f(x), f(x')) \in \xi_Y^*.$$

Thus we see that M is an input/output tolerance automaton if and only if each $M_q \in LP_\xi(X, Y)$.

There is a natural tolerance on $LP(X, Y)$,

$$(f, f') \in \xi_{LP} \quad \Leftrightarrow \quad (f(x), f'(x)) \in \xi_Y^* \text{ for all } x \in X^*.$$

We say that M is a *natural tolerance automaton* (with respect to input tolerance ξ_X and output tolerance ξ_Y) if for each $q \in Q$ and each $x \in X$

$$(M_q, M_q L_x) \in \xi_{LP}.$$

We could also define a tolerance $\bar{\xi}$ on X in terms of a tolerance ξ_Y on Y by

$$(x, x') \in \bar{\xi} \quad \Leftrightarrow \quad (\delta(q, x)\ \delta(q, x')) \in \xi_Y \text{ for all } q \in Q.$$

Note that we would normally require that a tolerance on X^* (or Y^*) be right invariant, that is,

$$(x, x') \in \xi_{X^*} \text{ and } x'' \in X^* \quad \Rightarrow \quad (xx'', x'x'') \in \xi_{X^*}.$$

All the tolerances explicitly introduced above share this tolerance property. In fact, we have always had the even stronger property that

$$(x, x') \in \xi_{X^*} \text{ and } (x'', x''') \in \xi_{X^*} \quad \Rightarrow \quad (xx'', x'x''') \in \xi_{X^*}.$$

Boundaries and optimality

(4.14) **Definition.** *C is called a* **cost space** *if*
(a) *C is a tolerance space with respect to ξ.*

(b) *C is an abelian group under $+$, partially ordered with respect to $\leq$.*

(c) *For each $c \in C$ there exist a, b, $\in C$ such that $a < c < b$, and $a < c' < b \Rightarrow (c, c') \in \xi$.*

In fact, we shall usually think of a cost space as the reals under a tolerance of the form $(c, c') \in \xi \Leftrightarrow |c - c'| < \epsilon$ for some fixed $\epsilon > 0$.

Given an automaton $M = (X, Y, Q, \lambda, \delta)$ and a cost space C, a cost function for M is a function $p: Q \times X \to C$. We extend p to $Q \times X^*$ by

$$p(q, xy) = p(q, x) + p(\lambda(q, x), y).$$

The *optimal control problem for automata* may be stated as follows:

(4.15)   **Optimal control problem.** *Let q_0 and q_1 be two states of Q, called the initial state and terminal state, respectively. We shall say that $u = (u_1, \ldots, u_n) \in X^*$ transfers M from q_0 to q_1 if*

$$\lambda(q_0, u) = q_1,$$

whereas

$$\lambda(q_0, u_1, \ldots, u_k) \neq q_1 \qquad \text{for all } k < n.$$

Among all sequences u in X^* which transfers M from q_0 to q_1, find that for which $p(q_0, u)$ is minimal.†

Let $M = (X, Y, Q, \lambda, \delta)$ be an automaton with cost function p and cost space C. We define the (usually infinite) automaton

$$(M, p) = (X, Y, Q \times C, \lambda_p, \delta_p)$$

by

$$\lambda_p(q, c, x) = (\lambda(q, x), c + p(q, x)),$$
$$\delta_p(q, c, x) = \delta(q, x).$$

We define the *reachable set R_{q_0}* in $Q \times C \times T$ to be

$$\{(\lambda_p(q_0, 0, u), l(u)): u \in X^*\},$$

where $\lambda_p(q_0, 0, \Lambda) = (q_0, 0)$ with $\Lambda = $ empty string.

(4.16)   **Definition.** *Let X and Y be tolerance spaces; then the **product tolerance space** is the cartesian product $X \times Y$ together with the tolerance*

$$((x, y), (x', y')) \in \xi \Leftrightarrow (x, x') \in \xi_X, (y, y') \in \xi_Y.$$

Now let M be a 1-tolerance automaton, and let us consider $Q \times C$ as a tolerance space with the product tolerance. For each n we consider

---

† A better formulation would be from q_0 to within tolerance of q_1. The present formulation is completely tentative.

the cross section of R_{q_0} at time n:

$$R_{q_0}^n = \{\lambda_p(q_0, 0, u): u \in X^n\} \subseteq Q \times C.$$

If u is optimal, then it has minimal C-coordinate of any point in $R_{q_0}^n$, and thus a *necessary* condition that u be optimal is that $\lambda_p(q_0, 0, u)$ be a point of the *boundary* of R_{q_0}.

(4.17) Theorem. *For every state q of a 1-tolerance automaton M and every $n \in T$, R_q^n is connected with respect to ξ_Q^2.†*

Proof. Let $q_1, q_2 \in R_q^n$. Then there exist $x = (x_1, \ldots, x_n)$ and $x' = (x_1', \ldots, x_n')$ in X^n such that $\lambda(q, x) = q_1$ and $\lambda(q, x') = q_2$. Let

$$\rho(m) = \lambda(q, x_1, \ldots, x_m, x_{m+1}', \ldots, x_n');$$

then $\rho(0) = q_1$, $\rho(n) = q_2$ and $(\rho(m), \rho(m + 1)) \in \xi_Q^2$.   □

At this stage a digression is necessary to specify "boundary" in terms of tolerance, rather than topological, spaces. Analogous to the usual definition for topological spaces, we have

(4.18) Definition. *Let S be a subset of a tolerance space X. Then we define:*
*The ξ-**closure** of S, $\bar{S} = \{x: (x, y) \in \xi$ for some $y \in S\}$.*
*The ξ-**interior** of S, int $S = \{x: (x, y) \in \xi \Rightarrow y \in S\}$.*
*The ξ-**boundary** of S, $\delta S = \bar{S} - $ int $S = \{x: (x, y) \in \xi$ for some $y \in S$ but $(x, z) \in \xi$ for some $z \notin S\}$.*

(4.19) Lemma. $X - \bar{S} = $ int $(X - S)$.

Proof.

$$x \in X - \bar{S} \Leftrightarrow (x, y) \in \xi \Rightarrow y \notin S,$$
$$\Leftrightarrow x \in \text{int } (X - S). \qquad \square$$

Thus we do indeed have, using Definition (4.14c), that if u is optimal, then $\lambda_c(q_0, 0, u) \in \delta R_{q_0}^n$.

(4.20) Theorem. *Let M be a 1-tolerance automaton and let*
$$u = (u_1, \ldots, u_n) \text{ be in } X^n. \text{ If } \lambda_c(q, u) \in \delta R_q^n, \text{ then}$$

$$(q_k, c_k) = \lambda_c(q_0, 0, u_1, \ldots, y_k) \in \delta R_{q_0}^k, \quad 1 \leq k \leq n.$$

Proof. Suppose there were some $k < n$ such that $(q_k, c_k) \in $ int R_q^k, whereas $(q_{k+1}, c_{k+1}) \in \delta R_q^{k+1}$. Then $(q_k, c') \in R_q^k$ for some $c' < c_k$. Let $(q_k, c') = \lambda_c(q_k, b)$, $v \in I^k$. Then $\lambda_c(q, vc_{k+1}) = (\lambda(q_k, c_{k+1}), c' + c(q_k, x_{k+1}))$, contradicting the optimality of u.   □

The control theorist will recognize this as the analog of the theorem on which rests the Pontryagin maximum principle of optimal control.

† The improvement from the ξ_Q^4 of [ARBIB, 1966] is due to [VERBEEK, personal communication].

7 Basic notions of automata and semigroups

The next three chapters are devoted to questions of the structure of the discrete systems we call automata. Because we are now turning from notions of calculus, with which most readers are familiar, to the notions of modern algebra, with which few readers are familiar, the next section sets forth in reasonable detail the algebraic background necessary for the remainder of the chapter.

We wish to understand the ways in which automata may be decomposed into a series and parallel composition of simpler machines. To this end we associate a semigroup with each machine. We then discover that a machine can always be decomposed into simpler machines unless it is a flip-flop, or has a simple group for its semigroup. Conversely, any decomposable machine M can be built up from flip-flops and simple group machines, where the simple groups required stand in a precise algebraic relation, called *divisibility*, to the semigroup of M.

7.1 Semigroups and congruences

This section is purely mathematical and provides the vocabulary in which our discussion of the structure of automata will be presented. The reader who is not familiar with all these topics should carefully read and reread this section until he has a feel for all the concepts involved before studying the subsequent structure theory.†

Let us first recall the relatively familiar notion of a group:

(1.1) Definition. *A* **group** *is a set G, together with a binary operation $m: G \times G \rightarrow G$ (we usually write ab for $m(a, b)$) subject to the conditions that*

(a)   *m is* **associative**; *that is, for all a, b, $c \in G$*

$$(ab)c = a(bc).$$

(b)   *G has an* **identity** *e; that is, there exists $e \in G$ such that for all $a \in G$*

$$ae = ea = a.$$

(c)   *Each element $a \in G$ has an* **inverse** *a^{-1}; that is, for all $a \in G$, there exists $a^{-1} \in G$ such that*

$$aa^{-1} = a^{-1}a = e.$$

For instance, let us consider the integers $\{\ldots, -2, -1, 0, 1, 2, \ldots\}$ under *addition*. Then we have a group, since

$$(a + b) + c = a + (b + c),$$
$$a + 0 = 0 + a = a,$$
$$a + (-a) = (-a) + a = 0.$$

However, the integers do not form a group under *multiplication*, for although

$$(ab)c = a(bc),$$
$$a1 = 1a = a,$$

the inverse is not defined as an integer, except for $a = \pm 1$.

(1.2) Note. We say a binary operation $m(a, b)$ is *commutative* if $m(a, b) = m(b, a)$ for all a and b. We call a group *abelian* or *commutative* if its operation is commutative.

Perhaps the easiest example of a *nonassociative* binary operation is subtraction:

$$a - (b - c) \neq (a - b) - c.$$

We shall not concern ourselves here with nonassociative operations, but we shall be interested in operations which do not *necessarily* have an inverse. This leads to the notion of a semigroup, which does not have

---

† Further material is contained in [CLIFFORD and PRESTON, 1961] and [HU, 1965, sec. 2].

to possess all the group properties:

(1.3) **Definition.** *A* **semigroup** *is a set* S, *together with a binary operation* $m: S \times S \to S$ *(we usually write* $a \cdot b$ *for* $m(a, b)$*) subject to the condition that*

(a) *m is associative; that is, for all* $a, b, c \in S$

$$(a \cdot b) \cdot c = a \cdot (b \cdot c).$$

If S *has an identity—that is, an element* e *such that*

$$e \cdot s = s \cdot e = s,$$

we sometimes call S *a* **monoid.**

Thus every group is a semigroup, but the converse is not true, for the integers under multiplication form a monoid (with identity 1) which is *not* a group. The integers *greater* than 1, form a semigroup which is *not* a monoid.

(1.4) **Definition.** *A* **subsemigroup** *of a semigroup* S *is a subset* $S' \subseteq S$ *which is closed under multiplication:*

$$s_1, s_2 \in S' \Rightarrow s_1 \cdot s_2 \in S'.$$

Consider a finite set Σ of symbols. Let Σ^* be the set of all finite sequences

$$\sigma_1 \cdots \sigma_n,$$

where each $\sigma_i \in \Sigma$. Let Λ denote the empty string with $n = 0$ symbols, and let us include it in Σ^*. We define a binary operation m on Σ^* by concatenation or juxtaposition; that is,

$$m(\sigma_1 \cdots \sigma_n, \tau_1 \cdots \tau_m) = \sigma_1 \cdots \sigma_n \tau_1 \cdots \tau_m.$$

We write xy for $m(x, y)$. Clearly, $(xy)z = x(yz)$. Λ is clearly a unit, $\Lambda x = x\Lambda = x$, and so Σ^* is a monoid. If $x = \sigma_1 \cdots \sigma_n$, we call n the length of σ, $l(\sigma)$. Then $l(\Lambda) = 0$, $l(xy) = l(x) + l(y)$, like a logarithm.

We refer to Σ^*, together with the operation of juxtaposition, as *the free monoid† generated by* Σ.

Let X be any set, and let f be any map of X into itself. We refer to f as a *transformation* of X. If we write the function symbol to the left of argument, the collection of such transformations becomes a semigroup $F_L(X)$ under the composition $(f \circ g)(x) = f(g(x))$. If we write the function symbol to the right of the argument, this collection becomes

---

† "Free" here means essentially that the elements are free of any restrictions except that imposed by associativity. No finite semigroup with more than one element is free. It suffices for our present purposes to merely use "free" as a label for Σ^*. The reader can find more details in [Hu, 1965, sec. 2.4].

a semigroup $F_R(X)$ under the composition $(x)(f \circ g) = [(x)f]g$. Thus the order of composition if reversed in going from $F_L(X)$ to $F_R(X)$.

(1.5) **Definition.** *A **homomorphism** of a semigroup S_1 into another S_2 is any map $p\colon S_1 \to S_2$ for which $p(s \cdot s') = p(s) \cdot p(s')$ for all s, $s' \in S_1$. If S_1 and S_2 are monoids and p maps the identity into the identity, we call p a **monoid homomorphism**.*

Thus multiplication by 2 is a homomorphism for the set of integers under addition but *not* for the set of integers under multiplication:

$$2(a + b) = 2a + 2b,$$

but

$$2(a \cdot b) \neq 2a \cdot 2b \qquad \text{(unless } a \cdot b = 0\text{).}$$

We say p is *onto* if for every $s_2 \in S_2$ we can find $s_1 \in S_1$ such that $p(s_1) = s_2$. We may abbreviate this to $p(S_1) = S_2$.

(1.6) **Definition.** *A **representation** of a semigroup S is a homomorphism*

$$T\colon S \to F_N(X)$$

for some space X, and $N = L$ or R; for example, if $N = L$ we represent an element s of S by a transformation $T(s)$ of the space X in such a way that

$$T(s \cdot s') = T(s)T(s').$$

Every semigroup S has two very simple representations, both of which represent an element s as a *transformation of S into itself*.

The *right-regular representation* $R\colon S \to F_R(S)$ represents s as the transformation R_s of right multiplication by s. In other words,

$$R(s) = R_s \in F_R(S),$$

where

$$(a)R_s = a \cdot s \qquad \text{for all } a \in S,$$
$$(a)R_{st} = ast = (as)R_t = (a)R_sR_t.$$

The *left-regular representation* $L\colon S \to F_L(S)$ represents s as the transformation L_s of left multiplication by s. That is,

$$L(s) = L_s \in F_L(S),$$

where

$$L_s(a) = s \cdot a \qquad \text{for all } a \in S,$$
$$L_{st}(a) = sta = L_s(ta) = L_sL_ta.$$

(1.7) **Definition.** *A **binary relation** R on a set S is simply a subset of $S \times S$. We say x **is in the relation** R **to** y (written xRy) if and only if $(x, y) \in R$.*

(1.8)   **Definition.**   *By a* **partially ordered set** *we mean a set P together with a binary relation $\leq$ which satisfies the conditions that*
(a) *For all $x \in P$, $x \leq x$ (reflexive).*
(b) *For all x and y in P, if $x \leq y$ and $y \leq x$, then $x = y$ (antisymmetric).*
(c) *For all x, y, and z in P, if $x \leq y$ and $y \leq z$, then $x \leq z$ (transitive).*

The most common example is that of the integers, where $x \leq y$ if and only if y is greater than or equal to x.

A more illuminating example, which points up that the ordering need only be partial, is the set of subsets of a set S under the subset relation, if X, $Y \subseteq S$, then $X \leq Y$ if and only if $X \subseteq Y$. Here we see that, given X and Y, it may well happen that neither $X \leq Y$ nor $Y \leq X$, unlike the situation for the integers. We then say that X and Y are *incomparable*.

A notion of fundamental importance in mathematics is that of equivalence:

(1.9)   **Definition.**   *A relation $\equiv$ on a set S is an* **equivalence relation** *if and only if it satisfies the conditions that*
(a) *For all $x \in S$, $x \equiv x$ (reflexivity).*
(b) *For all x, $y \in S$, $x \equiv y \Rightarrow y \equiv x$ (symmetry).*
(c) *For all x, y, $z \in S$, $x \equiv y$ and $y \equiv z \Rightarrow x \equiv z$ (transitivity).*

Familiar examples are equality in arithmetic, congruence in euclidean geometry, and similarity for nonsingular matrices.

If s is a typical element of S, then *"the equivalence class of s"* is $[s] = \{t : t \equiv s\}$.

Given a set S and an equivalence relation $\equiv$, we may define a new set $S/\equiv$, the set of equivalence classes of S modulo $\equiv$.

(1.10)   **Definition.**   *A* **partition** *P of S is a division of S into disjoint subsets S_1, S_2, . . . ; that is,*

$$S_i \cap S_j = \varnothing \quad \text{for } i \neq j$$

and

$$S = \bigcup_i S_i.$$

Each S_i is called a **block** *of P.*

Clearly, $[s] = [t]$ as elements of $S/\equiv$ if and only if $t \equiv s$. We call $S/\equiv$ *the partition of S induced by the equivalence $\equiv$.*

An equivalence relation R over the *semigroup S is right invariant* if whenever xRy, then $xzRyz$ for all z in S. Clearly, there is an analogous definition of *left-invariant* equivalence relations.   Then:

(1.11)   **Definition.**   *An equivalence relation over a semigroup S is a* **congruence relation** *if it is both right and left invariant.*

Define the map $N_R \colon S \to S/R$ by the equation

$$N_R(t) = [t]_R,$$

the equivalence class of t under R.

The importance of the notion of congruence resides in the following

(1.12)   **Lemma.**   *If R is an equivalence relation, we may make S/R a semigroup with the multiplication*

$$[x]_R \cdot [y]_R = [xy]_R$$

if and only if R is a congruence. N_R is then a homomorphism, and we call S/R a **factor semigroup.**

The easy verification is left to the reader.

(1.13)   **Definition.**   *An equivalence relation over a set S is of* **finite index** *if there are only finitely many equivalence classes under the relation.*

If, for a partition P, we define $\equiv_P$ on S by $x \equiv_P y$ if and only if x and y lie in the same block of P, then $\equiv_P$ is an equivalence and P is, essentially, the partition of S induced by the equivalence $\equiv_P$.

Finally, a short note on cardinal numbers: a set is *finite* if it has finitely many members; *denumerable* if it can be put in one-to-one correspondence with the set N of the integers (in which case we say the set has cardinality aleph-null, symbolized $\aleph_0$); and *nondenumerable* if none of the previous cases hold.

Given a finite set S with n elements, then there are 2^n subsets of S, since for each of the n elements we have two choices as to whether or not we include it in the subset.

By analogy, we denote by $2^{\aleph_0}$ the number of subsets of a denumerable set S; the one-to-one correspondence between S and N induces a one-to-one correspondence on their sets of subsets, which thus have the same cardinality.   With each subset T of S we may associate a real binary number:

$$0 \cdot t_1 t_2 t_3 \cdots t_n \cdots ,$$

where

$$t_n = \begin{cases} 1, & n \in T, \\ 0, & n \notin T. \end{cases}$$

Thus there is a one-to-one correspondence between the real numbers in the closed interval $(0, 1]$ and the subsets of T.† By the well-known Cantor diagonal argument, we know that these reals *cannot* be put in one-to-one correspondence with the integers:

$$2^{\aleph_0} > \aleph_0.$$

We shall make brief use of this fact in the next section.

7.2 Automata, reduced forms, and equivalence relations

Rather than consider continuous systems, we now restrict ourselves to stationary systems with finite input and output sets and with

$$T = \{0, 1, \ldots\}.$$

Formally, then, recalling the discussion from Section 6.1,

(2.1) **Definition.** *An* **automaton** (*or* **machine**) *is a quintuple*

$$M = (X, Y, Q, \lambda, \delta),$$

where X = *a finite set, the set of* **inputs,**
Y = *a finite set, the set of* **outputs,**
Q = *the set of* **states** (*not necessarily finite*),
$\lambda\colon Q \times X \to Q$ = *the* **next state function,**
$\delta\colon Q \times X \to Y$ = *the* **next output function.**

We shall not attempt to avoid confusing the abstract model and the machine it describes. The interpretation of an automaton is as a system such that if at time t it is in state q and receives input x, then at time $t + 1$ it will be in state $\lambda(q, x)$ and will emit output $\delta(q, x)$.

(2.2) **Definition.** *A machine M is a* **finite automaton** *if Q is finite.*

We immediately extend the applicability of λ and δ to X^*, the free monoid on X,

$$\lambda\colon Q \times X^* \to Q,$$
$$\delta\colon Q \times X^* \to Y,$$

by repeated application of the equalities

$$\lambda(q, x'x'') = \lambda(\lambda(q, x'), x''),$$
$$\delta(q, x'x'') = \delta(\lambda(q, x'), x''),$$

where the sequences x' and x'' are in X^*. We shall leave $\delta(q, \Lambda)$ undefined but set $\lambda(q, \Lambda) = q$.

---

† This is *not* quite correct. It is left as an exercise to the reader to patch it up.

The input-output behavior of the machine M starting in state q, $M_q: X^* \to Y$, is defined by $M_q(x) = \delta(q, x)$. Our general systems definitions reduce to

(2.3) **Definition.** q is **equivalent to** q' iff $M_q = M'_q$ as functions from X^* to Y.

(2.4) **Definition.** A machine M is in **reduced form** iff the map $q \mapsto M_q$ is one-to-one (the reduced machine cannot have one input-output function corresponding to more than one state).

(2.5) **Definition.** Given a machine $M = (X, Y, Q, \lambda, \delta)$ and a machine $M' = (X, Y, Q', \lambda', \delta')$, M is **strictly equivalent** to M' iff for each $q \in Q$ there exists a $q' \in Q'$ such that $M_q = M'_{q'}$, and vice versa.

(2.6) **Theorem.** Every machine M is strictly equivalent to a machine in reduced form, $M^0 = (X^0, Y^0, Q^0, \lambda^0, \delta^0)$.

Proof. The idea is to merge states with the same input-output behavior. Let $Q^0 = \{f: \text{there exists a } q \in Q \text{ for which } f = M_q\}$. Since several q, may yield the same M_q, we see that $\#(Q^0) < \#(Q)$ ($\#$ is cardinality) unless M is already in reduced form. Let $X^0 = X$ and $Y^0 = Y$. We must now define $\lambda^0(f, x)$ and $\delta^0(f, x)$. We introduce for each $x \in X^*$ the map $L_x: X^* \to X^*$, where $L_x(x') = xx'$ (left multiplication by x). Then we write $\delta(\lambda(q, x), x') = \delta(q, xx') = \delta(q, L_x(x')) = M_q L_x(x')$.

$M_{\lambda(q,x)} = M_q L_x$, a mapping so we may define λ^0 by $\lambda^0(f, x) = fL_x$ and $\delta^0(f, x) = f(x)$. Thus $\lambda^0(M_q, x) = M_{\lambda(q,x)}$ and $\delta^0(M_q, x) = \delta(q, x)$, and we easily check that M^0 is reduced and equivalent to M. $\qquad\square$

(2.7) **Exercise.** Show that M^0 is the *only* reduced form of M, that is, that any reduced form $\tilde{M}$ equivalent to M is obtained from M^0 simply by relabeling the states.

Consider a machine M started in some initial state q_0. Then every state that the machine will subsequently enter is of the form $\lambda(q_0, x)$ for some $x \in X^*$; that is, the $\lambda(q_0, x)$ are the only states that may be reached from the given initial state q_0.

Consider the reduced machine with only the states reachable from q_0. Call it $M(f) = (X, Y, Q_f, \lambda_f, \delta_f)$, where $f = M_{q_0}$. We may diagram its operation as shown in Figure 7.1. Then

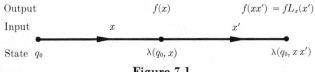

Output $f(x)$ $f(xx') = fL_x(x')$

Input x x'

State q_0 $\lambda(q_0, x)$ $\lambda(q_0, x\,x')$

Figure 7.1

$$Q_f = \{g: g = fL_x \text{ for some } x \in X^*\},$$
$$\lambda_f(g, x) = gL_x,$$
$$\delta_f(g, x) = g(x).$$

We shall hereafter mainly consider reduced machines, with all states reachable from a specified initial state—that is, machines of the form $M(f)$.

(2.8) **Definition.** *Given a function $f: X^* \to Y$, then f is* **realizable by a finite automaton** *if and only if Q_f is a finite set.*

(2.9) **Definition.** *The* **characteristic function** *χ_R of a set R is*

$$\chi_R(x) = \begin{cases} 1, & x \in R, \\ 0, & x \notin R. \end{cases}$$

(2.10) **Definition.** *A subset $R \subseteq X^*$ is* **realizable** *if and only if the characteristic function $\chi_R: X^* \to \{0, 1\}$ is realizable.*

A finite automaton is specified by a finite tableau. We can list these tableaux, one after another; hence the set of finite automata (and so, *a fortiori*, the set of realizable sets) is denumerable, that is, has cardinality $\aleph_0$. But X^*, having cardinality $\aleph_0$, must have $2^{\aleph_0}$ subsets, that is, non-denumerably many. Thus most subsets of X^* are *not* realizable.

(2.11) **Definition (Nerode equivalence).** *Given a function $f: X^* \to Y$, let E_f be the equivalence relation on X^*,*

$$xE_f x' \Leftrightarrow f(xz) = f(x'z) \text{ for all } z \in X^*.$$

Noting that $xE_f x' \Leftrightarrow fL_x = fL_{x'}$, we see that the *Nerode equivalence classes* $[x]_E$ may be thought of as elements of Q_f, the state set of $M(f)$, the minimal machine which realizes f. We may thus easily verify

(2.12) **Proposition.** *(a) E_f is a right-invariant equivalence relation, and (b) if f is realizable, E_f is of finite index, and vice versa.*

(2.13) **Example.** The set $R = \{0, 01, 01^2, \ldots\}$ is realizable.

Proof 1 (Nerode equivalence). Let $X = \{0, 1\}$. xEy implies that for all $z \in X^*$

$$xz \in R \Leftrightarrow yz \in R.$$

Observe that we have three equivalence classes: $\{\Lambda\}$, R, and $X^* - (R \cup \{\Lambda\})$. That is, E has *finite* index.

Proof 2. We may derive an automaton which realizes R as $M(\chi_R)$ (note that there will be three states, since states of the minimal machine correspond to Nerode equivalence classes), or else we may construct it directly.

Either way, we get

$$X = \{0, 1\},$$
$$Y = \{0, 1\},$$
$$Q = \{q_0, q_1, q_2\},$$

with q_0 the initial state and with λ and δ given by the tables

λ	0	1
q_0	q_1	q_2
q_1	q_2	q_1
q_2	q_2	q_2

δ	0	1
q_0	1	0
q_1	0	1
q_2	0	0

The output from state q_1 is 1, otherwise it is zero; that is, a sequence is accepted if it sends the machine from state q_0 to q_1. □

The reader should check that if a set is realizable, then so is any set that differs from it in only a finite number of elements.

(2.14) **Example.** The set $R = \{0^{n^2-1}1 : n > 0\}$ is *not* realizable.

Proof 1 (Nerode equivalence). $i \neq j \Rightarrow$ not $(0^i E 0^j)$, so E has nonfinite index.

Proof 2. Suppose $M(\chi_R)$ has only finitely many states. Since there are infinitely many strings 0^k, we must have a pair (k, l) such that

(2.15) $\chi_R L_{0^k} = \chi_R L_{0^l}$

Pick any $n^2 - 1 > k > 1$. Then for all t the positive integers

$$\chi_R(0^{n^2-1+t(k-l)}1) = \chi_R(0^{n^2-1}1),$$

by repeated application of (2.15), and R was not what we thought it. □

(2.16) **Definition (Myhill equivalence).** *Given a function* $f: X^* \to Y$, *let* $\equiv_f$ *be the* **Myhill equivalence relation** *on* X^* *defined by*

(2.17) $x \equiv_f x'$ $\Leftrightarrow$ $f(yxz) = f(yx'z)$ *for all* $y, z \in X^*$,
                    $\Leftrightarrow$ $f L_y L_x = f L_y L_{x'}$ *for all* $y \in X^*$,
                    $\Leftrightarrow$ *the input strings* x *and* x' *induce the same function on* Q_f, *that is,*

(2.18) $\mathfrak{M}_x = \mathfrak{M}_{x'}: Q_f \to Q_f$,

     where

$$\mathfrak{M}_x([y]_E) = [yx]_E.$$

Note, in particular, that $x \equiv_f x' \Rightarrow xEx'$, but the converse is not necessarily true.

(2.19) **Exercise.** Let $M(f)$ have n states and $|X^*/\equiv_f| = n'$. Show that $n \leq n' \leq n^n$. For arbitrary n give examples of functions which attain each bound.

Solution. $n' =$ the index (number of equivalence classes) of $\equiv_f$,
$\leq$ the index of $E_f = n$.
$n' \leq$ the number of functions from Q_f to $Q_f = n^n$.
Therefore $n \leq n' \leq n^n$.
   We now give examples whose (easy) verification is left to the reader.

(2.20) **Example.** $n' = n$. Let $M(f)$ have $n - 1$ reset inputs,

$$X = \{x_1, \ldots, x_{n-1}\},$$

with $f(zx_k) = x_k$ for all $z \in X^*$ $[f(\Lambda) = x_1,$ say$]$.

(2.21) **Example.** $n' = n^n$. Let Y have n elements $\{y_1, \ldots, y_n\}$. Let X have n^n elements $\{x_1, \ldots, x_{n^n}\}$, and let us label the n^n functions of Y into Y by elements of X as $g_{x_1}, \ldots, g_{x_{n^n}}$. Define $f(\Lambda) = y_1 \in Y$ and $f(zx_j) = g_{x_j}(f(x))$ for all $z \in X^*$. (In fact, we could choose X *much* smaller; it suffices that the g_x be a set of *generators* for the semigroup of maps of Y into Y.)
   In particular, we have

(2.22) **Proposition.**
   (a) $\equiv_f$ *of finite index implies that E_f is of finite index, which implies that $M(f)$ is finite.*
   (b) $M(f)$ *finite implies that Q_f is finite; this implies only finitely many functions from $Q_f \to Q_f$, which implies that $\equiv_f$ is of finite index.*

Notice that $x \equiv x'$ and $y \equiv y'$ imply that for all $z \in X^*$

$$fL_zL_{xy} = fL_zL_xL_y = fL_zL_xL_{y'} = fL_zL_{x'}L_{y'} = fL_zL_{x'y'} \Rightarrow xy \equiv_f x'y',$$

so the Myhill equivalence relation is a congruence, and we may form a factor monoid:

(2.23) **Definition.** *Given a function $f: X^* \to Y$, we define the* **monoid (semigroup) of** f *to be*

$$S_f = X^*/\equiv_f.$$

The elements of S_f correspond to input sequences regarded as state-transition functions of $M(f)$. Composition is consecutive state transformation.

(2.24) **Note (crucial).** If M_f is finite, S_f is a finite semigroup, and vice versa.

(2.25) **Example.** Compute S_{χ_R} for $R = \{0, 01, 01^2, \ldots \}$.

Construction. We shall use our knowledge of Q_f to compute $S_f(f = \chi_R)$ as a set of functions on Q_f. We saw that the states of $M(\chi_R)$ are $q_0 = [\Lambda]_E$, $q_1 = [0]_E$, and $q_2 = [1]_E$. We now have to subdivide these classes further into new sets such that all the strings in a given subset map Q_f identically. A little trial and error quickly yields the choice $s_0 = q_0, s_1 = q_1, s_2 = \{1^m\}$, and $s_3 = q_2 - s_1$. The functions on Q_f are given by the following table:

	s_0	s_1	s_2	s_3
q_0	q_0	q_1	q_2	q_2
q_1	q_1	q_2	q_1	q_2
q_2	q_2	q_2	q_2	q_2

We thus conclude that $S_{\chi_R} = \{s_0, s_1, s_2, s_3\} = \{[\Lambda], [0], [1], [10]\}$ and has multiplication table

	s_0	s_1	s_2	s_3
s_0	s_0	s_1	s_2	s_3
s_1	s_1	s_3	s_1	s_3
s_2	s_2	s_3	s_2	s_3
s_3	s_3	s_3	s_3	s_3

7.3 Machines and semigroups

We have seen how to go from a machine to a semigroup. We now want to turn our attention to how a semigroup characterizes a machine. We shall then devote the next two chapters to exploring the structure of machines via the structure of semigroups.

In the theory of semigroups a homomorphism is a map $h: S \rightarrow S'$ such that $h(ss_1) = h(s)h(s_1)$. The corresponding idea for a machine of the form $M(f)$ would thus be a homomorphism of the input semigroups $h: X^* \rightarrow (X')^*$, which in fact is fully defined by its values on X. If

$M(f')$ starts in state $f'L_t$ when $M(f)$ is in its initial state f, a map is induced by letting the action of x on f correspond to the action of $h(x)$ on $f'L_t$.

$$h_Q \colon Q_f \to Q_{f'},$$

defined by $h_Q(fL_x) = f'L_t L_{h(x)}$; we then have that if $q = fL_{x'}$, then

$$
\begin{aligned}
h_Q(\lambda_f(q, x)) = h_Q(fL_{x'x}) &= f'L_t L_{h(x'x)}, \\
&= f'L_t L_{h(x')} L_{h(x)}, \\
&= \lambda_{f'}(h_Q(q), h(x)),
\end{aligned}
$$

since h is a homomorphism.

Turning to general machines, this suggests

(3.1) Definition. *A* **homomorphism** $h \colon M \to M'$ **of machines** *is a map*

$$h \colon Q \cup X \to Q' \cup X'$$

with $h(Q) \subseteq Q'$ *and* $h(X) \subseteq X'$ *such that*

$$\lambda'(h(q), h(x)) = h(\lambda(q, x)) \qquad \text{for all } q \in Q,\, x \in X.$$

However, in terms of machines a more natural concept is that of *simulation*. We say that M simulates M' if, provided we encode and decode the input and output appropriately, M can process strings just as M' does when started in an appropriate state. We require the encoder and decoder to be memoryless (that is, to operate symbol by symbol) in order to make M do all the computational work involving memory (see Figure 7.2).

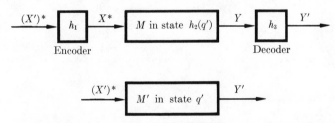

Figure 7.2 M simulates M' if h_1, h_2, and h_3 can be so chosen that the above systems have identical input/output behavior.

(3.2) Definition. M **simulates** M' *if there exists* (h_1, h_2, h_3) *where*
    (a) $h_1 \colon (X')^* \to X^*$ *is a monoid homomorphism,*
    (b) $h_2 \colon Q' \to Q,$
    (c) $h_3 \colon Y \to Y'$

are such that the following diagram is commutative:

(3.3)

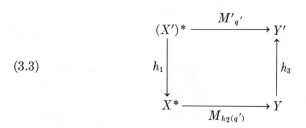

That is, the map from $(X')^*$ to Y' is independent of the route taken:

$$M'_q(x') = h_3(M_{h_2(q')}(h_1(x')))$$

or

$$\delta'[q', x'] = h_3(\delta(h_2(q'), h_1(x'))).$$

We say that M and M' are *weakly equivalent* if M simulates M' and M' simulates M. We write $M'|M$ if M simulates M', and read this as "M' *divides* M" (note change of order).

(3.4) **Lemma.** *Let M' be in reduced form. Then M simulates M' if and only if there exists (h_1, h_2, h_3), where*
   (a) $h_1: (X')^* \to X^*$ is a monoid homomorphism,
   (b) h'_2 maps a subset Q'' of Q onto Q': $Q \supseteq Q'' \xrightarrow{h'_2} Q'$,
   (c) $h_3: Y \to Y'$
are such that the following diagram is commutative:

(3.5)

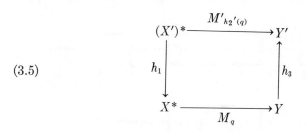

Proof. Keeping h_1 and h_3 fixed, we must show how to interchange h_2 of (3.3) with h'_2 of (3.5).

   (*a*) Starting from (3.3), since M' is reduced, no state of M can simulate two states of M', so that h_2 must be one-to-one. Thus we can take $Q'' = h_2(Q')$ and satisfy (3.5) by setting $h'_2 = h_2^{-1}$ on Q''.

   (*b*) Starting from (3.5), for $q' \in Q'$ choose $h_2(q')$ to be any q (many choices may be possible) such that $h'_2(q) = q'$; then (3.3) commutes. $\square$

(3.6)   **Remark.**   Criterion (3.5) for simulation of reduced machines is readily amenable to algebraic generalization, whereas (3.3) is not.

(3.7)   **Remark.**   If criterion (3.5) holds, then M simulates M' *whether or not M' is in reduced form.*

Given a machine $M(f)$, we obtain S_f, a monoid with the equivalence class of the empty string as identity.

Conversely, given a monoid S, we may obtain a machine, called the *machine of the semigroup S*,

$$M(S) \stackrel{\text{def}}{=} (S, S, S, \cdot, \cdot),$$

where $\cdot$ denotes semigroup multiplication.   If $M(S)$ is in state s at time t and receives input s', then its state and output at time $t + 1$ will both be $s \cdot s' \in S$.

If we go from the semigroup S_f to the machine $M(S_f)$, in what sense are this machine and the original $M(f)$ the same?   We realize they are not equal on recalling that if n is the number of states of $M(f)$ and n' is the number of states of $M(S_f)$, then $n \leq n' \leq n^n$, and the bounds are attainable.

We want to find some sense in which a machine and its semigroup are equivalent.

In our discussions of semigroups as representing machines, we are interested not only in the transformations from state to state, but also in what the output is.   So, given the semigroup S_f, we can also specify the function $i_f \colon S_f \to Y$, so that if x takes us from the initial state to the state represented by $s \in S_f$ (that is, if $s = [x]$), then we must have $i_f(s) = f(x)$. Note that i_f is well defined and depends only on s, and *not* the choice of representative x. $M(S_f, i_f)$ is then the machine $(S_f, Y, S_f, \cdot, \delta_f)$, where $\delta_f(s, s') = i_f(s \cdot s')$, that is, a state-output machine.   The next output may be determined completely from knowledge of the next state.

(3.8)   **Proposition.** $M(S_f, i_f)$ *is weakly equivalent to* $M(f)$.

Proof.   That $M(S_f, i_f)$ simulates $M(f)$ is immediate from Figure 7.3,

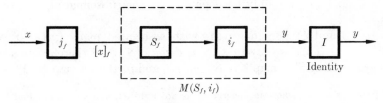

$$M(S_f, i_f)$$

Figure 7.3

where $j_f : X \to S_f$ is defined by $j_f(x) = [x]_f$ (the Myhill equivalence class).

To verify that $M(f)$ simulates $M(S_f, i_f)$ we have to find maps $h_1 : S_f \to X^*$, $h_2 : S_f \to Q_f$, and $h_3 : Y \to Y$ such that the following diagram

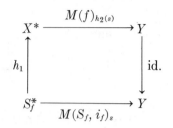

commutes.†

For each $s \in S$ choose $h_1(s)$ to be an $x \in X^*$ for which $s = [x]_=$. If $s = [x]_=$, set $h_2(s) = fL_x$. This does not depend on the choice of x, since $[x]_= = [x']_= \Rightarrow fL_x = fL_{x'}$. Finally, take $h_3(y) = y$.

There are many choices for h_1, but any one of them will make the diagram commute, for

$$
\begin{aligned}
M(S_f, i_f)_s(s') &= i_f(ss') = f(h_1(ss')), \\
&= f(h_1(s)h_1(s')), \\
&= M(f)_{[h_1(s)]_E}(h_1(s')), \\
&= M(f)_{h_2(s)}(h_1(s')). \qquad \square
\end{aligned}
$$

If we want to look at a machine in terms of algebra, we form S_f and i_f. Now we want to look at the semigroup and recapture machine concepts. The main concept so far is simulation. We now want to introduce the corresponding semigroup concept and check that it is suitable. Using Lemma (3.4b) as motivation, we introduce

(3.9) **Definition.** *We say a semigroup S* **divides** *a semigroup S' if there exists a subsemigroup $S'' \subset S'$ and a homomorphism Z of S'' onto S, or*

$$
S|S' \quad \Leftrightarrow \quad S' \supseteq S'' \xrightarrow{\ Z\ } S.
$$

Thus S' can simulate S in that the multiplication on S' induces that on S via Z, and we may only need part of S', the subsemigroup S'', to do this.

Let S and S' be semigroups with associated maps $i : S \to Y$ and $i' : S' \to Y'$. Then we say that (S, i) *divides* (S', i') if there exists a subsemigroup S'' of S', a map Z of S'' onto S, and further, a map $H : Y' \to Y$

---

† Here is a case where in general M' is not reduced, and we cannot find an h_2' to satisfy the criterion (3.5).

such that

$$i(Z(s)) = H(i(s)) \qquad \text{for all } s \in S'.$$

Thus   commutes.

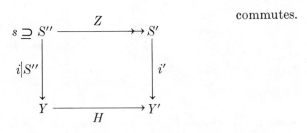

Let us write $g|f$ if $M(f)$ simulates $M(g)$ when the machines are started in their initial states, that is,

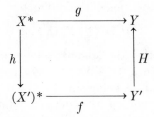

commutes, or $g = Hfh$.

Now we prove that if one semigroup divides another, the corresponding machines divide one another (and vice versa) when started in their initial states:

(3.10) Theorem. $g|f \Leftrightarrow (S_g, i_g)|(S_f, i_f)$.

Proof. Suppose $g = Hfh$. $S_g = (X)^*/\equiv_g$ and $S_f = (X')^*/\equiv_f$, so let $S'' = \{[h(x)]_f : x \in X^*\}$, and note that S'' is a subsemigroup of S_f. Define Z as the "inverse" of h reduced to equivalence classes; that is, $Z([h(x)]_f) = [x]_g$. Z will clearly be onto and a homomorphism as soon as we check that it is well defined; that is, $[h(x)]_f = [h(x_1)]_f \Rightarrow [x]_g = [x_1]_g$. But this follows, since for all y and z

$$\begin{aligned}
g(yxz) &= Hfh(yxz), \\
&= Hf(h(y)h(x)h(z)), \\
&= Hf(h(y)h(x_1)h(z)) \qquad \text{since } h(x) \equiv_f h(x_1), \\
&= Hfh(yx_1z).
\end{aligned}$$

We easily check that H will do to complete this half of the proof.

Converse.   If $(S_g, i_g) | (S_f, i_f)$, we have a commutative diagram

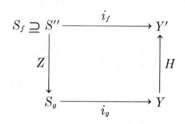

Given $x \in X$, choose $h(x)$ to be some $t \in (X')^* \ni Z([t]_f) = [x]_g$.   At least one such t exists, since Z is onto.   Then

$$Hfh(x) \;=\; H(i_f([h(x)]_f)) \;=\; i_g(Z([h(x)]_f)) \;=\; i_g([x]_g) \;=\; g(x). \qquad \Box$$

(3.11) **Important note.**   In the next two chapters we shall assume our machines to be *state-output* machines.   Thus

(3.12) $M = (X, Y, Q, \lambda, \beta)$

will henceforth denote that the present state of M determines the present output via the function $\beta : Q \to Y$ (we may recapture the next-output functions $\delta : Q \times X \to Y$ as being simply $\beta \circ \lambda$).

Given a machine denoted by some symbol, say N, we shall assume that it is represented by the quintuple whose entries are those of (3.12) subscripted by that symbol [for example, $(X_N, Y_N, Q_N, \lambda_N, \beta_N)$].

Given a machine denoted by M with some suitable modification, say M', we shall sometimes prefer to assume that it is represented by the quintuple whose entries are those of (3.12) with the same modification [for example, $(X', Y', Q', \lambda', \beta')$].   The two options should not cause any ambiguity.

8   Loop-free decomposition
of finite automata

It is well known (see, for example, [ARBIB, 1964, chap. 1]) that any finite automaton may be simulated by a network of modules with delays (*one-state* finite automata), provided we allow loops of arbitrary complexity in the network. In fact, we may use copies of just one module (the Sheffer-stroke module) to build such a network. In other words, a very simple set of components can be used to build up arbitrary finite automata *if we allow loops*. We shall see in the next two chapters that we cannot realize arbitrary automata from a finite set of components if we allow only loop-free synthesis. It is the purpose of this section to introduce the crucial definitions, state the main results of the algebraic theory of loop-free decomposition of finite automata, and to outline the strategy of our proofs.

8.1   An overview of the decomposition theorems

Let us first consider a portmanteau way of combining machines which subsumes series and parallel combinations.

(1.1) **Definition.** *Given state-output machines M' and M, a map $Z\colon \tilde{X} \times Y \to X'$, and a map $\eta\colon \tilde{X} \to X$, we define the* **cascade of M' and M with connecting map Z** *to be*

$$M' \times_Z^\eta M = (\tilde{X}, Y' \times Y, Q' \times Q, \lambda_Z, \beta_Z),$$

where we may read λ_Z and β_Z from Figure 8.1 *as*

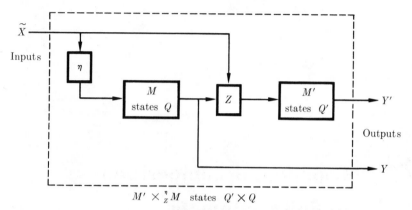

Figure 8.1

$$\lambda_Z((q', q), \tilde{x}) = (\lambda'(q', Z(\tilde{x}, \beta(q))), \lambda(q, \eta(\tilde{x}))),$$
$$\beta_Z(q', q) = (\beta'(q'), \beta(q)).$$

We usually omit η from explicit mention.

To get *series* composition (albeit preserving the output of M) we make Z independent of $\tilde{X}'$; to get *parallel* composition, we take $Z(\tilde{x}', y)$ to be just $\tilde{x}'$, as shown in Figure 8.2.

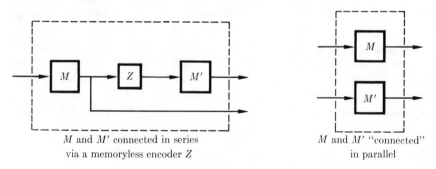

M and M' connected in series   M and M' "connected"
via a memoryless encoder Z   in parallel

Figure 8.2

The precise notion of loop-free composition we shall employ is then that of repeated formation of cascades, as well as memoryless encodings and decodings to obtain machines simulable by such combinations.   More precisely,

(1.2)   **Definition.**   *Given a collection $\mathfrak{M}$ of automata we call $C(\mathfrak{M})$†
the smallest collection of automata that contains $\mathfrak{M}$ and is closed under
cascades and simulations.*

We shall also speak of a machine obtained from more than two machines
in this way as being a *cascade* of these machines.

Before we state our main theorems we must introduce those machines
for which an input resets the machine to a state determined by that input,
or else the input leaves all states unchanged:

(1.3)   **Definition.**   *M is an **identity-reset machine** if for each $x \in X$
the map $\lambda(\cdot, x) \colon Q \to Q$ is either the identity map on Q, or else it is a
constant (reset) map, and the output of the machine is its state.*

Now let us consider the two-state identity-reset machine, which we
call the *flip-flop F*. F has states $\{q_0, q_1\}$ and inputs $\{e, x_0, x_1\}$ with the
actions $\lambda(q, e) = q$, where e is the identity input, and $\lambda(q, x_i) = q_i$, where
x_i is the "reset-to-q_i" input.

(1.4)   **Proposition.**   *Any identity-reset machine may be obtained as a
cascade of copies of the flip-flop F.*

Outline of construction.   In fact, we use only coding, decoding, and parallel
composition.   If M is an n-state identity-reset machine, we choose m so
that $n \leq 2^m$ and place m copies of F in parallel to obtain F^m.   We select
n of the 2^m states of F^m at whim and decode these as the outputs (states)
$\{q_1, \ldots, q_n\}$ of M.   If an input x to M acts as an identity, we code it
as the identity input to all m copies of F.   If x acts as a reset to state q_i,
we code it as that configuration of resets which will cause F^m to go into
the state which will be decoded as q_i.   See Figure 8.3, where the enclosed
region has the precise input/output behavior of M. $\square$

F has semigroup $U_3 = \{1, r_0, r_1\}$ whose elements are the Myhill
equivalence classes $1 = [e]$, $r_0 = [x_0]$, and $r_1 = [x_1]$,‡ and with multiplication $u \cdot 1 = u$ and $u \cdot r_i = r_i$ (1 is the identity and r_0 and r_1 are called
"right zeroes").

(1.5)   **Definition.**   UNITS *is the set of semigroups which divide U_3.*

Now U_3 has subsemigroups U_3, $U_2 = \{r_0, r_1\}$, $U_1 = \{1, r_0\}$ and its
isomorph $\{1, r_1\}$, and $U_0 = \{1\}$.

---

† [KROHN and RHODES, 1965] work with $SP(\mathfrak{M})$, based on series-parallel connections, rather than our $C(\mathfrak{M})$, based on cascades of machines.   The relationship
between these concepts is analyzed in [ARBIB, 1968*b*].

‡ The reader should make the easy verification.

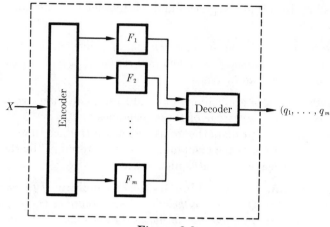

Figure 8.3

If $S|U_3$, we must have a diagram $U_3 \supseteq S' \twoheadrightarrow S$, and so it follows that, up to isomorphy,

$$\text{UNITS} = \{U_0,\ U_1,\ U_2,\ U_3\}.$$

We now want to capture the notion that a machine cannot be broken down into "smaller" machines with loop-free composition:

(1.6) Definition. *A machine N is* s-**irreducible** *if whenever it is simulated by a cascade of two machines, M_1 and M_2, with semigroups S_1 and S_2, then either $N|M(S_1)$ or $N|M(S_2)$.*

Thus if you claim that N can be broken up into simpler machines, you are fooling yourself in that actually the semigroup machine of either M_1 or M_2 alone suffices to simulate N. What machines are s-irreducible? The following answer will be our main target of proof in the remainder of the chapter:

(1.7) Proposition. *A machine M is* s-*irreducible if M has either an element of* UNITS *or a simple group† for its semigroup.*

This raises the question of what irreducible machines we need to build up a given machine. The answer, the proof of which will occupy us in the next chapter, is given by

---

† The reader unacquainted with group theory will find basic notions such as that of simple group expounded in the next section. The key theorem for the appreciation of the part played by simple groups in automata theory, the Jordan-Hölder theorem, will also be given there.

(1.8) **Proposition.** *Given a machine $M(f)$, we can simulate it by a cascade of flip-flops and machines of (not necessarily all) the simple groups which divide the semigroup S_f.*

Propositions (1.7) and (1.8) constitute the *Krohn-Rhodes theorem*, first established by [KROHN and RHODES, 1965]. They have expanded this material in [KROHN, RHODES, and TILSON, 1968], published in [ARBIB, 1968a] which contains a wealth of material on the algebraic theory of machines, languages, and semigroups. We conclude this section with an outline of our method of proof.

Following [ZEIGER, 1965], we start with

(1.9) **Definition.** *A* **PR-machine** *is one for which every input produces either a permutation or a reset.*

That is, given $x \in X$, either $\lambda(\cdot, x): Q \to Q$ is one-to-one (a permutation) or $\lambda(\cdot, x): Q \to Q$ is a constant map (a reset). The permutations generate a group, called the *group* of the *PR*-machine, which is not quite the semigroup of the machine, but is obtained from it by deleting resets.

The proof of the Krohn-Rhodes theorem result now breaks up into four pieces:

(1.10) *Any machine can be obtained by loop-free synthesis from PR-machines whose groups divide the semigroup of the original machine.*

This result is due to [ZEIGER, 1965], who emphasises that one might be satisfied to stop here and use *PR*-machines to synthesize arbitrary finite automata. However, we are pursuing the mathematical theory, and so require the remaining three pieces:

(1.11) *A PR-machine can be obtained as a cascade of a semigroup machine $M(G)$, G a group, and an identity-reset machine, with the group being the group of permutations of the PR-machine.*

(1.12) *Any group machine $M(G)$ can be simulated by a cascade of (not necessarily all) the machines of the simple groups which divide G.*

Divisibility is transitive, and so we see that any machine can be built up from flip-flops and from the machines of the simple groups which divide its semigroup. In fact, these machines are s-irreducible:

(1.13) *Flip-flops and the machines of the simple groups are s-irreducible.*

Statements (1.10) to (1.13) together give the Krohn-Rhodes result.

8.2   Results on groups and semigroups

We shall approach the conventional study of normal subgroups by the relatively novel route of asking what a congruence becomes when our semigroup is a group.

(2.1)   **Proposition.**   *Let G be a group and let $\equiv$ be a congruence relation on G, with $[g]$ denoting the equivalence class of an element g of G. Then $N = [1_G]$, where 1_G, the identity of G, is a subgroup of G.*

Proof.   The standard test for N to be a subgroup is just that for all $a, b \in N$ we have ab^{-1} in N. We prove this true:

$$ab^{-1} = ab^{-1} \cdot 1 \equiv ab^{-1} \cdot b = a \equiv 1. \qquad \square$$

(2.2)   **Proposition.**   *Let $aN = \{an: n \in N\}$ and $Na = \{na: n \in N\}$. Then $[a] = aN = Na$ for all $a \in G$.*

Proof.

$$\begin{aligned} b \in [a] &\Leftrightarrow a^{-1} \cdot b \equiv a^{-1} \cdot a = 1, \\ &\Leftrightarrow a^{-1}b \in N, \\ &\Leftrightarrow b = a \cdot a^{-1} \cdot b \in aN. \end{aligned}$$

Therefore

$$[a] = aN.$$

Similarly,

$$[a] = Na. \qquad \square$$

(2.3)   **Definition.**   *A subgroup N of a group G is called* **normal** *just in case*

$$gN = Ng \qquad for\ all\ g \in G.$$

We write $N \lhd G$. We note that a subgroup of a group need not be normal. For example, let A_n denote the alternating group on n letters (that is, the set of permutations of the set $\{0, 1, \ldots, n - 1\}$ which may be obtained by an even number of interchanged pairs of letters). It is well known (see, for example, [VAN DER WAERDEN, 1953, sec. 48]) that A_n $(n > 4)$ has *no* normal subgroups. Yet if $m < n$ we may consider A_m as a subset of A_n on identifying elements of A_m with permutations of $\{0, 1, \ldots, n - 1\}$ which leave $\{m, m + 1, \ldots, n - 1\}$ fixed.

We say a group is *abelian* if multiplication is commutative: ab always equals ba. It is immediate from the definition that *every* subgroup of an abelian group is normal.

(2.4)   **Proposition.**   *Given $H \leq G$ (read "a subgroup H of G"), the sets gH form a partition of G, called the* **left coset decomposition,** *that is, for all $g, g' \in G$, $gH \cap g'H \neq \phi \Rightarrow gH = g'H$.*

Proof. If $g'' \in gH \cap g'H$, we have $g'' = gh = g'h'$ for suitable elements $h, h' \in H$. But then
$$gH = g'h'h^{-1}H = g'H. \qquad \square$$

We have the corresponding result for right coset decomposition. We thus have the important result

(2.5) Lemma. *Given $H \leq G$, the left (right) coset decomposition of G is a congruence iff $H \lhd G$. Every congruence on G is the coset decomposition of some $N \lhd G$. If $N \lhd G$, we write G/N for the factor semigroup induced by the left (or right) coset congruence.*

(2.6) Lemma. *G/N is a group (we call it the factor group of G by N).*

Proof. We already know from our semigroup theory that it has an associative multiplication with

$$aN \cdot bN = [a] \cdot [b] = (ab)N.$$

Now we merely note that $N = [1]$ is the identity, and aN has inverse $a^{-1}N$. $\qquad \square$

The map $h: G \to G/N$ defined by $h(g) = gN$ is called the canonical epimorphism ("epi" means "onto," and "morphism" is short for "homomorphism"). We see that the following diagram is commutative, where $m_G: G \times G \to G$ is the group operation of G, etc.:

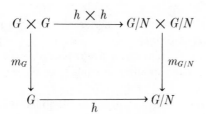

which is just a fancy way of emphasizing that $aN \cdot bN = (ab)N$.

Let $f: G_1 \to G_2$ be a homomorphism of G_1 onto G_2. Consider the kernel of f, which is defined to be the set of elements of G_1 which map onto 1_{G_2}; that is, $\ker f = f^{-1}(1_{G_2})$. Define $\sim_f$ on G_1 by $g \sim_f g' \Leftrightarrow f(g) = f(g')$. Then $\sim_f$ is a congruence, the equivalence classes are the sets $g \cdot \ker f$; and so $\ker f$ is a normal subgroup N of G_1, and we have $G_1/N \cong G_2$. (The easy details of proof are left to the reader.) This last result is referred to as the

(2.7) Homomorphism theorem. *If f is a homomorphism of G_1 onto G_2, then $G_1/\ker f \cong G_2$.*

Every group G has at least two normal subgroups, $\{1_G\}$ and G, and we have $G/\{1_G\} \cong G$ and $G/G \cong \{1\}$.

(2.8) **Definition.** *A group is* **simple** *if it has exactly two normal subgroups.*

Thus we observed above that every A_n is simple for $n > 4$. Given a group G, we call a series

(2.9) $G = G_0 \triangleright G_1 \triangleright G_2 \triangleright \cdots \triangleright G_k = \{1_G\}$

a normal series for G (note that $G \triangleright H \triangleright K \not\Rightarrow G \triangleright K$).

Any normal subgroup H of G is contained in a normal series, $G \triangleright H \triangleright \{1_G\}$. We can call the normal series

$$G = H_0 \triangleright H_1 \triangleright H_2 \triangleright \cdots \triangleright H_n = \{1_G\}$$

a *refinement* of (2.9) if every G_i is an H_j. The *factors* of (2.9) are the factor groups G_i/G_{i+1}.

Consider $Z_K = \{0, \ldots, K - 1\}$ under addition modulo K:

$$Z_6 \triangleright Z_3 \triangleright \{0\},$$
$$Z_6 \triangleright Z_2 \triangleright \{0\},$$
$$Z_6/Z_3 \cong Z_2/\{0\},$$
$$Z_6/Z_2 \cong Z_3/\{0\}.$$

This is an example of two normal series of a single group with the same factors but in different order.

We say that two normal series are *isomorphic* if their factors may be put into one-to-one correspondence such that corresponding factors are ismorphic.

A *composition series* is a normal series which has no proper refinements; that is, it is the only refinement of itself without repetitions. The crucial result we need (the proof is given in Section 8.4) is

(2.10) **Jordan-Hölder theorem.** *Any two composition series of a group are isomorphic.*

Of course, any *finite* group *must* have a composition series.

If we consider a group as being built up, in some sense, from the factors of its composition series, the Jordan-Hölder theorem states that G uniquely determines its building blocks. However, G may be built up from these blocks in several ways.

We now introduce some basic results from the theory of semigroups, and tie these in with our group theory and with our notion of divisibility. The great problem in dealing with semigroups is to refrain from attributing to them properties that they do *not* share with groups.

(2.11) **Definition.** *An element r of a semigroup S is called a* **left** *(or* **right**) **identity** *if $rs = s(sr = s)$ for all $s \in S$, a* **left** *(or* **right**) **zero** *if $rs = r$ $(sr = r)$ for all $s \in S$, and an* **idempotent** *if $r^2 = r$.*

Clearly, zeros and identities are always idempotents. Note that in a group G there are no zeros (unless G has only one element) and only one idempotent and right or left identity, all equal to 1_G.

(2.12) **Example.** Consider two groups, G_1 and G. Define S to be $G_1 \cup G_2 \cup \{0\}$, with the multiplication $s \circ s = ss'$ (group multiplication) if s and s' belong to the same G_k, and 0 otherwise. Then 0 is a left and right zero of S and 1_{G_1} and 1_{G_2} are idempotents of S, but S has no identities.

(2.13) **Definition.** *A subset A of S is a* **left** *(or* **right**) **ideal** *of S if $SA \subseteq A$ $(AS \subseteq A)$. We say that A is* **proper** *if $A \neq S$.*

Clearly, A is a subsemigroup of S: $AA \subset A$ $(SA = \{sa : s \in S, a \in A\}$, etc.).

If r is both a left and right identity (zero) of S, then it is the only left or right identity (zero), and we may then denote it unambiguously by $1_S(0_S)$. For example, if r is a right identity and r' is a left identity, then $r' = r'r = r$.

(2.14) **Proposition.** *A semigroup is a group if and only if it has no proper left or right ideals.*

Proof. If S is not a group, there are a, x, and y in S with $x \neq y$ but $ax = ay$.† This implies that $\{as : s \in S\}$ is a proper right ideal.

If $\hat{S}$ is a proper right ideal of S, then $a \in \hat{S} \Rightarrow aS \subseteq \hat{S}$, and so $ax = ay$ for some x and y in S, $x \neq y$, and S is not a group. $\qquad \square$

(2.15) **Proposition.** *Any finite semigroup S contains an idempotent. In fact, for any $a \in S$ there is an idempotent of the form a_k.*

This follows from the more careful analysis:

(2.16) **Theorem.** *Let a be an element of a semigroup S. Let $\langle a \rangle$ be the cyclic subsemigroup of S generated by a, that is $\{a, a^2, a^3, \ldots \}$. If $\langle a \rangle$ is infinite, then all the powers of a are distinct. If $\langle a \rangle$ is finite, there exist two positive integers, the* **index** *r and the* **period** *m of a, such that $a^r = a^{r+m}$, and*

$$\langle a \rangle = \{a, a^2, \ldots, a^{m+r-1}\},$$

† Or $xa = xy$, in which case a similar argument applies.

the order of $\langle a \rangle$ being $m + r - 1$. The set

$$K_a = \{a^r, a^{r+1}, \ldots, a^{r+m-1}\}$$

is a cyclic subgroup of S of order m. If n is the multiple of m satisfying $r \leq n \leq m + r - 1$, then a^n is idempotent and the identity of K_a.

Proof. If $a^r = a^s$ for some integers $r < s$, let s be the smallest such s and r be the corresponding r, and set $m = s - r$. The results all follow easily from Figure 8.4. □

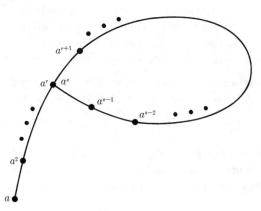

Figure 8.4

(2.17) **Lemma.** *If G' is a group dividing the finite semigroup S,*

$$S \supseteq S'' \xrightarrow{\;Z\;} G',$$

then there is a group $G \subseteq S$ such that $Z(G) = G'$ (that is, G' is isomorphic to a factor group of G).

Proof. Since S is finite, we may find a subsemigroup S_1 of S'' such that $Z(S_1) = G'$, and such that for any proper subsemigroup S_2 of S_1, $Z(S_2) \neq G'$.

We show that S_1 has no proper left or right ideals and is thus the desired group G. Suppose there were a right ideal $S_2 \subseteq S_1$. Then S_2 contains an idempotent e. Since $Z(e^2) = Z(e)$, we must have $Z(e) = 1_{G'}$.

Take any $g \in G'$ and an $s \in S_1$ with $Z(s) = g$. Then $es \in S_2$, since e is in the right ideal S_2. Thus $Z(es) = Z(e)Z(s) = 1_{G'}g = g$, and so $Z(S_2) = G'$, from which $S_1 = S_2$. Similarly, S_1 contains no proper left ideals. □

8.3   The irreducibility results

Suppose that in the definition of the cascade of two machines, we take $\tilde{X} = S_2 \times S_1$ and $\eta(s_2, s_1) = s_1$ and have Z independent of the s_1 of $\tilde{x}$. so that we may write $Z_{s_1}(s_2')$ for $Z((s_1', s_2'), s_1)$ (see Figure 8.5).

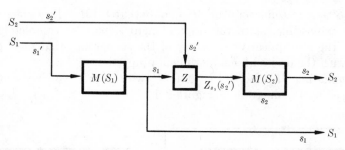

Figure 8.5

Further, let machine M be replaced by the semigroup machine $M(S_1)$ and M' be replaced by $M(S_2)$. We then have

$$\delta_Z[(s_2, s_1), (s_2', s_1')] = [s_2 Z_{s_1}(s_2'), s_1 s_1'].$$

The operation δ_Z becomes *associative* if we require that Z enjoy the properties

$$Z_{s_1 s_1'}(s_2) = Z_{s_1}(Z_{s_1'}(s_2)),$$

and $\qquad\qquad\qquad\qquad\qquad\qquad (s_1, s_1' \in S_1; s_2, s_2' \in S_2)$

$$Z_{s_1}(s_2 s_2') = Z_{s_1}(s_2) Z_{s_1}(s_2').$$

This leads us naturally to consider the semidirect product of S_1 and S_2:

(3.1)   **Definition.**   *Let S_1 and S_2 be semigroups, and let Z be a homomorphism of S_1 into* End S_2 *(the monoid of endomorphisms of S_2 under composition,)* $s_1 \mapsto Z_{s_1}(\cdot)$. *Then the* **semidirect product of S_1 and S_2 with connecting homomorphism Z** *is the semigroup $S_2 \times_Z S_1$ with elements the cartesian product set $S_2 \times S_1$ and multiplication*

$$(s_2, s_1)(s_2', s_1') = (s_2 Z_{s_1}(s_2'), s_1 s_1').$$

Similarly,

(3.2)   **Definition.**   *A semigroup S is* **irreducible** *if for all semidirect products $S_2 \times_Z S_1$ such that $S|S_2 \times_Z S_1$ we must have*

$$S|S_2 \qquad \text{or} \qquad S|S_1.$$

Our definitions have been so worded as to render highly plausible the result that a machine is s-irreducible if and only if its semigroup is irreducible. However, we shall see some proof is still required.

Given a cascade machine $M' \times_Z M$ with semigroup $\tilde{S}$, it is tempting to believe that we can always find a suitable homomorphism $\tilde{Z}$ such that

$$\tilde{S}|S' \times_{\tilde{z}} S,$$

where S' is the semigroup of M' and S that of M. However, this is usually impossible, since the original map Z may so completely "cut across" the multiplicative structure of the semigroups that no $\tilde{Z}$ can be found with the desired homomorphism properties.

Simulating M with $M(S)$ and M' with $M(S')$ we may represent $M' \times_Z M$ by the diagram in Figure 8.6.

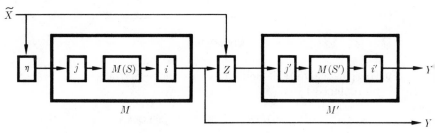

Figure 8.6

The (not necessarily reduced) state space of this cascade is $S' \times S$. Let us consider $\tilde{S}$, the semigroup of transformations of $S' \times S$ induced by input sequences. Let t be an input sequence, that is, an element of $(\tilde{X})^*$. The action of t on the $M(S)$ machine is simply multiplication by an appropriate element, $I_S(t)$, of S. However, the action of t on the *second* machine, $M(S')$, will depend on the state of $M(S)$ at the beginning of the operation; if the latter is s, then let us denote the former by $I_{S'}(t)(s)$. Thus $I_{S'}(t)$ is an element of $F(S, S')$, the semigroup of *maps* (not just homomorphisms) of S into S', with composition defined by $(f_1 \circ f_2)(s) = f_1(s) \cdot f_2(s)$.

Thus $\tilde{S}$ is obtained by replacing the element t of the input free semigroup by the element $(I_{S'}(t), I_S(t))$ of the *set* $F(S, S') \times S$. What is the semigroup multiplication induced on this set by the input action? The action of t_1 followed by t_2 yields

$$(3.3) \qquad (I_{S'}(t_1), I_S(t_1)) \cdot (I_{S'}(t_2), I_S(t_2)) = (I_{S'}(t_1 t_2), I_S(t_1 t_2)).$$

But the action of $t_1 t_2$ on $M(S)$ is simply that of t_1 multiplied by that of t_2,

$$(3.4) \qquad\qquad I_S(t_1 t_2) = I_S(t_1) I_S(t_2),$$

while the action of $t_1 t_2$ on $M(S')$ when $M(S)$ is started in state s is that of t_1 on $M(S')$ when $M(S)$ is started in state s, multiplied by that of t_2 on $M(S')$ when $M(S)$ is started in state $s \cdot I_S(t_1)$:

$$(3.5) \qquad I_{S'}(t_1 t_2)(s) = [I_{S'}(t_1)(s)] \cdot [I_{S'}(t_2)(s \cdot I_S(t_1))].$$

Refer back to the definition of the semidirect product and consider the map

$$W : S \to \text{End } F(S, S')$$

defined by

$$W_{s_1}(l)(s) = l(ss_1).$$

The multiplication defined on $F(S, S') \times S$ is that of the semidirect product $F(S, S') \times_W S$. This suggests that the latter semigroup deserves our special attention; we call it the *wreath product* of S and S' and denote it S' w S.

Note that W is completely defined when we give S and S' and so need not be mentioned explicitly.

We have $\hat{S}$ is a subsemigroup of the wreath product of S and S'. But the semigroup of $M' \times_Z M$ is a homomorphic image of $\hat{S}$, being simply the action of $\tilde{X}^*$ on the reduced state space. We thus have the crucial result:

(3.6) **Theorem.** *Let M have semigroup S and M' have semigroup S'. Then for any connecting map Z the semigroup $\tilde{S}$ of $M' \times_Z M$ divides the wreath product S' w S.*

Thus, although in general there is no semidirect product of S' and S which is divisible by $\tilde{S}$, it is always true that $\tilde{S}$ divides that semidirect product of $F(S, S')$ with S known as the wreath product.

We can now prove

(3.7) **Theorem.** *If M is s-irreducible as a machine, its semigroup S is irreducible as a semigroup.*

Proof. Suppose M is s-irreducible. To say that S divides the semidirect product $S_2 \times_Z S_1$ is just another way of saying that $M(S)$ is simulable by the cascade of machines $M(S_2) \times_Z M(S_1)$. Since $M|M(S)$, we have $M|M(S_2) \times_Z M(S_1)$, and by s-irreducibility, $M|M(S_1)$ or $M|M(S_2)$. But divisibility of machines implies divisibility of their semigroups, and so $S|S_1$ or $S|S_2$. Thus S is irreducible. $\qquad\qquad\square$

(3.8) **Theorem.** *If M is a machine whose semigroup S is irreducible as a semigroup, then M is s-irreducible.*

Proof. Let S be irreducible. Now, if $M|M_2 \times_Z M_1$, then S must divide S_2 w S_1, and so divide S_1 or $F(S_2, S_1)$. If S divides S_1, then $M(S)$

divides $M(S_1)$.   But $F(S_2, S_1) \cong S_2 \times \cdots \times S_2$ $|S_1|$ number of times. So if S divides $F(S_2, S_1)$, then S must divide S_2, by irreducibility, in which case $M(S)$ divides $M(S_2)$.   Thus $M|M(S_1)$ or $M|M(S_2)$. $\square$

Thus *a machine is s-irreducible if and only if its semigroup is irreducible.*† With this theorem, requirement (1.13) of our proof outline for the Krohn-Rhodes theorem is met by the two theorems which constitute the remainder of this section.

(3.9) **Theorem.** *Simple groups are irreducible.*

Proof.   Let G' be simple.   Given that $G'|S_2 \times_Z S_1$, we are to prove

$$G'|S_2 \quad \text{or} \quad G'|S_1.$$

We saw that we can choose a subgroup G of $S_2 \times_Z S_1$ mapping onto G',

$$S_2 \times_Z S_1 \geq G \xrightarrow{\varphi} G'.$$

G has elements (s_2, s_1) for $s_i \in S_i$.   Set $\pi(s_2, s_1) = s_1$; then π is a homomorphism from G to S_1.   Then $G_1 = \pi(G)$ is a subgroup of S_1, since it is a homomorphic image of G.

Let $1 = (1_2, 1_1)$ be the identity of G.   Set $G_2' = \{(s_2, 1_1) \in G\}$. G_2' is a subgroup of G.   In fact, since G_2' is the kernel of π, G_2' is, in fact, a normal subgroup of G.   Define $\psi\colon G_2' \to S_2$ by

$$\psi(s_2, 1_1) = Z_{1_1}(s_2).$$

First we show that ψ is a homomorphism.   For

$$(s_2, 1_1)(s_2', 1_1) = (s_2 Z_{1_1}(s_2'), 1_1),$$

and applying Z_{1_1}, we have

$$\begin{aligned} Z_{1_1}(s_2 Z_{1_1}(s_2'), 1_1) &= Z_{1_1}(s_2) Z_{1_1}(Z_{1_1}(s_2')), \\ &= Z_{1_1}(s_2) \cdot Z_{1_1}(s_2'), \end{aligned}$$

as was to be shown.

Next we show that ψ is one-to-one.   If

$$\psi(s_2, 1_1) = \psi(s_2', 1_1),$$

then

$$\begin{aligned} (s_2, 1_1) &= (1_2, 1_1)(s_2, 1_1), \\ &= (1_2 Z_{1_1}(s_2), 1_1), \\ &= (1_2 \psi(s_2, 1_1), 1_1), \\ &= (s_2', 1_1), \end{aligned}$$

as was to be proved.

---

† Our treatment thus far in Section 8.3 follows [ARBIB, 1968*b*], which contains an additional discussion of why our notion of machine irreducibility involves semigroups.   The remainder of the section follows [KROHN and RHODES, 1965].

Set $G_2 = \psi(G_2')$. Then $G_2 \leq S_2$.

Now G' is exactly the homomorphic image of G under φ, and G' is simple, by hypothesis:

$$G' \simeq G/\ker \varphi,$$

so that $K = \ker \varphi$ is a maximal normal subgroup of G. By (4.2), if A, B, and C are groups, and $A \lhd B$, and $C \lhd B$, then $A \cdot C \lhd B$. Hence we have $K \lhd K \cdot G_2' \lhd G$, with K maximal in G. Therefore either $K = K \cdot G_2'$ or $K \cdot G_2' = G$.

In case $K = K \cdot G_2'$, let us show that $G'|S_1$. Let $G_1' = \{(1_2, s_1) \in G\}$ and $G = G_2' \cdot G_1'$. It is obvious that $G_2' \cdot G_1' \subseteq G$. Thus we let (s_2, s_1) be a general element of G; then it is in $G_2' \cdot G_1'$ since

$$(s_2, 1_1)(1_2, s_1) = (s_2 Z_{1_1}(1_2), s_1),$$
$$= (s_2, s_1),$$

since Z is a homomorphism. If we define $P\colon G_1 \to G_1'$ by $P(s_1) = (1_2, s_1)$, then P is *onto* G_1' and is a homomorphism.

Now

$$\varphi(G_2' \cdot G_1') = \varphi(G) = G',$$
$$\varphi(G_2' \cdot G_1') = \varphi(G_2')\varphi(G_1').$$

By assumption, $K \cdot G_2' = K_1$, so $G_2' \leq K$, hence $\varphi(G_2') = \{1_G\}$, hence φ is onto G' from G_1', hence $\varphi \circ P$ is onto G' from $G_1 \leq S_1$; hence $G'|S_1$.

In case $K \cdot G_2' = G$, then $G'|S_2$. We have

$$G' = \varphi(K \cdot G_2') = \varphi(K)\varphi(G_2'),$$
$$= \varphi(G_2').$$

But ψ^{-1} is a homomorphism of G_2 *onto* G_2', and so the composite map $\varphi \cdot \psi^{-1}$ is a homomorphism of G_2 onto G'. $\qquad\square$

(3.10) Theorem. *Members of* UNITS *are irreducible.*

Proof. We prove U_3 is irreducible; the proofs for the remaining units are analogous and easier.

We first prove that if $U_3|S$, then $U_3 \subseteq S$. Suppose $S \supseteq S' \xrightarrow{\varphi} U_3$. Let $x = \varphi^{-1}(1)$. Then there is an idempotent e which is a power of x, and so $\varphi(e) = 1$ and $\varphi(eS'e) = U_3$, and e is an identity for $eS'e$.

Let S_1 be a subsemigroup of $eS'e$ of smallest order for which $\varphi(S_1) = U_1$. Then for each $s_1 \in S_1$ we have $\varphi(s_1 \cdot S_1) = \varphi(s_1)U = U_1$, since U_1 is right simple. Thus $s_1 \cdot S_1 = S_1$, and so S_1 is right simple.

By [CLIFFORD and PRESTON, 1961, Theorem 1.27], $S_1 \simeq G \times R_B$, where G is a group and R_B is the set B under right zero multiplication. B must contain at least two members, b_1 and b_2, since U_1 is not a group.

Then

$$U_3 \simeq \{e_1(1, b_1), (1, b_2)\} \subseteq S_1 \subseteq S.$$

Next we show that if $U_3|S_2 \times_Z S_1$, then U_3 divides S_2 or S_1. By the above,

$$U_3 \simeq \{(b, a), (b_0, a_0), (b_1, a_1)\} \subseteq S_2 \times_Z S_1.$$

As before, let $\pi_1(s_2, s_1) = s_1$:

$$\pi(U_3) = \{a, a_0, a_1\} \subseteq S_1.$$

If $a_0 \neq a_1$, then $\pi(U_3) \subseteq S_1$ is isomorphic to U_3, from which $U_3|S_1$; or

$$a = a_i(i = 0, 1) \Rightarrow Za = Za_i$$
$$\Rightarrow Z = a \qquad \text{for all } Z \in \{a, a_0, a_1\},$$

and we have $U_3 \simeq \{b, b_0, b_1\}$.

If $a_0 = a_1$, then $b_0 \neq b_1$. Let $p_2: U_3 \to S_2$, where $p_2(b', a') = Z_{a_0}(b')$. Noting that $Z_a Z_{a_1} = Z_{a_0} Z_a = Z_a$, we verify that p_2 is a homomorphism. Sheer enumeration of possibilities shows that p_2 is one-to-one. Thus in this case $p_2(U_3) \cong U_3$, and $p_2(U_3) \subseteq S_2$, and so $U_3|S_2$. □

8.4 Proof of the Jordan-Hölder theorem

Our aim now is to prove the Jordan-Hölder theorem, which states that the composition series of a group is unique up to isomorphism. We follow the proof of [KUROSH, 1956, pp. 77–78, 110–112].

(4.1) **Lemma.** *If $B \leq C$ and $A \lhd C$, then $AB \leq C$.*

Proof. Since A is normal, for all $c \in C$, $c^{-1}Ac = A$. Let $\alpha, \beta \in AB$. We must prove that $\alpha\beta^{-1} \in AB$. $\alpha = a_1 b_1$, $\beta = a_2 b_2$, $a_1 \in A$, and $b_1 \in B$.

$$\alpha\beta^{-1} = a_1(b_1 b_2^{-1})a_2^{-1}.$$

Choose $a_3 \in A$ such that $a_3(b_1 b_2{}^{-1}) = (b_1 b_2{}^{-1})a_2{}^{-1}$. Then

$$\alpha\beta^{-1} = (a_1 a_3)(b_1 b_2{}^{-1}) \in AB.$$ □

A similar proof yields

(4.2) **Lemma.** *If $B \lhd C$ and $A \lhd C$, then $AB \lhd C$.*

(4.3) **Zassenhaus' lemma.** *If $A, A', B,$ and B' are subgroups of a group G, $A' \leq A$, and $B' \leq B$, then*

$$A'(A \cap B') \lhd A'(A \cap B),$$
$$B'(B \cap A') \lhd B'(B \cap A),$$

and the corresponding factor groups are isomorphic:

$$A'(A \cap B)/A'(A \cap B') \simeq B'(B \cap A)/B'(B \cap A').$$

Proof. If we write $C = A \cap B$ and $D = (A \cap B')(B \cap A')$, then clearly $D \subseteq C$. Moreover, since B' is normal in B, and since C is a subgroup of B, we see that $C \cap B' = A \cap B \cap B' = A \cap B'$ is a normal subgroup of C.

By the symmetry of the assumptions on A and B, this also holds for the intersection $B \cap A'$. The same is true of D, since the product of normal subgroups is itself a normal subgroup. We can therefore speak of the factor group of D in C; we denote it by

$$H = C/D.$$

On the other hand, A' is a normal subgroup of A, so that the product $A'(A \cap B) = A'C$ is a subgroup.

Every element of $A'C$ has the form $a'c$, where $a' \in A$ and $c \in C$. If we associate with it the coset Dc (an element of H), and if $a'c$ has another representation in the same form

$$a'c = a_1'c_1,$$

then $a_1'^{-1}a' = c_1c^{-1} \in (A' \cap C) \subseteq A' \cap B \subseteq D$, and hence $c_1 = (a_1'^{-1}a')c \in Dc$. We thus obtain a single-valued mapping f of the group $A'C$ *onto* the group H, since every element $c \in C$ is mapped onto its coset Dc. f is homomorphic: since A' is normal in $A'C$, we have $a_1'c_1a_2'c_2 = a_3'(c_1c_2)$, where $a_3' \in A'$ (in fact, $a_3' = a_1a_4'$, where a_4' is chosen so $c_1^{-1}a_4'c_1 = a_2'$). ker f clearly must contain the subgroup $A'(A \cap B')$; we know that

$$A \cap B' = C \cap B' \subseteq D.$$

On the other hand, if an element $a'c$ is mapped by f into D, then $c \in D$; that is, $c = uv$, where $u \in (B \cap A')$ and $v \in (A \cap B')$. Then

$$a'c = a'uv = a_1'v \in A'(A \cap B').$$

The kernel of f is therefore the subgroup $A'(A \cap B')$. So, by the homomorphism theorem,

$$A'(A \cap B)/A'(A \cap B') \simeq H.$$

By symmetry, we also have the isomorphism

$$B'(B \cap A)/B'(B \cap A') \simeq H. \qquad \square$$

(4.4) **Schreier's theorem.** *Any two normal series of an arbitrary group have isomorphic refinements.*

Proof. Let

(4.5) $$G = G_0 \triangleright G_1 \triangleright G_2 \triangleright \cdots \triangleright G_k = \{1_G\},$$
(4.6) $$G = H_0 \triangleright H_1 \triangleright H_2 \triangleright \cdots \triangleright H_l = \{1_G\}$$

be two normal series of a group G. Put

$$G_{ij} = G_i(G_{i-1} \cap H_j),$$
$$H_{ij} = H_j(H_{j-1} \cap G_i).$$

Here G_{ij} and H_{ij} are groups, since, for example, G_i is a normal subgroup, and $G_{i-1} \cap H_j$ is a subgroup of G_{i-1}.

For $i = 1, 2, \ldots, k$ and $j = 1, 2, \ldots, l$ we now have

$$G_{i-1} = G_{i0} \geq G_{i,j-1} \geq G_{ij} \geq G_{ij} = G_i,$$
$$H_{j-1} = H_{0j} \geq H_{i-1,j} \geq H_{ij} \geq G_{kj} = H_j.$$

By Zassenhaus' lemma, $G_{ij} \triangleright G_{i,j-1}$ and $H_{ij} \triangleright H_{i-1,j}$, and

$$G_{i,j-1}/G_{ij} \simeq H_{i-1,j}/H_{ij}.$$

(Put $A = G_{i-1}$, $A' = G_i$, $B = H_{j-1}$, and $B' = H_j$.)

If we insert in (4.5) all the subgroups G_{ij}, $j = 1, 2, \ldots, l - 1$, between G_{i-1} and G_i, $i = 1, 2, \ldots, k$, then we obtain a refinement of (4.5) which is, in general, a normal series with repetitions, because some subgroups $G_{i,j-1}$ and G_{ij} may be equal. Similarly, we construct a refinement of (4.6). It is left as an exercise for the reader to verify that our two new normal series are isomorphic, even when we remove refinements. ☐

This immediately yields the

(4.7) **Jordan-Hölder theorem.** *Any two composition series of a group are isomorphic.*

Of course, any *finite* group *must* have a composition series.

(4.8) **Exercise.** Show that if a normal series is a composition series, each factor is simple, and vice versa.

Solution. The crucial point is to recall the following lemma:

$$A \triangleright B \triangleright C \qquad \text{and} \qquad A \triangleright C \Rightarrow A/C \triangleright B/C.$$

So if $G_k \triangleright G_{k+1}$ can be nontrivially refined by the insertion of H, $G_k \underset{\neq}{\triangleright} H \underset{\neq}{\triangleright} G_{k+1}$, we immediately have $G_k/G_{k+1} \underset{\neq}{\triangleright} H/G_{k+1} \underset{\neq}{\triangleright} 1$; hence G_k/G_{k+1} cannot be simple. Conversely, if $G_k/G_{k+1} \underset{\neq}{\triangleright} K \underset{\neq}{\triangleright} 1$, and $G_k \xrightarrow[\text{canonical}]{h} G_k/G_{k+1}$, we easily see that $G_k \underset{\neq}{\triangleright} h^{-1}(K) \underset{\neq}{\triangleright} G_{k+1}$.

(4.9)   **Exercise.**   Show that if P is a factor of a composition series of G, then P is simple and $P|G$.

Proof. $P = G_k/G_{k+1}$, say, is simple by Exercise (4.8).   But also $P|G$, since we have

$$G \geq G_k \xrightarrow{\;\eta\;} G_k/G_{k+1} = P,$$

where η is the canonical map $\eta(g) = gG_{k+1}$.

The converse is not true, since a *simple* group may have nontrivial (though not normal) subgroups.   That is, if A_n is the alternating group on n letters, both A_3 and A_5 are simple, yet $A_3|A_5$.

9   Proof of the decomposition results for finite automata

In this chapter we actually carry out the steps outlined previously to show that any machine may be obtained as a cascade of flip-flops and machines whose semigroups are simple groups which divide the semigroup of the original machine.

9.1   The decomposition of PR-machines

(1.1) **Lemma.** *Given a group G and a normal subgroup H, we can obtain $M(G)$ as a cascade of $M(H)$ and $M(G/H)$.*

Proof. Consider the cascade shown in Figure 9.1. For each g in G pick g' in G such that $Hg_1 = Hg_2 \Rightarrow g'_1 = g'_2$ and $Hg'_1 = Hg_1$, that is, so that g' is a coset representative. Define

$\beta_1 \colon G \to G/H$ by $g \mapsto Hg$,
$\beta_2 \colon G \times G/H \to H$ by $[g_2, Hg'_1] \mapsto g'_1 g_2 [(g'_1 g'_2)']^{-1}$,
$\beta_3 \colon G/H \times H \to G$ by $[Hg', h] \mapsto hg'$.

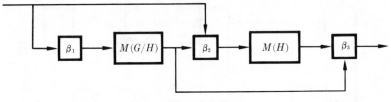

Figure 9.1

Then, if at time t the input is g_2 and the output is g_1, at time $t + 1$ the output will be $g_1 \cdot g_2$. □

(1.2) **Corollary.** *If G is a group, then $M(G)$ can be obtained as a cascade of machines of the composition factors of G, and so a fortiori of* PRIMES $(G) = \{P : P|G, P \text{ is a simple group}\}$.

Proof. This follows by induction from Lemma (1.1). Suppose

$$G = G_0 \rhd G_1 \rhd \cdots \rhd G_{k-1} \rhd G_k = \{1\}$$

is a composition series for G. Then the lemma lets us construct $M(G_{k-1})$ from $M(G_{k-1}/G_k)$ and $M(G_k)$, $M(G_{k-2})$ from $M(G_{k-2}/G_{k-1})$ and $M(G_{k-1})$, and so on, until finally we construct $M(G)$ from the machines of its factors. □

The next task is to decompose PR (permutation-reset) machines. Recall that an input x to a PR-machine acts either to permute the states or else to send every state to a single state q_x, depending only on the input ("reset" to q_x). The permutations generate the group of the PR-machine.

(1.3) **Lemma.** *If M is a PR-machine with permutation group G, then M can be obtained by loop-free synthesis from $M(G)$ and an identity-reset machine with the same state space as M.*

Proof. The construction realizing M is shown below in its two typical modes of operation, where we employ the same diagrammatic conventions as in the previous proof.

IR is an identity-reset machine with state set Q, the same as that of M. Its identity input is 1, and input $q \neq 1$ resets it to state q. $M(G)$ is the machine of group G. The action of $g \in G$ on $q \in Q$ is denoted by $(q)q$.

h_1 codes "reset to state q" as $(q, 1)$ and "permute by g" as $(1, g)$.

h_2 codes $(1, g)$ as $(1, g)$, but for $q \neq 1$ codes (q, g) as $((q)g^{-1}, g)$.

h_3 codes (q, g) as $(q)g$, and it is the output of h_3 that serves to simulate the state of M.

To reset to q', with $M(G)$ initially in state g and IR initially in state q (thus simulating M in state $(q)g$), we have the action shown in Figure 9.2.

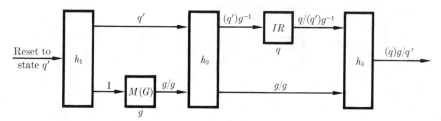

Figure 9.2

This rather complicated action is required because we cannot reset $M(G)$ back to 1 without using a loop.

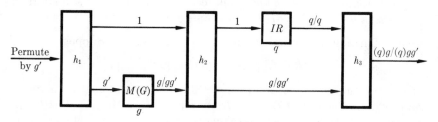

Figure 9.3

To permute the state by g', we have the action shown in Figure 9.3.

9.2   Semigroup-theoretic proofs of the decomposition theorem

The original proof that

(2.1)   *Any machine with semigroup S can be built as a cascade of flip-flops and machines of simple groups which divide S,*

given by [KROHN and RHODES, 1965] was based on a critical lemma, of which we provide our own elementary proof:

(2.2)   **Lemma.**   *Let S be a finite semigroup. Then one of the following conditions holds:*
   (a) *S is a cyclic semigroup.*
   (b) *S is left simple (that is, S has no proper left ideals).*
   (c) *There exists a proper left ideal T and a proper subsemigroup V of S for which $S = T \cup V$.*

Proof.   Either condition (b) holds, or we can find a maximal proper left ideal T of S. Suppose there exists $s \in S$ such that $Ts \subseteq T$. Then $Ts \cup T$

is a left ideal properly containing T, and so must equal S. But $V = Ts$ is a subsemigroup.

In the remaining case $Ts \subseteq T$ for all $s \in S$, and so T is a two-sided ideal of S. Then either $S - T$ is a subsemigroup, and we are done, or we can find $a, b \in S - T$ with $ab \in T$. Then $T' = Sb \cup T$ is a left ideal. It is proper, since $\#(T') \leq \#[(S - T - a)b] + \#T \leq \#S - 1$, and so must equal T. Thus $sb \subseteq T$, and so $b \cup T$ is a left ideal; hence $\{b\} = S - T$. Then either $\langle b \rangle = S$ and condition (a) holds, or we may take $V = \langle b \rangle$, and condition (c) holds. $\qquad\square$

This lemma provides the basis for the proof of (2.1) by induction on the number of elements of S.

To get the induction going Krohn and Rhodes note that (2.1) holds trivially if $\#S = 1$; they prove the following lemmas:

(2.3)   **Lemma.**   *If S is a cyclic semigroup, then (2.1) holds for S.*

(2.4)   **Lemma.**   *If S is left simple, then (2.1) holds for S.*

They then introduced two modifications of $M(f)$:

(2.5)   **Definition.**   *Let $f\colon X^* \to Y$ and $c \notin X \cup Y$. Then the **partial product of** f, PPf, is $M(f)$ augmented by having c as a "reset-to-initial-state" input. Precisely $PPf\colon (X \cup c)^* \to Y \cup c$ is defined by $x = x_1 c x_2$, $x_2 \in X^* \Rightarrow PPf(x) = f(x_2)$, and $PPf(x) = f(x)$ for $x \in X^*$.*

(2.6)   **Definition.**   *Let $f\colon X^* \to Y$ and $e \notin X \cup Y$. Then ef is $M(f)$ modified by having e as a "hold" input. Precisely, $ef\colon (X \cup e)^* \to Y \cup e$ is defined by*

$$x = x_1 e x_2 e \cdots e x_k, \text{ each } x_j \in X^* \Rightarrow ef(x) = f(x_1 x_2 \cdots x_k),$$

and $ef(x) = f(x)$ for $x \in X^*$.

Then they proved the following lemmas:

(2.7)   **Lemma.**   *If S is a finite semigroup with proper left ideal T and proper subsemigroup V such that $S = T \cup V$, then the machine of S can be built as a cascade of flip-flops and the machines ef_T and PPf_V, where f_T is the function of $M(T)$, etc.*

(2.8)   **Lemma.**   *If S satisfies (2.1), then both PPf_S and ef_S can be built as cascades of flip-flops and machines of simple groups dividing S.*

Now assume (2.1) holds for all semigroups with $\leq n$ elements. Let $\#S = n + 1$. If S is left simple or cyclic, we already know (2.1) holds for S. In the remaining case we may combine the last two lemmas to show that (2.1) holds for S on noting that $\#T \leq n$, $\#V \leq n$, and

$$\text{PRIMES}(T) \cup \text{PRIMES}(V) \subseteq \text{PRIMES}(S).$$

The proof of the last four lemmas, versions of which may be found in [KROHN and RHODES, 1965] are left as exercises!

Given a semigroup S, we may form $\hat{S}$ with multiplication reversed to get a new semigroup. This enables us to deduce from Lemma (2.2) its dual:

(2.9)   **Corollary.** *Let S be a finite semigroup. Then one of the following conditions holds:*
(a) S is a cyclic semigroup.
(b) S is right simple.
(c) There exists a proper right ideal T and a proper subsemigroup V of S for which $S = T \cup V$.

[ZEIGER, 1967] recently gave another proof of the decomposition theorem, the essential point of which is that he replaces Lemmas (2.7) and (2.8) by

(2.10)   **Theorem.** *Let M be a machine whose semigroup $S = T \cup V$, where T is a proper right ideal and V is a proper subsemigroup. Then, provided $Q \cdot T = Q$, M can be simulated by a cascade of machines M_1 with semigroup S_1 and M_2 with semigroup S_2, where*

$$S_1 | V \cup \{resets\},$$
$$S_2 | T \cup \{identity\}.$$

Proof.   Consider Figure 9.4. M_1 has state set T and M_2 has state set Q,

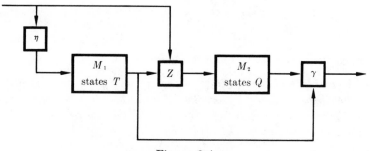

Figure 9.4

the state set of M.   If the cascade is in state $(p, r) \in T \times Q$, the output $\gamma(p, r) = r \cdot p$.   The input set is S.   If the input u is in T,

$$\eta(u) = \text{reset to } u, \quad Z_p(u) = p.$$
If u is not in T,
$$\eta(u) = u, \quad Z_p(u) = 1.$$

In either case the next state simulates $(r \cdot p) \cdot u$:

For u not in T we have $\gamma(p \cdot u, r) = r \cdot (p \cdot u)$.

For u in T, we have $\gamma(u, r \cdot p) = (r \cdot p) \cdot u$.

Thus $S_1|V \cup \{\text{resets}\}$ and $S_2|T \cup \{1\}$. ⃞

To apply this, and a related result in case $Q \cdot T \neq Q$, to prove the decomposition result now requires a double induction, just as we cannot induct on (2.7) without using (2.8). The thoughtful reader will see that (2.7) and (2.10) are virtually the same.

For a full proof of the decomposition result along these lines, see [ARBIB, 1968c]. Here, we now turn to another method of proof due to Zeiger, which uses the technique of "covers."

9.3 Decomposition via covers

Before we go into Zeiger's decomposition of a machine into PR machines, let us first discuss decomposition via "covers," a process most associated with the work of [HARTMANIS and STEARNS, 1966].

If C is a collection of subsets of a set Q, then *max* C denotes the set of all elements of C maximal with respect to set inclusion:

$$x, x' \in C,\ x \subsetneqq x' \implies x \notin \max C.$$

E_A denotes the identity map on a set A.

If S is a semigroup of transformations of a set Q, then $\tilde{S}$ is the *augmented semigroup* got by adjoining to S both E_Q and the reset maps $w_q(q \in Q)$ such that $w_q(Q) = \{q\}$.

For a machine N and $x \in X_N^*$, let us introduce the notation

$$x_N(\cdot) = \lambda_N(\cdot, x): Q_N \to Q_N.$$

We still use

$$\beta_N: Q_N \to Y_N$$

to denote the state-output function. For a machine M, S_M is the semigroup of functions $x_M: Q_M \to Q_M$ with $x \in X_M^*$.

(3.1) **Definition.** *A **cover** C for a machine M is a nonempty collection of nonempty subsets of Q_M such that for each $w \in \tilde{S}_M$ and $R \in C$, $w(R) \subseteq \in C$ [that is, there exists an R' such that $w(R) \subseteq R' \in C$]. (Note that C does cover the whole set, since $\tilde{S}_M$ contains all the resets.)*

The condition says that, given $R \in C$ and $x \in X^*$, we can find $R' \in C$ (depending on x and R) such that $q \in R \implies x_M(q) \in R'$, *independent* of the choice of q within R.

If C is a cover for M and N is a machine, then "N tells where M is in C" means:

$X_N = X_M$ and β_N maps Q_N onto C.

For each $x \in X_M$ and each $q \in Q_N$, $x_M(\beta_N(q)) \subseteq \beta_N(x_M(q))$.

That is, N keeps track of how inputs to M move around the blocks of C considered as subsets of Q_M, but N need not tell us anything about how individual states move within the blocks of C.

Given a machine M and a cover C, we may define a machine N, which tells us where M is in C simply by taking

$$X_M = X_N,$$
$$C = Q_N = Y_N,$$
$$\beta_N(R) = R,$$

and, for each $R \in C$ and $x \in X_N = X_M$, taking $x_N(R)$ to be some element of C which contains $x_M(R) = \{x_M(q): q \in R\}$. Such a choice is always possible by the condition in the definition of a cover.

Now, suppose C' is a refinement of C; that is, $R' \in C'$ implies that there exists $R \in C$ with $R \supseteq R'$. We say C' is a proper refinement if C' has fewer members than C, or at least one element of C' is a proper subset of an element of C. As before, we may find a machine N' which tells us where M is in C'. The crucial point is to note that N' can be realized as a loop-free combination of N with a new machine L.

The job of L is to supply the additional information which specifies, given which set of C we are in (say R), the additional information as to which C' subset of R we are in (see Figure 9.5).

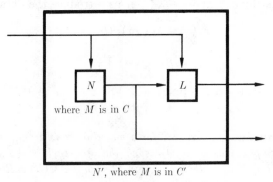

Figure 9.5

Given an element R of C, we number its C'-subsets in some arbitrary order as $R_1, R_2, \ldots, R_{j(R)}$, say, each $R_i \in C'$.

Let
$$\rho = \max \{j(R) \colon R \in C\} \geq 1.$$
Let
$$X_L = X_M \times Y_N,$$
$$Q_L = Y_L = \{1, \ldots, \rho\},$$
$$\beta_L(i) = i.$$

The composition of N and L is to tell us where M is in C', and so if the output of N is R, and the output of L is i, this should mean that the current state of M is an element R_i in C'. It should now be clear that we can always define $x_L(\cdot) = \lambda_L(\cdot, x)$ to meet this requirement.

In particular, if C' is the cover of Q_M by singletons (one-element sets), we conclude that, given any cover C of M and a machine K which tells where M is in C, we may loop-free-augment K by a machine $M_?$ to obtain a new machine which simulates M (see Figure 9.6). If we now refine C

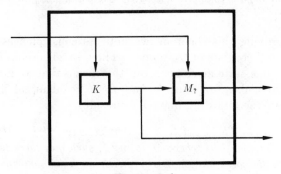

Figure 9.6

to obtain C', we may break $M_?$ up into two boxes, L and $M_{??}$, such that K combines with L to tell us where M is in C' and $M_{??}$ contains the remaining information for a complete simulation of M (see Figure 9.7).

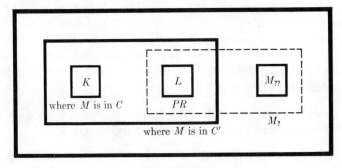

Figure 9.7

9.4   Decomposition into PR-machines

The next theorem, due to [ZEIGER, 1965], shows that we may always choose C' so that we may take L to actually be a PR machine whose group divides the semigroup of M:

(4.1)   **Theorem.**   *Let M be a machine, C a cover for M not consisting entirely of singletons, and K a machine which tells where M is in C. Then there exist machines N and L and a cover C' for M for which*

(a) *C' is a proper refinement of C.*

(b) *N tells where M is in C'.*

(c) *N is a cascade of K and L.*

(d) *L is a PR-machine.*

(e) *There is an $R \subseteq Q_M$ such that the permutation group in S_L is a homomorphic image of $\{wE_R : w \in S_M$ and wE_R is a permutation of $R\}$.*

Before proving this theorem, we note its consequences. Consider the case where C consists of Q_M alone. Then, in Theorem (4.1) we may take K to be a one-state machine. But this means that we may disregard K and, in particular, consider N and L to be identical in this case. Hence we have the following immediate result:

(4.2)   **Corollary.**   *There exists a nontrivial cover C' for M and a PR-machine L which tells where M is in C', such that there is an $R \subseteq Q_M$ for which the permutation group in S_L is a homomorphic image of $\{wE_R : w \in S_M$ and wE_R is a permutation$\}$.*

Thus, applying the corollary once and then repeatedly applying the theorem, we get a sequence of covers C', C'', . . . , $C^{(k)}$, . . . and a corresponding sequence of PR-machines $L^{(k)}$ such that the group part of the semigroup of $L^{(k)}$ always divides the semigroup of M, and such that we may obtain $M^{(k)}$, a machine which tells us where M is in $C^{(k)}$, by loop-free synthesis from L', L'', . . . , $L^{(k)}$. Now, the process of refinement of covers can terminate only when we reach an n with $C^{(n)} = \{$singletons$\}$, but this must happen for some n. The corresponding $M^{(n)}$ tells us where M is in $C^{(n)}$, but since $C^{(n)}$ is made up of singletons, $M^{(n)}$ actually simulates M. We have thus proved

(4.3)   **Proposition.**   *Any machine M can be obtained as a cascade of PR-machines whose groups divide the semigroup of M.*

We say elements R_1 and R_2 of a cover C of M are *similar*, $R_1 \sim R_2$, if there exist w_1, $w_2 \in \tilde{S}_M$ such that $R_1 = w_1(R_2)$ and $R_2 = w_2(R_1)$.

Similarity is clearly an equivalence relation, and so we may split C up into similarity classes.

We call $R \in C$ *initial* if, whenever $R = w(R')$ for some $R' \in C$ and $w \in \tilde{S}_M$, we have $R \sim R'$ and if, further, no block of C has more elements. We easily deduce that if $R \sim R'$ and R is initial, then R' is also initial. Let us call an "initial class" any similarity class composed of initial elements. Since C is finite, at least one similarity class is initial (if we assumed this false, we would end up with an infinite regression of "ancestors," and a contradiction.)

Notice that if an initial element R of C contains only one element, then C consists only of singletons, for a reset will map *any* element of C onto R, but R is initial and so must be similar to all elements; and $w \in \tilde{S}_M$ can map R *onto* a set only if that set is a singleton.

So, given a cover C, we may, assuming C is not composed entirely of singletons, choose an initial class D of C, all of whose members have at least two elements. We then form

$$C' = C_1 \cup C_2,$$

where $C_1 = C - D$, and

$$C_2 = \bigcup_{R \in D} \max \left[\{ Hw : H \in C, w \in \tilde{S}_M, Hw \subset R \} - \{R\} \right].$$

C' is a cover for M, since for every block H of C' and every $w \in \tilde{S}_M$ $w(H)$ is included either in a block of $C - D$ or in some subset of a block in D which will be included in a block of C_2.† Note that $R \in D$ can be an image (onto) only of a block $R' \in D$, but *all* these blocks are deleted. C' is a proper refinement of C because the blocks in D of C are "replaced" in C' by smaller ones. This discussion shows why in defining C' we did not take D to be an arbitrary subset of C', but asked that it comprise initial elements. We now have our refinement C' of C. It remains to be seen that we can choose the "correction-term" machine L so that it has the properties advertised in conditions (d) and (e) of the theorem. We first prove a technical

(4.4) Lemma. *If P, $R \in D$, then there exist maps v_P^R and $v_R^P \in \tilde{S}_M$ for which $v_P^R E_P$, and $v_R^P E_R$ are inverses.*

Proof. Since P and R are similar, we can find w and $y \in \tilde{S}_M$ for which $w(P) = R$ and $y(R) = P$. Hence $yw(P) = P$ and $wy(R) = R$. Thus yw and wy are permutations, but we have no reason to expect them to be inverses. However, we can find integers n and m for which $(yw)^n E_P = E_P$

---

† This definition of C' is from [ZEIGER, 1968], correcting an error in [ZEIGER, 1965]. See also [A. GINSBURG, 1966].

and $(wy)^{nm}E_R = E_R$, since every permutation has a power which is the identity. Let $v_P^R = w$ and $v_R^P = (yw)^{nm-1}y$ and observe that they are inverses. □

We have now reached the home stretch, where we construct N and L. Let $D = \{R_0, R_1, \ldots, R_{k-1}\}$. For each $R_j \in D$ we select maps $v_{R_0}^{R_j}$ and $v_{R_j}^{R_0}$ as in the lemma, and stick to this selection henceforth. Let

$$B_j = \{H \in C' : H \subset R_j\}.$$

Our choice for the state space of L is now $Q_L = B_0$, which is just the set of "fragments" of D comprising R_0. We now show that our choice of $v_{R_0}^{R_j}$ and $v_{R_j}^{R_0}$ allows us to assign these fragments to other elements R_j of D as well; that is, our choice of $v_{R_0}^{R_j}$ and $v_{R_j}^{R_0}$ imposes within each element of D a common coordinate system, the state space of L. The lemma is from [GINSBURG, 1966].

(4.5) **Lemma.** *Each B_j has the same cardinality, say α. Further, the blocks of B_j may be so labeled as $H_{j1}, \ldots, H_{j\alpha}$ that*

$$v_{R_0}^{R_j}(H_{0p}) = H_{jp} \qquad and \qquad v_{R_j}^{R_0}(H_{jp}) = H_{0p}$$

for $1 \leq p \leq \alpha$ and $0 \leq j \leq k - 1$.

Proof. Let H_{0p} be in B_0. Then $v_{R_0}^{R_j}(H_{0p})$ lies in R_j and so must be contained in some block, H_{jg}, say, of B_j. Now

$$H_{0p} = v_{R_j}^{R_0} v_{R_0}^{R_j}(H_{0p}) \subseteq v_{R_j}^{R_0}(H_{jg}).$$

Since $v_{R_j}^{R_0}$ maps R_j onto R_0, there is a block, H_{0r}, say, of B_0 such that $v_{R_j}^{R_0}(H_{jg}) \subseteq H_{0r}$. The maximality of the blocks of B_0 implies that

$$H_{0p} = H_{0r} = v_{R_j}^{R_0}(H_{jg}).$$

Since $v_{R_j}^{R_0}$ is a one-to-one mapping, $|H_{0p}| = |H_{jg}|$, and since $v_{R_0}^{R_j}$ is also one-to-one, $v_{R_0}^{R_j}(H_{0p}) = H_{jg}$. For $H_{0p_1} \neq H_{0p}$ the same reasoning gives $v_{R_0}^{R_j}(H_{0p_1}) = H_{jg_1}$, with $H_{jg_1} \neq H_{jg}$, because otherwise $v_{R_0}^{R_j}$ would take $H_{0p} \cup H_{0p_1}$ onto H_{jg}, while $|H_{0p} \cup H_{0p_1}| > |H_{0p}| = |H_{jg}|$.

Thus $v_{R_0}^{R_j}$ maps distinct blocks of B_0 onto distinct blocks of B_j. The roles of B_0 and B_j can be reversed—hence the conclusion that each B_j has the same cardinality.

Enumerate the blocks of B_0 arbitrarily as $H_{11}, \ldots, H_{1\alpha}$, and then label $v_{R_0}^{R_j}(H_{0p})$ as H_{jp} to obtain the complete result. □

We make the following choices:

$$X_N = X_K = X_M,$$
$$X_L = X_K \times Q_K \qquad \text{(we identify the state and output of } K),$$
$$Q_N = Q_K \times Q_L.$$

We have to define $\beta_N\colon Q_N \xrightarrow{\text{onto}} C'$ so that N tells us where M is in C'. So, given $(p, r) \in Q_N$:

   (i) If $p \notin D$, then p is an element of C which is also an element of C', and K already tells where M is in C', without any extra help from L. That is, we take $\beta_N(p, r) = p$.

   (ii) If $p \in D$, then we need to use the information in L to go from the element p of C to the fragment in C'. In line with our choice of "coordinate system" above, we take $\beta_N(p, r)$ to be some element of C' including $v_{R_0}^P(r)$.

It only remains to specify our state transitions consistently with our choice of the output map:

For $x \in X_N$, $x_N(p, r) = (x_K(p), x_L(p, r))$ with x_K given, whereas x_L is to be specified. So if $(p, r) \in Q_N$, we have

1. If $x_K(p) \notin D$, then by the definition of β_N, $x_L(p, r)$ is irrelevant, and so we may satisfy condition (d) as we please by letting $x_L(p, \cdot)$ be a reset or identity map.

2. If $x_K(p) \in D$, then $p \in D$, since D is initial. We define $x_N(p, r)$ to be (s, t), where $s = x_K(p)$, of course, and t is an element of Q_L containing $v_s^{R_0} x_M v_{R_0}^p(r)$. To complete our proof, we must check that $y = v_s^{R_0} x_M v_{R_0}^p$ is a reset or permutation on $Q_L = \max \{R\colon R \subseteq R_0,\ R \subset \in C'\}$.
Clearly, $y(R_0) \subseteq R_0$, since $R_0 \xrightarrow{v_{R_0}^p} p \xrightarrow{x_M} x_M(p) \subseteq x_K(p) = s \xrightarrow{v_s^{R_0}} R_0$. If $y(R_0)$ is a *proper* subset of R_0, then it is actually an element of C', and so all states of Q_L are mapped to this state, and x_L is a reset. If $y(R_0) = R_0$, then y is a permutation on R_0, and it follows that x_L must be a permutation on Q_L. This completes the proof of Zeiger's theorem and thus assures us of the truth of the crucial theorem: *Any machine M can be obtained by cascade synthesis from PR machines whose groups divide the semigroup of M.*

This completes our discussion of automata theory. We have shown that the structure of machines is best elucidated by studying their representation by algebraic systems—in this case, semigroups. We may expect that as we place mathematical restrictions on our class of machines, so shall we need new algebraic systems to exploit them. We shall see in Part Four that if we restrict our attention to linear machines, then we may turn from semigroups to modules to make use of the extra information provided by linearity.

PART FOUR

Advanced theory of linear systems

R. E. Kalman

"Dies ist der Jugend edelster Beruf:
die Welt, sie war nicht, eh ich sie erschuf."

Goethe: *Faust,* 6793-4 (Part II)

10 Algebraic theory of linear systems

This chapter has the format of a short textbook. We shall be concerned with linear systems, with the foundations of the theory as well as with the mathematical machinery. We begin with first principles and then build up the theory along strictly algebraic lines. This treatment is new. It was influenced primarily by the development of automata theory (see Chapters 6 and 7) and secondarily by the classical machinery of modules over a principal-ideal domain, as applied to the determination of canonical forms of matrices.

We shall use a rigidly algebraic style of exposition. It may *seem* needlessly abstract and even eccentric to readers who were brought up on the Laplace-transform or state-variable type of linear system theory. In due course, however, all the familiar concepts (impulse-response function, transfer function, state-transition equations, and so on) will make their appearance, frequently in a sharper or more general form. The new theory has many practical advantages: the Laplace-transform and state-variable approaches are *merged* into a single framework; linear systems over a finite field become a *special case* of the general theory; new methods are obtained for the *effective computation* of realizations, and so on.

This chapter may be viewed also as a natural generalization of the elementary investigations in Chapter 2. As we saw there, a dynamical system must satisfy two conditions for the regulator problem to have a solution: the system must be completely controllable and it must be completely constructible. For a continuous-time, constant system Σ (Equations (3.2) of Chapter 2), these conditions take the "dual" forms

(*) $$\text{rank } [G, FG, \ldots, F^{n-1}G] = \dim X_\Sigma,$$
(†) $$\text{rank } [H', F'H', \ldots, (F')^{n-1}H'] = \dim X_\Sigma$$

(Theorems (3.4) and (6.15) of Chapter 2). We have already noted (Remark (3.19) of Chapter 2) that the first condition remains essentially unchanged if "continuous-time" is replaced by "discrete-time." The same is true also for the second condition. We emphasize once more: *the entire theory of the regulator problem in Chapter 2 depends on algebraic properties of the matrices F, G, and H satisfying these two conditions.*

It is tempting to ask whether the algebraic theory can be built up solely from these conditions. The answer is very much in the affirmative. We interpret (*) algebraically by viewing it as a condition for generating a module over polynomials in the matrix F. Thus module theory enters into system theory in a wholly natural way.

A crucial theoretical advantage of the module viewpoint is that it leads to a complete and (hopefully) final clarification of the problem of realization, especially the uniqueness theorem, first published in [KALMAN, 1963c, theorem 7(ii)]. This theorem, doubted by many at the beginning, has now become a trivial consequence of the fact that a linear system may be viewed as a module. Because of the central importance of the realization problem in system theory, our entire exposition is built around this theme. We examine the problem repeatedly from different sides in Sections 10.6, 10.10, 10.11, and 10.13.

In the initial development of realization theory after 1963, too much attention was paid to the notion of a *minimal* realization, which is one whose state space has minimal dimension. This formulation is not really satisfactory, however, because "dimension" is defined only in the linear case. It is more fruitful to return to the original formulation of [KALMAN, 1963c, theorem 7(ii)] and characterize a minimal realization by the property that it is completely reachable and completely observable; we call such a realization *canonical*. This formulation is completely general (not tied to linearity) because it corresponds to the canonical factorization of an input/output map into an onto and a one-to-one map. Many of the current difficulties of realization theory are resolved by strict adherence to this setup. In particular, in Section 10.13 we give a new and (hopefully) definitive account of the realization theory of continuous-time linear systems.

The mathematical level of this chapter is quite elementary, but of course we have to require some familiarity with the concepts of modern algebra (abelian groups, rings, modules, abstract constructions). For convenience of the reader, we collected in Appendix A all the essential facts used in the text.

Historical and bibliographical note. The evolution of this material may be seen quite directly through the following chain of papers: [KAL-MAN, 1963c, 1965a–b, 1966b, 1967]. As far as pure mathematics is concerned, [VAN DER WAERDEN, vol. 2, chapter on linear algebra] has used module theory consistently since the early 1930s for the determination of canonical forms and related advanced questions in linear algebra. This point of view has now become encyclopedic [BOURBAKI, 1962, 1964]. That it has not yet fully penetrated university instruction may be attributed to pedagogical constraints. For instance, it is amusing to find in [GREUB, 1967, chap. XIII] an account based directly on (*), which develops all facts and machinery for the elementary theory of $\mathbf{R}[z]$-modules without actually defining the concept of the module itself. Our present exposition has many innovations and new results which go beyond [KAL-MAN 1963–1966]. For example, the first proof of B. L. Ho's fundamental realization algorithm (Section 10.11) succeeds with almost no reference to module theory (though the latter is decidedly helpful in understanding what is going on), while the second proof gives strong hints of a still more general machinery beyond the usual confines of linear algebra.

10.1 Basic definitions

Our program for this chapter is the following: we shall give precise and sharp definitions of various properties of linear systems, and then we examine the consequences of the definitions by purely algebraic methods. In other words, we are interested in knowing just what kind of an *algebraic structure* a linear dynamical system is. The exposition is self-contained, and after this section it can be read independently of the rest of the book. Our purpose here is to make contact with the basic definitions of Chapter 1, thereby relating to the preceding parts of the book everything that will follow.

Standing assumptions. Until Section 10.13 we shall consider only systems Σ which are:

1. discrete-time,
2. constant,
3. linear,
4. equipped with finitely many input and output terminals,
5. constructed with numbers from an arbitrary but fixed field K.

In Section 10.13 we relax the first three assumptions and apply the basic algebraic ideas of the theory to much more general problems.

It may be convenient (and intuitively appealing) to think of such systems in terms of Fig. 10.1. The system Σ has m input terminals, at

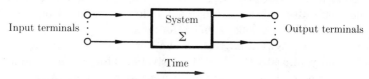

Figure 10.1 Schematic representation of a system.

each of which the system accepts a number from the field K at each instant $t \in \mathbf{Z}$ = integers. There are p output terminals, at each of which the system emits a number in K at each instant in $\mathbf{Z}$. The internal behavior of the system is governed by some given linear law of state transitions, one state transition taking place at each $t \in \mathbf{Z}$. The states and state transitions are all described in terms of numbers belonging to K.

To be entirely precise, let us formally transliterate the preceding assumptions into the language of Definition (1.1) of Chapter 1. We have:

T = time set = $\mathbf{Z}$ = (ordered abelian group of) integers;

U = input values (input alphabet) = K^m = vector space of m-tuples over the field K;

Y = output values (output alphabet) = K^p;

X = state space = K^n;

Ω = input space = arbitrary functions $T \to U$, that is, arbitrary sequences . . . , $\omega(-1)$, $\omega(0)$, $\omega(1)$, . . . , with $\omega(t) \in U$;

Γ = output space = arbitrary functions $T \to Y$;

φ = state-transition map $T \times T \times X \times \Omega \to X$ given by

(1.1) $(t + 1; t, x, \omega) \mapsto \varphi(t + 1; t, x, \omega) = Fx(t) + G\omega(t)$,

where F and G are $n \times n$ and $n \times m$ matrices over K;

η = readout map $T \times X \to Y$ given by

(1.2) $(t, x) \mapsto \eta(t, x) = Hx$,

where H is a $p \times n$ matrix over K.

(1.3) **Comment.** The choice of U and Y as the concrete vector space of m-tuples and p-tuples over K simply expresses the fact that there is a certain fixed way in which the system interacts with its environment. So this is a realistic assumption. On the other hand, to think of the state space X as K^n is merely a convention which enables us to specify the internal description of the system in a numerical form via the matrices F, G, H. As has often been emphasized before in this book, the state of a system is to be regarded in general as an abstract entity; properties related to state should be independent of the particular coordinate system set up in X. Hence it is better to view X as an *abstract n-dimensional vector space over K.*

In view of this discussion, we shall adopt the following conventions:

(1.4) **Definition.** *A* **discrete-time, constant, linear, m-input, p-output dynamical system** Σ **over a field** K *is a composite concept* (F, G, H) *where*

$$F: X \to X,$$
$$G: K^m \to X,$$
$$H: X \to K^p$$

are abstract K-homomorphisms, and X is an abstract vector space over K. We define dim Σ *by* dim X.

The dynamical interpretation of Σ is as given in the preceding paragraphs. To be explicit, we think of the triple (F, G, H) as defining the equations

(1.5)
$$\begin{aligned} (a) \quad & x(t + 1) = Fx(t) + Gu(t), \\ (b) \quad & y(t) = Hx(t), \end{aligned}$$

with $t \in \mathbf{Z}$, $x \in X$, $u(t) \in K^m$, and $y(t) \in K^p$.

The notation $u(t)$ for a *point* in $U = K^m$ will be used interchangeably with the notation $\omega(t)$ for *values of a function* $\omega: T \to U$. We shall usually not make a distinction between (F, G, H) as a triple of K-homomorphisms or as a triple of matrices representing these homomorphisms with respect to a given basis in X. The qualifications "discrete-time," "constant," "m-input," and "p-output" will be understood (and generally not explicitly mentioned) throughout Sections 10.2 to 10.12. Almost always, we assume also that dim $\Sigma < \infty$.

We emphasize that (1.4) is an axiomatic definition, which is to be viewed as the anchor point of the following mathematical investigations. Whenever the word "system" is used informally in the following discussions, its precise meaning is always understood to be in the sense of Definition (1.4).

In view of our insistence that X be an abstract vector space, it is natural to agree that two systems are equivalent when their input/output and state-transition properties are identical. This means that the two systems are equal up to an isomorphism of their state spaces. The precise rule is given by the following

(1.6) **Definition.** *Two linear dynamical systems* $\Sigma = (F, G, H)$ *and* $\hat{\Sigma} = (\hat{F}, \hat{G}, \hat{H})$ *are* **isomorphic** *iff there is a K-isomorphism* $\alpha: X \xrightarrow{\sim} \hat{X}$ *such that the diagram of K-homomorphisms*

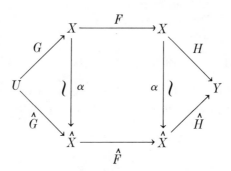

is commutative. If A is the matrix over K representing $\alpha: X \xrightarrow{\sim} \hat{X}$, then system isomorphism is expressed by the following matrix relations:

$$
\begin{array}{ll}
(a) & \hat{F} = AFA^{-1}, \\
(1.7) \qquad (b) & \hat{G} = AG, \\
(c) & \hat{H} = HA^{-1}.
\end{array}
$$

10.2 Input/output maps of a linear system

We shall adopt in this chapter an unconventional definition of an input/output map. The choice is dictated by our desire to have a very simple relationship between the external and internal properties of a system. Of course, this requires a certain amount of foresight. Eventually the definition will be fully justified by the mathematical setup which we shall develop.

Our input/output maps are intended to express the outcome of an "experiment" of the following sort:

1. We apply an input sequence of finite duration to the system. We terminate the input at $t = t_0$; in other words, for all $t > t_0$ the input is identically zero.

2. We observe the output sequence of the system only after the input is terminated, that is, for $t > t_0$. We assume that the output is known for every value of t, no matter how large.

3. Since we shall only consider constant systems (except in Section 10.13), we may choose any suitable integer for the "present" instant $t = t_0$. The obvious choice is $t_0 = 0$.

4. We note that, because of linearity, the special conventions 1 to 3 above do not imply any restriction on generality.

We formalize the preceding system-theoretic picture in the following way:

(2.1) **Definition.** *A* **linear, zero-state, input/output map over** *K is a map $f\colon \Omega \to \Gamma$ defined as follows:*

(a) *$\Omega = \{all \; K\text{-vector sequences } \omega\colon \mathbf{Z} \to K^m \text{ such that } \omega(t) = 0 \text{ for every } t > 0 \text{ and every } t < t_{-1} \leq 0, \text{ where } t_{-1} \text{ is some integer possibly dependent on } \omega\}.$*

(b) *$\Gamma = \{all \; K\text{-vector sequences } \gamma\colon \mathbf{Z} \to K^p \text{ such that } \gamma(t) = 0 \text{ for all } t \leq 0\}.$*

(c) *f is invariant under translation with respect to time, in the following sense: the diagram*

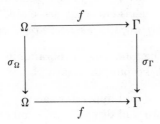

commutes with respect to the shift operators σ_Ω and σ_Γ defined as

$$\sigma_\Omega\colon (0, \ldots, \omega(-1), \omega(0); 0, \ldots)$$
$$\mapsto (0, \ldots, \omega(0), 0; 0, \ldots) \; (\text{``prefix a zero''}),$$
$$\sigma_\Gamma\colon (0, \ldots, 0; \gamma(1), \gamma(2), \ldots)$$
$$\mapsto (0, \ldots, 0; \gamma(2), \gamma(3), \ldots) \; (\text{``discard } \gamma(1)\text{''}).$$

(d) *Ω and Γ are K-vector spaces (in the usual sense, see below) and f is a K-homomorphism relative to this structure on Ω and Γ.*

(2.2) **Comment.** The fact that Ω and Γ may be endowed with the structure of a K-vector space should be quite clear. For instance, consider Ω. Addition is defined as

$$(\omega + \hat{\omega})(t) = \omega(t) + \hat{\omega}(t), \qquad \omega, \hat{\omega} \in \Omega$$

(addition on the right-hand side takes place in the K-vector space K^m). Similarly, scalar multiplication is defined as

$$(\alpha \cdot \omega)(t) = \alpha \cdot \omega(t), \qquad \alpha \in K, \omega \in \Omega$$

(scalar multiplication on the right-hand side takes place in K^m).

(2.3) **Comment.** The sequences $[f(e_i)]_j$ ($= j$-th components of the vector sequence $f(e_i)$, where $e_i \in \Omega$ is the sequence $\omega_k(0) = \delta_{ik}$, $\omega_k(t) = 0$ if $t \neq 0$) provide the same information as the impulse-response map of a continuous-time, constant linear system. $[f(e_i)]_j$ is sometimes called the *pulse response sequence at the j-th terminal due to excitation at the i-th terminal.* It is well known that knowledge of all these sequences suffices to determine the zero-state input/output behavior of a constant linear system. We see again that our definition does not entail any loss of generality.

(2.4) **Comment.** Definition (2.1) is formulated in such a way that causality is automatically "built in," because there is no way to express the dependence of the present output (namely $\gamma(1)$) on future values of the input (namely $\omega(1)$, $\omega(2)$, . . .). The very fact that Ω and Γ consist of functions defined on the disjoint subsets $(-\infty, 0]$ and $[1, \infty)$ of $\mathbf{Z}$ shows that past $\Rightarrow$ future. All this follows already from (2.1a) to (2.1c), without any need for linearity. It is this assumption that will imply the existence of a "state" which is naturally associated with any input/output map. In short, *causality implies the existence of state.*

(2.5) **Comment.** The following "programming" interpretation of the shift operators is worth keeping in mind:

σ_Ω = shift backward and append a zero.
σ_Γ = shift backward and discard first symbol.

(2.6) **Computing technique.** Here is how f is used to compute the zero-state response of a linear system to an arbitrary input sequence $(\omega(0), \omega(1), \omega(2), . . .)$, $\omega(t) \in K^m$. The output sequence $(\gamma(1), \gamma(2), \gamma(3), . . .)$ is clearly given by

$$\gamma(1) = f(0, . . . , 0, \omega(0); 0, . . .)(1),$$
$$\gamma(2) = f(0, . . . , \omega(0), \omega(1); 0, . . .)(1),$$
$$\gamma(3) = f(0, . . . , \omega(0), \omega(1), \omega(2); 0, . . .)(1),$$
$$\cdot \; \cdot$$

where $f(\;)(1)$ is the first term of the sequence $f(\;)$.

(2.7) **Notation convention.** Remember that output symbols are produced with a delay of one unit of time after the application of the input symbol.

To relate the concept of an input/output map as defined here with the concept of a dynamical system defined in Section 10.1, we need the following definition, which will occupy our attention throughout the chapter.

(2.8)   **Definition.**   *A linear dynamical system* Σ *defined by* (1.6) *is a* **realization** *of an input/output map f defined by* (2.1) *iff the input/ output map f_Σ of* Σ *is equal to f.*

Comparing (1.5) with the verbal discussion preceding (2.1), we obtain the

(2.9)   **Proposition.**   Σ *realizes f if and only if*

(2.10) $[f(e_i)]_j = ([HG]_{ji}, [HFG]_{ji}, [HF^2G]_{ji}, \ldots).$

(*The square brackets refer, temporarily, to components of vectors or matrices.*)

To avoid any possible ambiguity concerning the definition of f_Σ, we write down the explicit formula:

(2.11) $f_\Sigma: \omega \mapsto \left(\sum_t HF^{-t}G\omega(t), \sum_t HF^{-t+1}G\omega(t), \ldots \right).$

10.3   *K[z]*-module structure on Ω and Γ

For the moment, let $K = \mathbf{R}$.   In the analysis of discrete-time linear systems it has become traditional to replace sequences in Ω and Γ by their *discrete Laplace transforms*, also called *z-transforms* (see [FREEMAN, 1965]), which are defined as

(3.1) $\hat{\omega}(z) = \sum_{t \in \mathbf{Z}} \omega(t)z^{-t},$

(3.2) $\hat{\gamma}(z) = \sum_{t \in \mathbf{Z}} \gamma(t)z^{-t}.$

The "transformations" $\omega \mapsto \hat{\omega}$ and $\gamma \mapsto \hat{\gamma}$ must be viewed, of course, as mappings

$$\Omega \to \{\text{maps } \mathbf{C} \to \mathbf{C}^m\},$$
$$\Gamma \to \{\text{maps } \mathbf{C} \to \mathbf{C}^p\},$$

since $z \in \mathbf{C}$ = field of complex numbers. Because the transform is regarded as a *function* of the complex argument z, the definition is meaningful only if the sums in (3.1) and (3.2) are *convergent* (for some z).

Under our assumptions on Ω, $\hat{\omega}(z)$ is a polynomial and hence defined for all z. By contrast, the sum (3.2) converges only under special conditions; for instance, if

(3.3) $|z| > 1$ and $|\gamma(t)| < C < \infty$ for all t.

Although convergence conditions are essential to lend mathematical respectability to transform methods, in the engineering literature they are treated *very loosely and yet with apparent impunity.*

Why?

The reason is that, in the vast majority of applications of linear system theory, we may restrict our attention to finite-dimensional systems; then convergence plays no role whatever, since everything may be expressed in algebraic terms.

We shall now develop a rigorous and yet very simple machinery to exploit this observation. Appropriate pedagogical reforms in engineering education will undoubtedly follow in due time. Our attention here will be directed primarily toward the mathematical issues.

We shall adopt the following setup, to be used throughout the rest of this chapter.

We assume once more: K = arbitrary field.

Instead of regarding (3.1) as defining a function of z, we regard it as expressing an isomorphism between Ω and the K-vector space $K^m[z]$ of *polynomials in the indeterminate z with coefficients in K^m.* So we have

$$(3.1^*) \qquad \omega \approx \sum_{t \leq 0} \omega(t)z^{-t} \qquad \text{(finite sum!)},$$

$$= \begin{bmatrix} \omega_1 \\ \cdot \\ \cdot \\ \cdot \\ \omega_m \end{bmatrix}, \qquad \omega_k \in K[z], \quad k = 1, \ldots, m,$$

where the symbol ω_k, the k-th component of the vector ω, denotes the polynomial

$$\omega_k(z) = \omega_k(0) + \omega_k(-1)z + \cdots + \omega_k(-t_1)z^{t_1}.$$

(3.4) **Comment.** We emphasize once more: *For us a polynomial is an algebraic object, not a function of a complex variable.* A polynomial is simply an alternate way of viewing a sequence in Ω. The indeterminate z may be viewed as a time-marker, with z^k corresponding to $t = -k$. Of course, this convention does nothing that is basically new. We are not really concerned with "transforms." The main advantage is notational: manipulations of polynomials are presumably familiar to everyone. (Advice: This is a good time to review Appendix A.)

We shall not distinguish sharply between ω = finite vector sequence and ω = vector polynomial.

The critical algebraic observation, which *is new* (as far as system theory is concerned), turns out to be the following

(3.5) Proposition. Ω, *viewed as the K-vector space* $K^m[z]$, *admits the structure of a finite free module, as follows*:

(a) *Abelian group of Ω: abelian group of Ω, regarded as a K-vector space (see Section 2; in particular, Comment (2.2)).*

(b) *Ring acting on Ω: $K[z]$, regarded as a ring with the usual multiplication of polynomials.*

(c) *Scalar multiplication in Ω: $\pi \cdot \omega$ (for $\pi \in K[z]$, $\omega \in \Omega$), the ordinary componentwise product of a polynomial vector by a scalar polynomial.*

(d) *Generators of Ω: the vectors*

$$e_1 = \begin{bmatrix} 1 \\ \cdot \\ \cdot \\ \cdot \\ 0 \end{bmatrix}, \ldots, e_m = \begin{bmatrix} 0 \\ \cdot \\ \cdot \\ \cdot \\ 1 \end{bmatrix}$$

in K^m.

Proof. Trivial (and well-known) verification of the module axioms (A.4) to (A.7). $\square$

(3.6) Interpretation. Important: The multiplication in $K[z]$ (the familiar, well-known, usual, ordinary multiplication of polynomials) is equivalent to convolution of scalar sequences. To see this, write

$$\pi(z) = \sum_t \pi_t z^t;$$

then

$$(\pi\pi')(z) = \sum_{r+s=t} \pi_r \pi'_s z^t,$$
$$= \sum_{t,s} \pi_s \pi'_{t-s} z^t = \sum_{t,s} \pi_{t-s} \pi'_s z^t.$$

(3.7) Remark. It is also important to note that we have not defined a *ring* structure (multiplication) in Ω, even when $m = 1$. Rather, (3.5) gives an *operator structure on Ω*, where operators in the ring $K[z]$ act on Ω via scalar multiplication. Note also that scalar multiplication by z corresponds to the shift operator σ_Ω, because

$$z \cdot \omega = \sigma_\Omega(\omega),$$

as may be seen immediately from (3.1*). Hence *multiplication by z is a representation of the shift operator* σ_Ω. Speaking somewhat loosely, the module structure put on Ω by (3.5) allows us to express the shift operator (which is crucial in the study of dynamics) in a purely algebraic way.

In view of (3.2), we define an isomorphism between Γ and the *K*-vector space $K^p[[z^{-1}]]$ *of all formal power series in the indeterminate* z^{-1} *with coefficients in* K^p. In other words, we replace (3.2) by

(3.2*) $$\gamma \approx \sum_{t>0} \gamma(t)z^{-t}.$$

"Formal power series" is just another name for "infinite sequence." There is no question of convergence; we are *not* interested in computing the number $\gamma(z)$, $z \in \mathbf{C}$, even for those values of z for which (3.2) may happen to converge. The indeterminate z again serves as a time-marker, and z^{-k} corresponds to $t = k$, in strict agreement with the notation convention adopted in (3.1*). Note also that the *zeroth* term of the power series (3.2*) is always zero. This is a result of our notation convention (2.7).

We shall not distinguish sharply between $\gamma =$ infinite sequence and $\gamma =$ vector formal power series.

Now consider the diagram

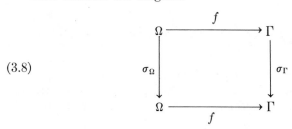

(3.8)

By Definition (2.1*c*), this diagram must be commutative for Ω and Γ regarded as sets and f, σ_Ω, and σ_Γ regarded as set-maps. In the linear case, Ω and Γ are *K*-vector spaces; the shift operators σ_Ω and σ_Γ are then automatically *K*-homomorphisms; the map f is required to be a *K*-homomorphism by (2.1*d*). Thus, in the linear case, (3.8) is a commutative diagram of *K*-homomorphisms. We already know that σ_Ω may be written as $\omega \mapsto z \cdot \omega$, because of the possibility of putting a *K[z]*-module structure on Ω. We want the same representation for σ_Γ. If we can find a *K[z]*-module structure on Γ such that $\sigma_\Omega : \gamma \mapsto z \cdot \gamma$, then f becomes a *K[z]*-homomorphism (because the commutativity of the diagram

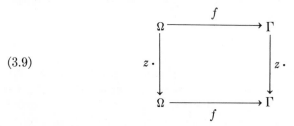

(3.9)

is the *definition* of a *K[z]*-homomorphism Ω → Γ).

So, to endow Γ with the structure of a $K[z]$-module, we simply examine the various necessary conditions implied by the fact that f is required to be a $K[z]$-homomorphism. We have

$$f(z \cdot \omega) = z \,\square\, f(\omega) = \sigma_\Gamma(f(\omega))$$

where $\square$ is the (as yet unknown) scalar multiplication in Γ. The definition of σ_Γ (see also Comment (2.5)) shows that we must define

(3.10) $\pi \cdot \gamma$ = *ordinary product of a formal power series* $\gamma \in K^p[[z^{-1}]]$ *by a polynomial* $\pi \in K[z]$, *followed by deletion of all terms corresponding to nonnegative powers of* z.

With this important point settled, routine verification of the module axioms establishes the desired

(3.11)   **Proposition.** Γ, *viewed as the K-vector space of all formal power series in* z^{-1} *with coefficients in* K^p *and* $\gamma_0 = 0$, *admits the structure of a* $K[z]$-*module with scalar multiplication defined by* (3.10). *Relative to this module structure the shift operator* σ_Γ *is represented by* $\gamma \mapsto z \cdot \gamma$.

Note that Γ, as a $K[z]$-module, is neither free nor finite.

10.4   Module and Nerode equivalence

We have now established that f, defined by (2.1), is a $K[z]$-homomorphism. It is this *mathematical* consequence of the *system-theoretic* definition (2.1) which makes linear system theory interesting (and successful).

As a first application of the "module" point of view, we note that there is a module *naturally associated with* or *naturally induced by* the input/output map f: the quotient $K[z]$-module $\Omega/\text{kernel } f$. Although we assume that the reader is familiar with the notion of a quotient module, we shall review this concept here because of its system-theoretic interpretation.

Given f, a $K[z]$-homomorphism, we have an equivalence relation $\underset{f}{\approx}$ defined by

(4.1) $$\omega \underset{f}{\approx} \hat{\omega} \quad \text{iff} \quad f(\omega) = f(\hat{\omega})$$

or, equivalently, by

(4.2) $$\omega \underset{f}{\approx} \hat{\omega} \quad \text{iff} \quad \omega - \hat{\omega} \in \text{kernel } f.$$

To make contact with the material of Chapter 7, let us note the

(4.3)   **Proposition.** $\underset{f}{\approx}$ *is a congruence on* Ω *regarded as a* $K[z]$-*module.*

Proof. Let $[\omega]_f$ denote the equivalence class of ω under $\underset{f}{\approx}$.

We must prove that the equivalence class of the sum $\omega + \omega'$ contains all sums $\hat{\omega} + \hat{\omega}'$ where $\hat{\omega}$ and $\hat{\omega}'$ are arbitrary representatives of $[\omega]_f$ and $[\omega']_f$. Then $\omega - \hat{\omega}$ and $\omega' - \hat{\omega}'$ belong to kernel f by (4.2). So $\omega + \omega' - (\hat{\omega} + \hat{\omega}') \in$ kernel f, since kernel f is a subspace of Ω. Hence, again by (4.2), $\hat{\omega} + \hat{\omega}' \in [\omega + \omega']_f$.

Similarly, we can prove that if $\omega' \in [\omega]_f$ then $\alpha\omega' \in [\alpha\omega]_f$ (here, of course, $\alpha \in K[z]$). $\square$

Suppose the $[\omega]_f$ are regarded as abstract entities, not as subsets of Ω. Then the proof of (4.3) shows that the abstract set $\{[\omega]_f : \omega \in \Omega\} = X_f$ admits the structure of a $K[z]$-module, and that in this module the operations $+$ and $\cdot$ are *defined* as

$$(4.4) \qquad \begin{aligned} [\omega]_f + [\omega']_f &= [\omega + \omega']_f, \\ z \cdot [\omega]_f &= [z \cdot \omega]_f. \end{aligned}$$

By definition, $\Omega/$kernel $f = X_f$. So Proposition (4.3) is equivalent to

(4.5) **Proposition.** *The equivalence classes X_f under the (module) equivalence relation $\underset{f}{\approx}$ admit the structure of a $K[z]$-module, which is called the* **quotient module** $\Omega/$kernel f.

Let us recall now the discussion in Section 7.2 of the Nerode equivalence relation $\underset{f}{\sim}$ induced by f. The essential point is the following: Two inputs, ω and $\hat{\omega}$, are *Nerode equivalent* iff the output sequences $f(\omega)$ and $f(\hat{\omega})$ are the same and remain the same whenever both ω and $\hat{\omega}$ are followed by an arbitrary input $\nu \in \Omega$. "Followed by" is the concatenation multiplication, which is defined in the present context by

$$\omega \circ \nu = (0, \ldots, 0, \omega_q, \omega_{q-1}, \ldots, \omega_0, \nu_{q'}, \nu_{q'-1}, \ldots, \nu_0; 0, \ldots).$$

In the polynomial notation, we write this as

$$(4.6) \qquad \omega \circ \nu = z^{1+q'}\omega + \nu,$$

where

$$q' = \deg \nu = \max_k \{\deg \nu_k : k = 1, \ldots, m\},$$

and "deg" is the ordinary degree of a polynomial. So the precise definition of the *Nerode equivalence relation induced by f* is

$$(4.7) \qquad \omega \underset{f}{\sim} \hat{\omega} \quad \text{iff} \quad f(\omega \circ \nu) = f(\hat{\omega} \circ \nu) \text{ for all } \nu \in \Omega,$$

where ν may be the empty sequence $\varnothing$ for which $\omega \circ \varnothing = \omega$. We denote the *Nerode equivalence classes* by $(\omega)_f$.

Now we have an all-important but easy result:

(4.8) **Proposition.** $[\omega]_f = (\omega)_f$ *for all* $\omega \in \Omega$.

Proof. That (4.7) implies (4.1) is immediately clear from the definitions. By linearity of f and (4.4)

$$f(\omega \circ \nu) = f(z^{1+\deg \nu}\omega) + f(\nu),$$
$$= z^{1+\deg \nu}f(\omega) + f(\nu).$$

Hence (4.1) implies (4.7). $\qquad\qquad\qquad\qquad\qquad\square$

We saw in Section 7.2 that the Nerode equivalence classes constitute the natural definition of the state set associated with an input/output map. Hence we have from (4.8) the following result, which, for many reasons yet to be elucidated, we regard as the

(4.9) **Fundamental theorem of linear system theory.** *The natural state set* X_f *associated with a linear, constant, zero-state, discrete-time input/output map* f *over* K *admits the structure of a* $K[z]$-*module.*

We have already employed and shall continue to employ the special

(4.10) **Notation convention.** A set is always denoted by the same symbol, regardless of the structure the set may carry. Similarly, elements of a set are denoted consistently by the same letters, regardless of structure. Thus Ω is either an abstract set, or a K-vector space, or a $K[z]$-module, and so on. Elements of Ω are written as ω, and ω may be either a sequence or a polynomial. Whenever necessary, the assumed structure will be specially emphasized. The reader should make an effort to get used to the abstractions involved until he can "put on" or "take off" structures without conscious mental effort. *Warning:* The first time a new structure is used, its existence must be carefully checked, as in Proposition (3.11).

10.5 The state space as a module

Now we turn our attention to a linear dynamical system Σ described via its state-transition equations (1.5a). More precisely, we view Σ as the pair $(F, G, -)$, where $F: K^n \to K^n$ and $G: K^m \to K^n$ are K-homomorphisms (and the map H is immaterial, as suggested by the notation). We want to associate with Σ a $K[z]$-module X_Σ.

A standard recipe for the construction of $K[z]$-modules is given by the following

(5.1) **Proposition.** *Let* $A: V \to V$ *be an arbitrary* K-*endomorphism of*

*the K-vector space V. Then V admits the structure of a K[z]-module
with scalar multiplication*

$$(\pi, v) \mapsto \pi \cdot v = \pi(A)v, \qquad \pi \in K[z], \, v \in V,$$

*where $\pi(A)$ is "A substituted into π," that is, $\pi(A)$ is the endomorphism
$\pi_0 I + \cdots + \pi_n A^n$ and $\pi(A)v$ is the usual action of the endomorphism
$\pi(A)$ on v.*

*Conversely, given any K[z]-module V, the K[z]-endomorphism
$v \mapsto z \cdot v$ induces the K-endomorphism $v \mapsto Av = z \cdot v$.*

We call the module X_Σ associated with Σ via this process (with
$V = X_\Sigma = $ state space of Σ and $A = F_\Sigma$) the *K[z]-module induced by* Σ.
This module has a special set of elements, given by

$$\{Ge_1, \ldots, Ge_m\} = \{g_1, \ldots, g_m\},$$

with e_k as defined by (3.5d). For reasons soon to be explained, we call
this set the *accessible generators* of X_Σ.

The definition of Σ tells us explicitly how inputs act on states in the
dynamical sense. This action may be regarded as a kind of multiplication. (We write it on the right in order to distinguish it from scalar
multiplication, and we denote it by ∘ because dynamical action is closely
related to concatenation.) Thus the map

$$\varphi_\omega: \text{initial state} \mapsto \begin{array}{l} \text{state resulting one unit of time} \\ \text{after the end of the input sequence } \omega \end{array}$$

is denoted by

$$x \mapsto x \circ \omega,$$

and is defined explicitly as

$$(5.2) \qquad x \circ \omega = F^{1 + \deg \, \omega} x + \sum_{t = -\deg \, \omega}^{0} F^{-t} G \omega(t).†$$

As one of the first facts justifying the name of Theorem (4.9), we
prove the easy

(5.3) **Proposition.** *Dynamical action may be expressed via module
action.*

Proof. By definition, the map $x \mapsto Fx$ (vector-space notation) is equivalent to the map $x \mapsto z \cdot x$ (module notation). Moreover, since our

---

† While this expression is certainly well defined, it does not lend itself to describing the effect of an "input" consisting of a string of zeros, because $\omega = 0$ is a polynomial of degree 0, that is, a string of *exactly one* zero. The modifications necessary
to cover that case may be found in Section 10.12.

notations are

$$\omega(t) = \sum_{k=1}^{m} \omega_k(t)e_k,$$

$$\omega_k(z) = \sum_{t} \omega_k(t)z^{-t},$$

and

$$Ge_k = g_k,$$

it is clear that

(5.4) $$x \circ \omega = z^{1+\deg \omega} \cdot x + \sum_{k=1}^{m} \omega_k \cdot g_k. \qquad \square$$

(5.5) **Interpretation.** In particular, we have

(5.6) $$0 \circ \omega = \sum_{k=0}^{m} \omega_k \cdot g_k.$$

The left-hand side of the formula denotes a state x which is *reachable from the state 0 by the application of the input ω.* (Review Definition (2.13) of Chapter 2.)   The right-hand side expresses this state (regarded as a module element) as a linear combination of the accessible generators $g_k = Ge_k$.   In particular, ω_k is the "signal" which must be applied to the kth input terminal of Σ to bring about $x = 0 \circ \omega$.   In other words, *linear combinations of module elements correspond to generating a state via suitable inputs.*   Module scalar multiplication is therefore a very compact notation for writing down the effect of inputs applied to specific input terminals.   Thus $g_k \in X_\Sigma$ corresponds to the kth input terminal; "accessible" means that such input terminals are actually available to the external user of Σ; that is, the state transition $0 \mapsto \omega_k \cdot g_k$ can be achieved without opening up the system and injecting special signals into it.

In view of this discussion, we have immediately the

(5.7) **Proposition.** *The reachable states of Σ are precisely those lying in that submodule of X_Σ which is generated by the accessible generators of X_Σ.*

(5.8) **Warning.** Since all states are not necessarily reachable, in general, the *accessible* generators do not necessarily generate all of X_Σ. But X_Σ is finitely generated whenever X_Σ (the K-vector space of states of Σ) is finite dimensional, because then the elements of a basis of X_Σ (vector space) serve as generators of X_Σ (module).

In summary, a big advantage of the module setup is that the act of "injecting" an input sequence (of arbitrary but finite length) into a system initially in the zero state can be represented by a single algebraic operation. We express this operation by the new notation

$$(5.9) \qquad \bar{G}_\Sigma: \Omega \to X_\Sigma: \omega \mapsto 0 \circ \omega = \sum_{k=1}^{m} \omega_k \cdot g_k.$$

The restriction of $\bar{G}_\Sigma$ to K^m (regarded as isomorphic to the polynomials of degree 0 and hence a submodule of $K^m[z]$) is just the map $G_\Sigma = G: K^m \to K^n$ occurring in the definition of Σ. Note that $\bar{G}_\Sigma$ is a $K[z]$-module homomorphism (immediate from the definition).

The map $\bar{G}$ may also be defined for the module $X_f = \Omega/\text{kernel } f$ induced by f:

$$(5.10) \qquad \bar{G}_f: \Omega \to X_\Sigma: \omega \mapsto [\omega]_f.$$

As before, $\bar{G}_f$ is a $K[z]$-homomorphism, because

$$\bar{G}_f(z \cdot \omega) = [z \cdot \omega]_f = z \cdot [\omega]_f = z \cdot \bar{G}_f(\omega)$$

by definition of scalar multiplication in X_f (see (4.5)). This fact could have been used to guess the "right" module structure on X_f, which we simply pulled out of the air before. In other words, it is a remarkable accident that dynamical considerations are nicely dovetailed with algebraic necessity, as in the construction of the quotient module. This is another reason for the name of (4.9).

Let us consider now the output map H_Σ of a dynamical system Σ. Clearly H induces the map

state $\mapsto$ output sequence resulting from state.

We write

$$(5.11) \qquad \bar{H}_\Sigma: X \to \Gamma,$$
$$: x \mapsto (Hx, HFx, HF^2x, \ldots)$$
$$= (Hx, H(z \cdot x), H(z^2 \cdot x), \ldots),$$

and call $\bar{H}_\Sigma$ the *extension* of $H_\Sigma = H$. $\bar{H}_\Sigma$ is trivially a $K[z]$-homomorphism.

Similarly f induces the map

$$(5.12) \qquad \bar{H}_f: X_f \to \Gamma: [\omega]_f \mapsto f(\omega).$$

We must check that $\bar{H}_f$ is well-defined, but this is trivial: if $[\omega]_f = [\hat{\omega}]_f$, then $f(\omega) = f(\hat{\omega})$, by (4.1).

Now we have seen how to translate the most important concepts associated with f and Σ into $K[z]$-modules and $K[z]$-module homomorphisms.

(5.13) **Notations.** Let us review the terminology set up so far. The dynamical system $\Sigma = (F, G, H)$ induces the triple $(X_\Sigma, \{g_1, \ldots, g_m\},$

$\bar{H}_\Sigma$), where

$X_\Sigma = K[z]$-module induced by F (Proposition (5.1)),

$g_k = Ge_k, k = 1, \ldots, m$, are the accessible generators of X_Σ,

$\bar{H}_\Sigma$ = map (5.11) induced by H.

It is clear that, given $(X, \{g_1, \ldots, g_m\}, \bar{H})$, we can induce from it a system Σ simply by reversing the previous steps. For convenience we shall call $(X_\Sigma, \{g_1, \ldots, g_m\}, \bar{H}_\Sigma)$ the *module of* Σ and shall quite often refer to it simply as X_Σ. In short, Σ *is equivalent to* X_Σ. (The subscript Σ may be dropped as a matter of convenience.)

Similarly the input/output map f induces a triple $(X_f, \{g_1, \ldots, g_m\}, \bar{H}_f)$, where

$X_f = \Omega/\text{kernel } f$ (as $K[z]$-module),

$g_k = [e_k]_f$,

$\bar{H}_f$ = map (5.12) induced by f.

By reversing this process, we see immediately that $f = \bar{H}_f \circ \bar{G}_f$, where

$$\bar{G}_f: \omega \mapsto [\omega]_f = \sum_{k=1}^m \omega_k \cdot g_k.$$

We shall call $(X_f, \{g_1, \ldots, g_m\}, \bar{H}_f)$ simply the *module of* f and abbreviate it as X_f. Again we have that f *is equivalent to* X_f.

Sometimes it is convenient to consider the triple $(X, \bar{G}, \bar{H})$ instead of $(X, \{g_1, \ldots, g_m\}, \bar{H})$. The two are obviously equivalent.

10.6 Abstract realization theory

We recall the fundamental definition (2.8): Σ *realizes* f iff $f_\Sigma = f$. Realizations certainly exist for any f: take a dynamical system Σ with $X_\Sigma = \Omega$ (vector space), $F = \sigma_\Omega$, G = identity, and $\bar{H} = f$. This realization is, of course, both trivial and useless. The module machinery set up in the preceding sections provides another realization, which is natural, useful, nontrivial, and therefore interesting: we associate with f its module $(X_f, \{g_1, \ldots, g_m\}, \bar{H}_f)$ and then view the triple as a dynamical system Σ. It is easy to verify that this is a realization.

The problem is to show that the realization just constructed (only abstractly, of course) is in some sense the only realization we need to consider. We view the realization problem as "guessing the equations of motion of a dynamical system from its input/output behavior," or "setting up a physical model which explains the experimental data," or "drawing a circuit diagram which *may* correspond to the interior of a black box whose input/output map is f."

In 1962, well before the development of the present formalism, Kalman obtained what were perhaps the first definitive results in this direction.   Paraphrasing slightly [KALMAN, 1963c, theorem 7] to suit the present context, we can state these results as follows:

1. *If a dynamical system Σ is tested via causal experiments and its (causal) input/output map f determined from the equations of motion, then f depends only on a subsystem Σ_0 of Σ which is both completely reachable and completely observable.   The other parts of Σ have no effects on f and may be chosen completely arbitrarily without altering f.*

2. *If any two systems, Σ and $\hat{\Sigma}$, are both completely reachable and completely observable and have the same input/output map f, then they differ only in the coordinatization of their state spaces.*

As immediate consequences of this basic result (also discussed at length in [KALMAN, 1963c]) we have also:

3. *If a realization Σ of f is not completely reachable or not completely observable, then Σ contains certain parts which have no relation to the experimental data as represented by f but arise in a completely arbitrary way determined only by the particular algorithm used to construct Σ.*   (Note: there are many instances of such a situation in the literature prior to 1965.)

4. *If a realization Σ of f is completely reachable and completely observable, then it is essentially uniquely determined by f, since the coordinatization of states can never be inferred from input/output experiments.   In any case the coordinatization of states is irrelevant, since the labeling of internal states in Σ has no intrinsic physical significance.*

We shall see that the statement "essentially uniquely determined by f" is a direct consequence of Theorem (4.9).

We shall now adopt the following basic

(6.1)   **Definition.**   *A realization Σ of f is* **canonical** *(or* **natural***) iff it is both completely reachable and completely observable.*   (For the last two definitions, refer to (2.12) and (6.4) of Chapter 2.)

We wish to develop realizability theory in the simplest and most elementary way.   For the moment we abandon linearity to state a trivial but useful lemma on the set-theoretic level, based on the work of [ZEIGER, 1967b]:

(6.2)   **Zeiger fill-in lemma.**   *Let A, B, C, D be arbitrary sets.   Consider the commutative diagram*

(6.3)

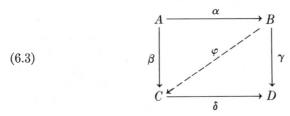

and suppose that α = onto and δ = one-to-one.

Then there is a unique map φ (shown by the dashed arrow) such that the diagram remains commutative.

Proof. ("Diagram chasing"). Take any element $b \in B$. Since α is onto, the set $\alpha^{-1}(b)$ is not empty. For any $a \in \alpha^{-1}(b)$ commutativity implies that

$$(\delta \circ \beta)(a) = d = (\gamma \circ \alpha)(a) = \gamma(b).$$

Since δ is one-to-one, there is a unique c such that $c = \delta^{-1}(d)$. We define

$$\varphi: b \mapsto c = (\delta^{-1} \circ \gamma)(b).$$

Then the northwest triangle commutes, because

$$\beta(a) = (\delta^{-1} \circ (\gamma \circ \alpha)(a) = ((\delta^{-1} \circ \gamma) \circ \alpha)(a) = (\varphi \circ \alpha)(a)$$

by commutativity (and associativity of maps under composition). Similarly, the southeast triangle commutes, because

$$(\delta \circ \varphi)(b) = (\delta \circ (\delta^{-1} \circ \gamma))(b) = \gamma(b).$$

No other prescription can work; hence φ is unique. $\qquad\square$

(6.4) **Corollary.** *If the commutative diagram (6.3) involves K-vector spaces and K-homomorphisms, then φ is a K-homomorphism.*

Proof. φ is a composition of the K-homomorphisms γ and δ^{-1}: image $\gamma \circ \alpha \to C$. $\qquad\square$

The same proof gives also the

(6.5) **Corollary.** *If the commutative diagram (6.3) involves $K[z]$-modules and homomorphisms, then φ is also a $K[z]$-homomorphism.*

We emphasize these trivial facts for important reasons. The basic properties of a realization depend only on the facts expressed by Zeiger's lemma, that is, on very elementary properties of maps. But the reason that Zeiger's lemma can be *applied* to dynamical systems is that the structural properties of linearity are compatible with the diagram (6.3), not only in the sense of K-vector spaces, but also in the sense of $K[z]$-modules. Roughly speaking, *the power of the algebraic method in dynamics depends on the possibility of putting additional structure on diagrams such as (6.3) in such a way that this structure mirrors certain important dynamical properties.*

On the elementary set-theoretic level of Zeiger's lemma, the counterpart of Definition (6.1) is the following

(6.6) **Definition.** *Let A and B be arbitrary sets and $f: A \to B$ an arbitrary map. We say that f is **factored through** (a set) C iff there is*

a commutative diagram

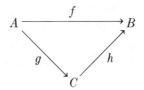

The factorization is **canonical** (*or* **natural**) *iff* $g = onto$ *and* $h = one$-*to*-*one*.

The first application of Zeiger's lemma is the

(6.7) **Proposition.** *Any two canonical factorizations* $f = h \circ g = \hat{h} \circ \hat{g}$ *of* f *through* C *and* $\hat{C}$ *are equivalent in the following sense: there is a unique set isomorphism* $\alpha\colon C \xrightarrow{\sim} \hat{C}$ *such that* $\hat{g} = \alpha \circ g$ *and* $h = \hat{h} \circ \alpha$.

Note, in particular, that this implies that the number of elements of C is equal to the number of elements of $\hat{C}$.

Proof. Consider the commutative diagram

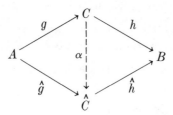

Since g is onto and $\hat{h}$ is one-to-one, a unique α exists, by Zeiger's lemma. Since $\hat{g}$ is onto and h is one-to-one, there is also a unique β going in the opposite direction. By commutativity, $\beta \circ \alpha$ is the identity on C and $\alpha \circ \beta$ is the identity on $\hat{C}$. So $\beta = \alpha^{-1}$, and α is an isomorphism of sets. ☐

The corollaries of Zeiger's lemma imply the

(6.8) **Corollary.** *If* f *is a* K-*homomorphism* [$K[z]$-*homomorphism*], *then any two canonical factorizations are equivalent via a* K-*homomorphism* [$K[z]$-*homomorphism*] α. *In particular,* C *and* $\hat{C}$ *are then isomorphic* K-*vector spaces* [$K[z]$-*modules*].

We can now prove the

(6.9) **Fundamental theorem of linear realization theory.** *Any two canonical realizations of an input/output map* f *are isomorphic as* $K[z]$-*modules and therefore also as dynamical systems.*

  In summary: All canonical realizations of any fixed f *are essen-*

tially the same; the equivalence classes of canonical realizations are in one-to-one correspondence with the class of input/output maps.

Proof. All we really have to do is to translate the definition of "canonical realization" into the more abstract context of factorization of a map.

Let $X = (X, \bar{G}, \bar{H})$ be any canonical realization of f. Then X is completely reachable, which means that $\bar{G}: \Omega \to X$ is onto. X is also completely observable, which means that $\bar{H}: X \to \Gamma$ is one-to-one. (The reader should verify this by reviewing the definitions in Section 2.6.) By definition of a realization, $f = \bar{H} \circ \bar{G}$. Hence a canonical realization of f induces a canonical factorization of f.

Suppose $\hat{X} = (\hat{X}, \hat{\bar{G}}, \hat{\bar{H}})$ is another canonical realization. Then $f = \hat{\bar{H}} \circ \hat{\bar{G}}$. By (6.8), X is isomorphic to $\hat{X}$ as a $K[z]$-module; moreover, $\hat{\bar{G}} = \alpha \circ \bar{G}$ and $\bar{H} = \hat{\bar{H}} \circ \alpha$.

In view of the facts explained at the end of Section 10.5, this implies immediately that the dynamical systems Σ and $\hat{\Sigma}$ corresponding to X and $\hat{X}$ are also isomorphic in the sense of Definition (1.6).

More explicit argument: Notice that the matrix representation of α has elements over K (*not* $K[z]$); this follows immediately from the fact that α is an isomorphism $X \xrightarrow{\sim} \hat{X}$ with respect to the K-module (vector-space) structure of these sets, since K may be viewed as a subring of $K[z]$. Then $\hat{\bar{G}} = \alpha \circ \bar{G}$ implies (1.7b), $\bar{H} = \hat{\bar{H}} \circ \alpha$ implies (1.7c), and

$$\alpha(Fx) = \alpha(z \cdot x) = z \cdot \alpha(x) = z \cdot \hat{x} = \hat{F}\hat{x} = \hat{F}\alpha(x)$$

implies (1.7a). □

(6.10) **Proposition.** dim Σ *is minimal over the class of all realizations of f if and only if Σ is a canonical realization of f.*

Proof. Let $\hat{\Sigma}$ be any realization and Σ a canonical realization. Then $f = \hat{h} \circ \hat{g}$ and therefore $\hat{h}(\hat{X}) \supset \text{range } f$. But $X \approx \text{range } f$ by (6.8), since f may be factored canonically as $\Omega \to \text{range } f \to \Gamma$ (the second map is the natural injection). So

$$\text{dim } \hat{\Sigma} \underset{\triangle}{} \text{dim } \hat{X} \geqq \text{dim range } f = \text{dim } X \underset{\triangle}{} \text{dim } \Sigma.$$

If the equality sign holds in the middle then $\hat{h}(\hat{X}) \approx \text{range } f$, that is, $\hat{h}$ is one-to-one; moreover, range $\hat{g} \approx \hat{X}$ (since otherwise $f \neq \hat{h} \circ \hat{g}$), which implies that $\hat{g} = \text{onto}$. Hence dim $\hat{X} = \text{minimum}$ implies that the factorization is canonical. □

(6.11) **Terminology.** From here on we shall say interchangeably and equivalently that a realization is *canonical* or *minimal*. (Only at the end of Section 10.11 will we encounter a situation where "minimal" is not necessarily the same as "canonical".) In view of (6.10), we define dim f as the *dimension* of a canonical realization of f; this terminology is fully justified by the fundamental theorem (6.9). We say that a linear map f is *finite-dimensional* iff f may be factored through a finite-dimensional space. (The standard terminology in this case is to say that f has *finite rank;* for us here, however, the term "finite-dimensional" is more suggestive.)

Alternate approach (*Zeiger*). We may, of course, prove Theorem (6.9) without any overt use of Theorem (4.9) and without even mentioning modules, keeping the machinery of proof entirely within the usual K-vector-space structure. This type of attack also has considerable power, as will become clear in Section 10.11.

Define a *realization* to be any factorization $f = \bar{H} \circ \bar{G}$ of f subject to the condition

$$\bar{G}(\omega) = \bar{G}(\hat{\omega}) \qquad \text{implies} \qquad \bar{G}(\sigma_\Omega \omega) = \bar{G}(\sigma_\Omega \hat{\omega}).$$

Since f is constant, Definition (2.1c) implies that we have the following commutative diagram of K-homomorphisms (the existence of F will be proved later):

(6.12)

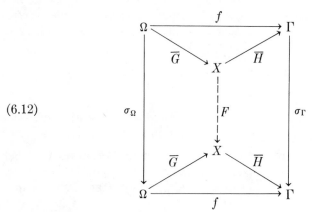

This notion of a realization agrees with the standard one and has the additional advantage of being quite near to the intuitive notion of state. It says that the input sequence ω can be replaced by the state $x = \bar{G}(\omega)$ which has the property that the output sequence $\gamma = f(\omega)$ can be computed from it as $\gamma = \bar{H}(x)$. That is, the state space X incorporates all information about the past which *may be* necessary and *is* sufficient to compute the future. In addition, the states must be defined in a "constant" way: if we agree that ω and $\hat{\omega}$ define the same state (that is, $\bar{G}(\omega) = \bar{G}(\hat{\omega})$), then $\sigma_\Omega \omega$ and $\sigma_\Omega \hat{\omega}$ must also define the same state.

Let us verify that this notion of a realization implies the existence of a triple (F, G, H) specifying a dynamical system Σ. The existence of H and G is immediate from the definition of a factorization (and from the standing definition of Ω and Γ). To define F it suffices to fill in the dashed arrow in the commutative diagram (6.12), for then the diagram becomes equivalent to equations (1.5). The existence of F is easy:

(6.13) $\qquad F: x \mapsto x' = \begin{cases} \bar{G}(\sigma_\Omega \omega) & \text{if } x = \bar{G}(\omega), \\ \in \bar{H}^{-1}(\sigma_\Gamma(\bar{H}(x))) & \text{if } x \notin \text{range } \bar{G}. \end{cases}$

Note that F is not unique unless $\bar{G}$ is onto. Note that F is well defined by hypothesis on $\bar{G}$.

A realization is *canonical* iff the factorization of f through X is canonical. This implies immediately that F is uniquely determined by (6.13). Of course, canonical realizations do exist: for instance, let $X = \Omega/\text{kernel } f$ (in the vector-space sense).

Now let $f = \hat{\hat{H}} \circ \hat{\bar{G}}$ be another canonical factorization of F. By (6.4), we have the commutative diagram

(6.14)

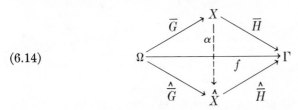

where $\alpha: X \xrightarrow{\sim} \hat{X}$ is the *unique* vector-space isomorphism which is compatible with the commutativity of the diagram. This proves the last two equivalence relations (1.7). To prove the first, we combine each of the commutative diagrams

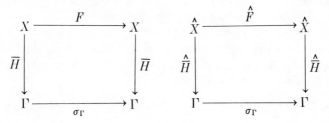

and two of the commutative diagrams

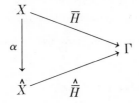

into the diagram

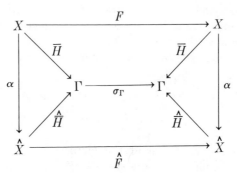

We *must show* that the square commutes. By commutativity of the component diagrams, we see that

$$\hat{\hat{H}} \circ \alpha \circ F = \bar{H} \circ F,$$
$$= \sigma_\Gamma \circ \bar{H},$$
$$= \sigma_\Gamma \circ \hat{\hat{H}} \circ \alpha,$$
$$= \hat{\hat{H}} \circ \hat{F} \circ \alpha.$$

Since $\hat{\hat{H}}$ is one-to-one, $\alpha \circ F = F \circ \alpha$, and the square commutes. This implies (1.7a), since α is an isomorphism.

(6.15) **Remark.** This alternate procedure differs from the "standard" proof given before only in a technical sense. For instance, here the existence of F is deduced from σ_Γ; in module theory σ_Γ is used to put the "right" module structure on Γ, and then F may be represented as the map induced from the scalar product on Γ (see below). Moreover, the existence of the various diagrams we use to prove all the needed relations follows, fundamentally speaking, from the existence of the module structure guaranteed by Theorem (4.9). It is largely a matter of personal psychology whether one is more successful with one method than with another. The diagram method can be very effective in certain contexts; see, for instance, the second proof of (11.14).

(6.16) **Remark.** There is no profound reason for taking

$$X_f = \Omega/\text{kernel}\, f$$

as the fundamental definition of the state space. We could take equally well $\Xi_f = f(\Omega)$. The two are *naturally* isomorphic, since

$$[\omega]_f \ (\text{subspace of } \Omega) \approx f(\omega) \ (\text{element of } \Gamma).$$

Instead of defining state transitions as

$$F\colon [\omega]_f \mapsto [z \cdot \omega]_f = z \cdot [\omega]_f \qquad (\cdot \text{ in } \Omega),$$

we can also define them as

$$F: f(\omega) \longmapsto f(z \cdot \omega) = z \cdot f(\omega) \qquad (\cdot \text{ in } \Gamma).$$

Since a completely detailed exposition of the situation has not been published before, it is not surprising that about half the literature uses one formalism and the remaining half uses the other.

It may seem that the abstract attack employed in this section is rather remote from conventional mathematics or concrete numerical algorithms. This is a decidedly false impression. Important applications to polynomial matrices and structure theory of linear systems will be developed in Section 10.10. Then we show in Section 10.11 that Definition (6.1) leads quite directly to a highly ingenious algorithm for numerical computation of a realization.

(6.17) **Historical notes.** The initial proof of these results, obtained in 1962, was clumsy, and only the results were published in [KALMAN, 1963c]. The proof was based on direct manipulations of the integrals in Theorems (2.24) and (6.6) of Chapter 2 and is reproduced in Appendix C of this chapter: See also the discussion of Theorem (13.19).

Shortly after the publication of [KALMAN, 1963c] it became apparent that the whole apparatus could be sharpened by adopting some of the notions of automata theory, especially the Nerode equivalence relation. This is explored in Chapter 6, based on work [ARBIB, 1965] completed in mid-1964. A pertinent comment may be found also in [KALMAN, 1963c, p. 169, immediately following theorem 7]. The module approach was worked out in the spring and summer of 1965 and first published in [KALMAN, 1965b].

After 1963, under pressure of public opinion, it became fashionable to call the "natural" realization "minimal," because the state space of a natural realization has minimal *dimension* and because this characterizes the natural realization completely (see [YOULA, 1966, p. 534, after equation (45)]). The name "irreducible" has been used also, because a realization that is not natural can be "reduced" by throwing away superfluous states (see [KALMAN, 1965a]). It seems, in retrospect, that neither terminology is a satisfactory alternative to "natural" or "canonical." "Minimal" dimension makes sense only in the vector-space context, but the theory is actually of much wider applicability. "Reduction" of realizations is an interesting problem, but its theory is deeper than the questions connected in a narrow sense with properties or construction of natural realizations.

The exposition of realization theory adopted here (and continued in Sections 10.10, 10.11, and 10.13) is one of the main fruits (so far?) of the interaction of linear system theory and automata theory through modern algebra. From a fundamental scientific point of view, it is not very likely that the picture presented here can be substantially improved or simplified.

For additional motivation, development of ideas, suggestions for further investigations, and so on, the original sources may still be read with profit.

10.7 Cyclic modules

The preceding developments were directed toward illuminating the relationship between module structure and dynamics. We were mainly concerned with establishing equivalences between primary definitions. Now we begin to exploit mathematical properties of modules.

For the moment, let R be an arbitrary (not necessarily commutative) ring and X an arbitrary R-module. Let Y be any subset (not necessarily a submodule) of X. We recall: the *annihilator* A_Y of Y is the subset

$$(7.1) \qquad A_Y = \{r: r \in R \text{ and } r \cdot y = 0 \text{ for all } y \in Y\}$$

of R. The definition shows that A_Y is a (left) ideal.

If there is an $r_x \neq 0$ for each x in X such that $r_x \cdot x = 0$, we say that X is a *torsion* module. (This terminology comes from algebraic topology and has no intuitive significance in the present context.)

We are interested mainly in the ring $R = K[z]$, in which case R is a principal-ideal domain. We have then the basic

(7.3) **Definition.** *Let X be an arbitrary torsion module over $K[z]$. Write $A_X = K[z]\psi_X$; if $\psi_X \neq 0$, we call it the* **minimal** (*or* **annihilating**) *polynomial of X.*†

In Section 10.8 we shall show that this terminology is in strict agreement with the definition of the minimal polynomial of a matrix. (The latter notion was first used in a system-theoretic context in Section 2.3.)

(7.4) **Proposition.** *Let X be a finite $K[z]$-module generated by m elements and let $\psi_X \neq 0$. Then* $\dim X$ *(as a vector space)* $\leq m \cdot \deg \psi_X$.

Proof. Every element x of X is a sum of terms $\pi_k \cdot g_k$ ($k = 1, \ldots, m$). We can reduce each π_k modulo ψ_X without altering the sum. Hence $x \in X$ is determined by m polynomials of degree $< \deg \psi_X$. ☐

Later we shall develop an exact formula for $\dim X$. For the time

---

† It is not yet clear that "X = finite torsion module" implies "$\psi_X \neq 0$" when $R = K[z]$, or, more generally, when R = principal-ideal domain. We do not need this result here; but it does in fact follow from the fundamental structure theorem (8.1).

being, however, we are interested in exploring those modules whose properties are *wholly* determined by their minimal polynomial.

We recall that X is a *cyclic*† R-module iff it is generated by a single element g. The structure of cyclic modules is very simple:

(7.5) **Proposition.** *Let X be a cyclic R-module with generator g over an arbitrary ring R. Then X is isomorphic with the quotient ring R/A_g.*

Proof. Consider the R-homomorphism $\rho: R \to X: r \mapsto r \cdot g$ (here R is viewed as an R-module over itself). "Cyclic" means that ρ is onto. Moreover, kernel $\rho = A_g$. Hence the conclusion follows at once by the homomorphism theorem. □

This easy result has very interesting and basic consequences in the dynamical context.

Suppose $m = 1$. Then $\Omega = K[z]$ is a free cyclic module which is generated by 1 (the polynomial which is identically 1). Then, for any input/output map f the $K[z]$-module $X_f = \Omega/\text{kernel } f$ is also cyclic and is generated by $g = [1]_f$. In view of (7.5), $A_g = \text{kernel } f = K[z]\psi_f$ consists of all polynomials ω such that $f(\omega) = 0$; $\psi_f \neq 0$ is the polynomial of least degree in this set. So we have the following successively more explicit equivalent descriptions of X_f:

$$X_f = K[z]/\text{kernel } f,$$
$$= K[z]/K[z]\psi_f,$$
$$= \{\{\omega + \pi\psi_f: \pi \in K[z]\}: \omega \in K^m[z]\}.$$

Each equivalence class $\{\omega + \pi\psi_f\}$ has a *unique* representative $\bar{\omega}$ of least degree, which is the remainder of ω after division by ψ_f. Clearly, $\deg \bar{\omega} < \deg \psi_f$. We summarize our observations as the

(7.6) **Representation theorem.** *If X is a cyclic $K[z]$-module with annihilating polynomial ψ, then $X \approx \mathcal{P}_{\deg \psi} = \{$all polynomials $\pi \in K[z]$ with $\deg \pi < \deg \psi\}$. In particular, this is always true if $m = 1$ and $X = X_f$.*

(7.7) **Notation.** To simplify certain formulas we may sometimes be forced to write $\omega \bmod \psi$ instead of $\bar{\omega}$, even though the former notation should be used only for the equivalence *class* mod ψ, not for a special representative of that class.

(7.8) **Remark.** The dynamical behavior of a cyclic X_f is described essentially by the map: inputs $\to$ state, which is written as

$$\omega \mapsto [\omega]_f = \omega \bmod \psi_f = \bar{\omega}.$$

---

† This terminology is archaic, misleading, but entrenched. It would be better to use a literal translation of Bourbaki, who calls cyclic modules "monogenic."

We may view X_f intuitively as a "pattern-recognition" device: an input ω presented to X_f is stored in the form of its remainder $\bar{\omega}$ after division by ψ_f. Since ψ_f may be quite complicated, the stored pattern $\bar{\omega}$ may bear no *obvious* resemblance to the input pattern ω, but in the algebraic sense the operation of the system is very simple.

Some interesting special cases of minimal polynomials are provided by the following five examples.

(7.9)   **Example.** $\psi_f(z) = z$. Then $[\omega]_f = \bar{\omega} = \omega_0$, the last received input value. Such a system has a "memory" extending over just one unit of time.

(7.10)   **Example.** $\psi_f(z) = z^k$. Then $[\omega]_f = \bar{\omega} = \omega_0 + \cdots + \omega_{k-1}z^{k-1}$. This system "remembers" the last k input values.

(7.11)   **Example.** $\psi_f(z) = z - 1$. This means that $z = 1 \bmod \psi_f$. Consequently

$$[\omega]_f = \bar{\omega} = \omega_0 + \cdots + \omega_r \qquad (r = \deg \omega).$$

This system is an "averagor" or "integrator."

(7.12)   **Example.** $\psi_f(z) = z^k - 1$. Then $z^{kl}\omega = \omega \bmod \psi_f$ for any $l \geq 0$. This means that the system is sensitive to periodic patterns of period k: such patterns are reinforced, whereas nonperiodic patterns tend to be averaged out. (*Exercise:* Write down the explicit formula for $[\omega]_f$.)

(7.13)   **Example.** $\psi_f(z) = z^k - \alpha$. Here $z^{kl}\omega = \alpha^l\omega \bmod \psi_f$. This means that the system is still sensitive to periodic inputs of period k, but past inputs can be enhanced or deemphasized depending on whether $\alpha > 1$ or $\alpha < 1$.

Let us now consider some typical problems which can be formulated and solved very nicely with the present machinery. We assume in each case that X is a cyclic $K[z]$-module with generator g and annihilator ψ. In view of (7.6) we can write

$$x = \xi \cdot g = \bar{\xi} \cdot g$$

if $\xi = \bar{\xi} \bmod \psi$.

(7.14)   **Problem.**   *Compute a sequence ω which transforms x to 0.* That is, ω is the solution of $x \circ \omega = 0$. In module notation (see (5.4))

$$x \circ \omega = (z^{1+d}\xi + \omega) \cdot g = 0,$$

where $x = \xi \cdot g$ and d is an arbitrary integer satisfying $\geq \deg \omega$.†   Since

---

† The possibility that $d > \deg \omega$ corresponds to taking the first $d - \deg \omega$ values of the input equal to zero. We must consider this case, and there is no other way of expressing it via the module notation.

$1 \cdot g \neq 0$ (otherwise everything is trivial), it is clear that ω must satisfy

(7.15) $$z^{1+d}\xi + \omega = 0 \bmod \psi, \qquad d \geqq \deg \omega.$$

This may appear to be a difficult equation to solve since $\deg \omega$ enters nonlinearly (in the exponent). But, since the equation is required to hold only mod ψ, we may choose $\deg \omega = r = d$ arbitrarily subject to $r \geqq n - 1$. Then a solution is

$$\omega = -z^{1+r}\xi + \nu\psi_f,$$

with $\nu \in K[z]$ chosen so that $\deg \omega$ is indeed r. The inequality $d \geqq \deg \omega$ turns out to be irrelevant.

If we want a solution of (7.15) such that $\deg \omega =$ minimum, the situation is much more difficult. (This corresponds to the simplest discrete-time time-optimal control problem.) We must take

$$\deg \omega = 1 \qquad \text{iff} \qquad \deg \widetilde{z^2\xi} = 1,$$
$$\deg \omega = 2 \qquad \text{iff} \qquad \deg \widetilde{z^3\xi} = 2,$$

and so on. (*Exercise:* Work out all details for various interesting ψ's.)

(7.16) **Problem.** *Find all periodic motions through x of period q.* We seek all ω's of degree $\leqq q - 1$ such that

$$x \circ \omega = x.$$

In module notation, with $x = \xi \cdot g$, this condition is equivalent to

$$z^q\xi + \omega = \xi \bmod \psi.$$

This implies

(7.17) $$\omega = (1 - z^q)\xi \bmod \psi.$$

If $q \geqq \deg \psi$, the solution is easy, as before. If $q < n$ and ξ is fixed, a solution may fail to exist, unless $z^q = 1 \bmod \psi$, in which case all states have period q under the free motion.

It is interesting to consider also the converse

(7.18) **Problem.** *Find all states periodic under an input ω.* This problem is equivalent to solving (7.17) for ξ given ω. For this we require that $(z^q - 1) \neq 0$ be a unit (invertible element) in $K[z]/K[z]\psi$. We recall: in this ring an element ν is a unit iff it is not a divisor of zero and ν is a divisor of zero if $\nu|\psi$ (see (A.13)). If ω is a unit and $z^q - 1$ is a nonunit, ξ cannot exist. Otherwise the situation is a little more complicated; the complete solution is left as an exercise.

(7.19) **Problem.** *Build a machine that recognizes an input φ but does* *not respond to an input θ.* This is a pattern-discrimination problem. In module language, we seek a ψ such that $\psi \nmid \varphi$ but $\psi | \theta$. A solution is possible iff $\theta \nmid \varphi$.

(7.20) **Problem.** *Find "the" machine that recognizes φ as the state $\bar{\varphi} \cdot g$.* Here ψ must satisfy $\varphi = \bar{\varphi}$ mod ψ, which is equivalent to

$$\psi | (\varphi - \bar{\varphi}), \qquad \deg \psi > \deg \bar{\varphi}.$$

A solution certainly exists, but in general it will not be unique.

(7.21) **Remark.** There is a close connection between cyclicity and the "control canonical form" (5.8) of Chapter 2. Consider a linear dynamical system Σ and assume that X_Σ is a cyclic $K[z]$-module with generator g_Σ and annihilator ψ_Σ. Then $X \approx \mathcal{O}_{\deg \psi_\Sigma}$. Hence X_Σ, as a vector space, admits the basis $\{g, z \cdot g, \ldots, z^{n-1} \cdot g\}$, with $n = \deg \psi_\Sigma$. Corresponding to this basis, the operator $x \mapsto z \cdot x$ is represented by the matrix

$$(7.22) \qquad F = \begin{bmatrix} 0 & 0 & \cdots & 0 & -\alpha_n \\ 1 & 0 & \cdots & 0 & -\alpha_{n-1} \\ \cdot & \cdot & & \cdot & \cdot \\ \cdot & \cdot & & \cdot & \cdot \\ \cdot & \cdot & & \cdot & \cdot \\ 0 & 0 & \cdots & 1 & -\alpha_1 \end{bmatrix}$$

where the elements of the last column are minus the coefficients of the polynomial ψ. This follows immediately by writing ψ in the form

$$z^n = -\alpha_1 z^{n-1} - \cdots - \alpha_n \text{ mod } \psi.$$

Observe that the "mod ψ" operation is the module version of the Cayley-Hamilton theorem. (The "control canonical form" (5.8) of Chapter 2 is deduced by a similar reasoning.)

Closely related to all this is the classical

(7.23) **Criterion.** *The characteristic polynomial χ_F of a square matrix F over K is equal to its minimal polynomial ψ_F if and only if there is a basis in K^n such that F has the form (7.22); that is, if and only if F is similar to (7.22).*

Proof. Sufficiency. Let F be a matrix equal to or similar to (7.22). The polynomial ψ defined via the last column of (7.22) is clearly the polynomial of least degree such that $\psi \cdot x = 0$ for all x (or, equivalently, such that $\psi(F) = 0$). Hence $\psi = \psi_F$, the minimal polynomial of F. (See also (3.6), Chapter 2.) A direct computation based on (7.22) shows that

$$\psi(z) = \det (zI - F) = \chi_F(z),$$

the characteristic polynomial of F (recall that χ_F is independent of the basis in which F is exhibited).

Necessity. Instead of a fixed matrix F, we can consider, equivalently, the $K[z]$-module X_F induced by F (see Proposition (5.1)). Suppose a basis of the required type does not exist. Then every set $\{x, z \cdot x, \ldots, z^{n-1} \cdot x\}$ is linearly dependent. That is, every x is annihilated by polynomials in $A_x = K[z]\pi_x$, where deg $\pi_x < n$. Clearly X_F is generated by a finite set $\{g_1, \ldots, g_l\}$ where $l \leq n$. We claim that the annihilator of X_F, $A_{X_F} = K[z]\psi_F$, is generated by a polynomial ψ_F which is the least common multiple of the polynomials $\{\pi_{g_1}, \ldots, \pi_{g_l}\}$. This is immediate from the fact that in this situation all annihilators are principal ideals: since ψ_F annihilates g_i, $\psi_F \in A_{g_i}$ and hence $\pi_{g_i}|\psi_F$ for all $i = 1, \ldots, l$; on the other hand, every common multiple of $\{\pi_{g_1}, \ldots, \pi_{g_l}\}$ annihilates all of X_F. Now consider the vectors

$$\hat{g}_1 = g_1, \; \hat{g}_2 = g_1 + \alpha_2 g_2, \; \ldots, \; \hat{g}_l = g_1 + \alpha_2 g_2 + \cdots + \alpha_l g_l,$$

where the $\alpha_k \in K$ are chosen in such a way that each vector $\hat{g}_k$ is nonzero. (This is obviously always possible.) By induction, it is easily proved that $\pi_{\hat{g}_k} = $ least common multiple of $\{\pi_{g_1}, \ldots, \pi_{g_k}\}$. Hence $\pi_{\hat{g}_l} = \psi_F$, and, by hypothesis, deg $\psi_F > \pi = \deg \chi_F$. This contradiction proves the necessity of our criterion. ☐

(7.24) **Exercise.** Find a matrix which fails to satisfy criterion (7.23).

(7.25) **Remark.** Henceforth we shall call a matrix F *cyclic* iff F satisfies (7.23). So F is cyclic iff the $K[z]$-module X_F induced by F via $z \cdot x = Fx$ (see Proposition (5.1)) is cyclic for some $g \in K^n$.

(7.26) **Remark.** In view of the representation theorem (7.6), *the states of a dynamical system may be described in the same language as the inputs, namely as polynomials.* In this sense, the present approach to linear system theory eliminates the apparent incompatibility between the Laplace-transform and the state-variable methods of system analysis.

(7.27) **Remark.** It is embarrassing to have to observe that the state space X_f has even more structure than is claimed in (7.6). Not only is X_f an abelian group, but, at least if $m = 1$, it is even a ring with multiplication

(7.28) $$[\omega]_f \cdot [\omega']_f = [\omega\omega']_f,$$

where $\omega\omega'$ is the ordinary product in $K[z]$. The product (7.28) is well defined, because $\omega \cdot [\omega']_f = [\omega\omega']_f$ is a well-defined scalar product on X_f. No system-theoretic interpretation of (7.28) is known at present, although the same phenomenon has also been observed in automata theory [DAY and WALLACE, 1967].

In the next section we shall see that the preceding results may be generalized from the cyclic case to arbitrary finite torsion modules.

10.8   The structure of finite $K[z]$-modules

The central result in the theory of modules over a principal-ideal domain is that *every* such module is a simple combination, namely a sum, of cyclic modules. In more intuitive terms, we may say that linear systems are built up as sums of the simplest building blocks, which are cyclic systems.

The conventional approach to linear system theory has always stressed the importance of scalar linear systems (those with $m = 1$, $p = 1$), which are necessarily cyclic. What was missing in this theory is a systematic way of treating the most general case. Some attempts in this direction go back to the early 1940s, but no complete theory of linear systems was available prior to the introduction of the present module-theoretic viewpoint. In other words, *the full complexity of linear systems is most clearly understood via the module machinery; conversely, an understanding of this complexity is equivalent to an implicit knowledge of the module machinery.*

Let us now give a precise statement of the main

(8.1) **Structure theorem for finite modules over a principal-ideal domain R.** *Let X be such an R-module. Then X is isomorphic to a direct sum of cyclic modules*

$$X \approx R/R\psi_1 \oplus \cdots \oplus R/R\psi_q \oplus R \oplus \cdots \oplus R,$$

where the number of pieces in the direct sum is at most equal to the number of generators of X, the $\psi_i \in R$ are unique modulo units in R and satisfy $\psi_{i+1}|\psi_i$, $i = 1, \ldots, q - 1$.

(8.2) **Definition.** *The $\psi_i \in R$ are known as the* **invariant factors** *of X.*

For instructive proofs of this basic result we refer the reader to [CURTIS and REINER, 1962, sec. 16; BOURBAKI, Algèbre, chap. 7; LANG, 1965, chap. XV, sec. 2; JACOBSON, 1953, chap. 3]. These proofs are given in very different styles; reading *each* of them is truly a worthwhile exercise. The theorem itself is a direct transliteration into module language of the classical invariant-factor theorem (see (A.16)). In Section 10.10 we shall apply the algorithm used in proving this theorem to the computation of the invariant factors of X and the isomorphism (8.2). These computations will constitute an indirect proof of Theorem (8.1).

From a purely mathematical standpoint (8.1) might seem to be a dull exercise in abstraction, since the entire proof (see especially the exposition of [CURTIS and REINER, 1962]) depends critically on the invariant-factor algorithm. In *our* context, however, the module version of the theorem is essential; the algorithm is of interest only as a computing device. So the mathematical trend in the direction of greater abstraction (vector spaces to modules) is also motivated by its applications to concrete problems in dynamical systems.

(8.3) Interpretation. The general form of the structure theorem as given above includes the case when X is a "mixture" of torsion cyclic modules (the $R/R\psi_i$) and free cyclic modules (the R's). If X is a torsion module, the latter terms are missing, and, by the divisibility condition, the annihilating polynomial ψ of X is the same as ψ_1. Clearly X *is cyclic iff* $q = 1$.

(8.4) Remark. "Finite" corresponds to "finitely many input terminals"; this restriction does not significantly limit the applicability of the theorem. But the assumption R = principal-ideal domain is rather undesirable, since it rules out, for instance, $R = \mathbf{Z}[z]$ (see Exercise (A.15)). Although more general structure theorems are not yet available in the mathematical literature, there are strong indications that such theorems can be deduced via the system-theoretic approach.

Let us note an immediate and highly useful consequence of (8.1), which is an improvement over (7.4):

(8.5) Corollary. *Let* X *be a torsion module over* $K[z]$ *with annihilating polynomial* ψ. *Then*

$$\dim X \ (\textit{as a K-vector space}) = \sum_{i=1}^{q} \deg \psi_i \leqq q \cdot \deg \psi.$$

Proof. Immediate from (7.6). $\square$

There is a second structure theorem which generalizes the method of partial-fraction expansions in the theory of the Laplace transform.

We write, for any module X over a *commutative* ring R,

$$X_r = \{x \colon x \in X,\ r^q \cdot x = 0 \text{ for fixed } r \in R,\ \text{some } q > 0\}.$$

It is easily verified (using commutativity) that X_r is an R-module. For instance, if $X \approx R/Rr^s$, then $X = X_r$.

(8.6) Structure theorem for finite torsion modules over a principal-ideal domain R. *Suppose* X *is such an R-module with annihilating polynomial* ψ. *Let*

$$\psi = \epsilon \rho_1^{n_1} \cdots \rho_s^{n_s} \qquad (\epsilon = \textit{unit} \in R)$$

be a representation of ψ *by prime factors* ρ_i.
Then X *is the direct sum of the torsion modules* X_{ρ_i}.

If $R = K[z]$, then a prime element of R is an irreducible polynomial in $K[z]$. In particular, if $K = \mathbf{R}$, then the ρ_i are polynomials of degree one or two depending on whether the roots of ρ_i are real or complex; if $K = \mathbf{C}$, then each ρ_i is a polynomial of degree one.

The proof of (8.5) is straightforward. The reader is referred to [BOURBAKI, Algèbre, chap. 7, sec. 2, N° 1; LANG, 1965, chap. XV, sec. 2].

Theorems (8.1) and (8.6) are essentially all that we have to know to reduce the study of general linear systems to that of simple linear systems. For instance, applying first (8.6) and then (8.1) gives the abstract counterpart of a partial-fraction expansion with repeated roots. See also [VAN DER WAERDEN, 1931, §88].

The invariant factors of a $K[z]$-module can be used to define corresponding invariants for matrices over K. (Note: historically, exactly the opposite has happened!) This is just a generalization of Remark (7.25):

(8.7) **Definition.** *Let $A : V \to V$ be an arbitrary endomorphism of the K-vector space V (or its matrix representation) and let V_A be the $K[z]$-module induced over V by A (see Proposition (5.1)). The* **invariant factors** *of A are the invariant factors of V_A.*

As is well known [VAN DER WAERDEN, 1931, vol. 2], the invariant factors of A determine the Jordan canonical form of A. A large part of linear system theory could be carried through by reducing everything to the computation of the Jordan forms of matrices; with the module-theoretic approach we can use the invariant factors *directly*, rather than through the intermediary of the Jordan canonical form. The very terminology "invariant" (independent of basis) is clearest when the ψ_i are defined through modules.

10.9   Transfer functions

One of the most important results in the conventional theory of linear systems is the existence of a transfer function associated with the input/output map f (actually, the "transform" of $f(1)$). The usual discussion of transfer functions relies on convergence arguments. As a result, the student is sometimes given the false impression that only stable systems can be studied by transform methods.

We shall see, however, that the existence of transfer functions in the finite-dimensional case is a purely algebraic fact, having nothing whatever to do with questions of convergence. The essential point is the existence of a $K[z]$-module structure on Γ for which the input/output map f is a module homomorphism (see Section 10.3). More explicitly:

since f is a $K[z]$-homomorphism, we have

$$f(\omega) = \sum_{k=1}^{m} \omega_k f(e_k).$$

This shows that f is completely described by giving the elements $\epsilon_i = f(e_i)$ in Γ. Each ϵ_i is a formal power series. The idea of "transfer functions" is equivalent to describing each of these power series as the formal expansion of a ratio of a polynomial vector by a scalar polynomial.

Our basic assumption is:

(9.1) X_f *is a torsion module with annihilating polynomial* ψ_f.

From the constructions developed earlier, we know that f admits the factorization expressed by the commutative diagram

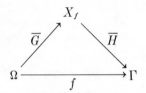

of $K[z]$-homomorphisms. Let us consider a generator e_k $(k = 1, \ldots, m)$ of the free $K[z]$-module Ω. From the diagram we get

$$\begin{aligned}
f(\psi_f \cdot e_k) &= (\bar{H}_f \circ \bar{G}_f)(\psi_f \cdot e_k) && \text{(factorization of } f\text{),} \\
&= \bar{H}_f(\psi_f \cdot \bar{G}_f(e_k)) && (\bar{G} \text{ is } K[z]\text{-homomorphism),} \\
&= 0 && (\psi_f \text{ annihilates } X_f).
\end{aligned}$$

On the other hand,

$$f(\psi_f \cdot e_k) = \psi_f \cdot f(e_k) \qquad (f \text{ is } K[z]\text{-homomorphism).}$$

Hence $\psi_f \cdot f(e_k) = 0$ as an element of the $K[z]$-module Γ. By the definition of scalar multiplication in Γ, we conclude that $\psi_f f(e_k)$ (the ordinary product of a formal power series by a polynomial) belongs to $K^p[z]$. We write $(\psi_f f(e_k))(z) = \theta_k(z)$, $\theta_k \in K^p[z]$.

The desired result will follow easily from the following simple

(9.2) **Lemma.** *Let* $\gamma \in \Gamma \subset K^p[[z^{-1}]]$, $\pi \in K[z]$, $\delta \in K^p[z]$, *and* $\deg \delta$ $< \deg \pi$. *Then the identity*

$$\pi\gamma = \delta$$

is equivalent to

$$\gamma = \delta/\pi,$$

where / *means formal long division into positive powers of* z^{-1}.

Proof. Write out both identities in detail and verify their equivalence term by term. ☐

So we can identify $f(e_k) = \epsilon_k$ with the ratio $w_k = \theta_k/\psi_f$. Notice that, in view of $w_k \in \Gamma$, we must have deg $\theta_k <$ deg ψ_f; that is, the vector w_k is a proper rational function (it takes the value 0 at $z = \infty$). Further, if ψ is the least common denominator of all the w_k, then $\psi_f = \psi$.

Conversely, a finite family of proper rational functions $w_k \in \Gamma$, $k = 1, \ldots, m$, induces a $K[z]$-homomorphism

(9.3) $$f: \Omega \to \Gamma: e_k \mapsto w_k.$$

Summarizing, we have the following important

(9.4) **Representation theorem.** *Let* $f: \Omega \to \Gamma$ *be any* $K[z]$-*homomorphism with annihilating polynomial* ψ_f. *Then* f *is uniquely determined by its* $p \times m$ **transfer-function matrix**† W_f *whose columns are the p-vector proper rational functions* $w_k = f(e_k), k = 1, \ldots, m$, *in the indeterminate z.* *Conversely, any matrix* W *with proper rational elements induces a unique* $K[z]$-*homomorphism* f_W *according to rule* (9.3), *and the least common denominator* ψ_W *of* W *is the annihilating polynomial* ψ_f *of the* $K[z]$-*module* X_f *induced by* f_W.

In short: f *and* W *are equivalent objects.*‡

To compute $f(\omega)$ we may henceforth proceed according to the following rules:

1. *Multiply* W_f *by* ω (*ordinary multiplication of rational matrix by polynomial vector*).

2. *Expand by long division into ascending powers of* z^{-1}.

3. *Discard all terms involving nonpositive powers of* z^{-1}; *the result is just* $f(\omega)$.

The terms thrown away in step 3 correspond to the output of the system up to and including $t = 0$. More precisely, these terms are given by Δ in

(9.5) $$W_f\omega = \Theta_f\omega/\psi_f = \Delta + \widetilde{\Theta_f\omega}/\psi_f,$$

where $\Theta_f = (f(e_1)\psi_f, \ldots, f(e_m)\psi_f)$ is a $p \times m$ matrix of polynomials and $\sim$ is the operation of picking the representative of minimal degree in the equivalence class of polynomials modulo ψ_f. (Recall (7.6).) This shows that $f(\omega)$ depends only on $\Theta_f\omega$ modulo ψ_f. Hence we may simplify the above rules for computing $f(\omega)$ by replacing step 1 by

1'. *Multiply* $W_f = \Theta_f/\psi_f$ *by* ω, *the product being computed in the ring* $K[z]/K[z]\psi_f$, *always working with representatives of least degree in each equivalence class.*

---

† The classical terminology "transfer function" (sometimes also known as "system function") is not good but it is very entrenched.

‡ When talking of transfer-function matrices it is always assumed (if the question arises at all) that elements of W are ratios of *relatively prime* polynomials.

Then step 2 is the same as before, but step 3 is no longer necessary.

(9.6)   **Conclusion.**   *"Transfer functions" and "modulo ψ operations" are intimately related.*

(9.7)   **Notation.**   By analogy with the notation conventions for Γ, let us agree that

$W\omega$ = ordinary multiplication of power series by polynomial,

$W \cdot \omega$ = the preceding product viewed as a power series in which all terms with nonpositive powers of z^{-1} are deleted.

The computing rules just explained are equivalent to those arising in the conventional theory of the z-transform. All the usual results [FREEMAN, 1961] are therefore valid for our setup as well. Let us mention an interesting problem illustrating such calculations.

(9.8)   **Problem.**   *Given an input ω (deg $\omega = -t_{-1}$) in Ω, find an initial state x in X_f (X_f torsion module with annihilator ψ_f) at $t = t_{-1}$ such that the output of the system is identically zero for all $t > t_{-1}$.* Since X_f is completely reachable, any $x \in X_f$ may be represented as $x = [\xi]_f$ for some $\xi \in \Omega$. So the condition of the problem may be expressed as

$$z^{-\deg \omega} W_f\omega + W_f \cdot \xi = 0.$$

In words: $z^{-\deg \omega} W_f\omega$ must be an output function of the system; that is, it must belong to $f(\Omega)$. A more explicit equivalent condition is

(9.9) $$\Theta_f\omega = -z^{\deg \omega}\Theta_f\xi.$$

To simplify matters, we consider only the scalar case of this equation; that is, we assume that $m = p = 1$, and we write $\Theta_f = \theta_f \in K[z]$. (The general case requires the theory developed in the last section and is left as a nontrivial exercise for the reader.) An obvious necessary condition for the existence of ξ is that $z^{\deg \omega} | \theta_f\omega$. Then $\sigma = \theta_f\omega/z^{\deg \omega} \in K[z]$ and, of course, deg $\sigma <$ deg ψ_f. We must solve

$$\sigma = -\theta_f\xi \bmod \psi_f.$$

Now, θ_f is relatively prime to ψ_f, because otherwise ψ_f is not the *minimal* polynomial of the module X_f. Hence (see (A.13)) θ_f is a unit in the ring $K[z]/K[z]\psi_f$. It follows that θ_f^{-1} exists in this ring and is represented by a polynomial of degree less than deg ψ_f. So

$$\begin{aligned} \xi &= \theta_f^{-1}\sigma \bmod \psi_f, \\ &= \theta_f^{-1}\sigma + \alpha\psi_f, \qquad \alpha \in K[z]. \end{aligned}$$

Hence

$$\theta_f\xi = \theta_f(\theta_f^{-1}\sigma + \alpha\psi_f) = \sigma + \beta\psi_f;$$

this implies that $\theta_f\xi = \sigma$. So we have proved: *our problem has a solu-*

tion iff $\theta_f\omega$ *contains the factor* $z^{\deg \omega}$.   This is true, in particular, whenever $\omega(z) = \kappa z^q$, $\kappa \in K$, $q \geqq 0$.

10.10   Applications of the invariant-factor algorithm

The structure theorem (8.1) for finite modules over a principal-ideal domain first appeared in mathematics in the form of the so-called invariant-factor theorem for polynomial matrices (see (A.16)). For us the importance of this result lies in the fact that its proof provides an *algorithm* for computing the invariant factors, and knowledge of the invariant factors will lead to the determination of the structure of linear systems.

We shall now develop the connections between the abstract theory of Section 10.8 and the invariant-factor algorithm. We shall be interested mainly in questions related to the theory of realization. The reader should compare the present treatment with the abstract proofs in Section 10.6 and also with the linear-algebraic algorithm of the next section, which does *not* require knowledge of the invariant factors.

We begin with an elementary

(10.1)   **Remark.**   Recall that $\psi = \psi_W$ is the least common denominator of the proper rational matrix W (Theorem (9.4)). So ψW is a *polynomial* matrix. By the invariant-factor theorem (A.16), we have the representation

(10.2) $\psi W = A\Lambda B$,   (det A, det B = units in $K[z]$; Λ = diagonal and unique up to units in $K[z]$.)

A forteriori, we also have a representation

(10.3) $\psi W = PLQ \bmod \psi$ (det P, det Q = units in $K[z]/K[z]\psi$; L = diagonal and unique up to units in $K[z]/K[z]\psi$.)

In fact, (10.3) follows from (10.2) trivially, by replacing each polynomial π in (10.2) by its canonical representative of least degree $\tilde{\pi}$ in the equivalence class $\pi \bmod \psi_W$. So

(10.4) $\psi W = \psi\widetilde{W} = \tilde{A}\tilde{\Lambda}\tilde{B} + \psi V = PLQ + \psi V$,

where V is a suitable polynomial matrix. In the sequel we shall need mainly the weaker representation (10.3). This affords a certain flexibility in our computations. For instance, the invariant-factor algorithm may be carried out over the ring $K[z]/K[z]\psi$ rather than over the ring $K[z]$. (*Exercise:* investigate the programming modifications and savings in computer time resulting from this observation.)

From the representation (10.3) we can quickly deduce all the invariants of the module X_{f_W} associated with W. The steps are actually just rearrangements of the terms in (10.3).   The basic idea is to compute the equivalence class $[0]_{f_W}$ of the input/output map $f_W: \omega \mapsto W \cdot \omega$ corresponding to W.

(10.5) Computation.

1. An input $\omega \in K^m[z]$ to X_{f_W} sets up the state 0 iff $\bar{G}(\omega) = 0$; or iff $W \cdot \omega = 0$; or iff $W\omega \in K^p[z]$; or iff $(\psi W)\omega = 0 \bmod \psi$.

2. By (10.2), $(\psi W)\omega = 0 \bmod \psi$ iff $\Lambda Q \omega = 0 \bmod \psi$, since $\det P$ is a unit in $K[z]$, hence a unit in $K[z]/K[z]\psi$.

3. Change variables by setting $\omega \mapsto \hat{\omega} = Q\omega$; this map is one-to-one, since $\det Q$ is a unit in $K[z]/K[z]\psi$.   More precisely: $\omega \mapsto Q\omega$ is a $K[z]$-module isomorphism.

4. Now $\bar{G}(\omega) = 0$ is equivalent to $\tilde{\Lambda}\hat{\omega} = 0 \bmod \psi$.   Write $\tilde{\Lambda} = \mathrm{diag}$ $\{\tilde{\lambda}_1, \ldots, \tilde{\lambda}_q, 0, \ldots, 0\}$.   By the invariant-factor theorem, the λ_i are uniquely determined by W up to units in $K[z]$ and $\lambda_i | \lambda_{i+1}$, $i = 1, \ldots,$ $q - 1$.   By a simple calculation $\tilde{\lambda}_i | \tilde{\lambda}_{i+1}$, $i = 1, \ldots, q - 1$.   Further, let $\mu_i = (\tilde{\lambda}_i, \psi)$ (largest common factor) and let $\psi_i = \psi/\mu_i$.   Then $\tilde{\lambda}_i/\mu_i$ is a unit in $K[z]/K[z]\psi$ iff $\mu_i \neq \psi$ and 0 iff $\mu_i = \psi$.   So there are three cases to consider: (i) $\tilde{\lambda}_i \neq 0$, $i = 1, \ldots, q$; (ii) $\tilde{\lambda}_i = 0$, $i = q + 1, \ldots,$ $r = \mathrm{rank}\ \psi W$; (iii) $\tilde{\lambda}_i = \lambda_i = 0$, $i = r + 1, \ldots, \min(m, p)$.   Consequently $\tilde{\Lambda}\hat{\omega} = 0 \bmod \psi$ iff

$$\hat{\omega}_i = \begin{cases} 0 \bmod \psi_i, & i = 1, \ldots, q; \\ \text{arbitrary}, & i = q + 1, \ldots, \min(m, p). \end{cases}$$

5. The ψ_i satisfy a divisibility condition of the *opposite type* as the $\tilde{\lambda}_i$; thus $\psi_{i+1} | \psi_i$, $i = 1, \ldots, q - 1$.   Moreover, $\psi_1 = \psi$, since ψ is the *least common denominator* of W.

Let us summarize these calculations.   From the fact that $X_f \approx \Omega/[0]_f$ and from (10.5d), we have the following

(10.6) Proposition.   *The elements of X_{f_W} are isomorphic to the q-vectors*

$$\begin{bmatrix} \hat{\omega}_1 & \bmod & \psi_1 \\ & \cdot & \\ & \cdot & \\ & \cdot & \\ \hat{\omega}_q & \bmod & \psi_q \end{bmatrix}.$$

From Section 10.7, we know that for each $\psi \in K[z]$ the set of residue classes $\{\pi \bmod \psi: \pi \in K[z]\}$ is the cyclic $K[z]$-module $K[z]/K[z]\psi$.   Hence it follows that

$$X_f \approx K[z]/K[z]\psi_1 \oplus \cdots \oplus K[z]/K[z]\psi_q$$

In view of the divisibility properties of the ψ_i (see step 5 of (10.5)), *this is just the structure theorem (8.1) for the torsion module X_f.*

We can immediately generalize this result to arbitrary finite $K[z]$-modules with torsion. (If there is no torsion, then W is not rational, and Remark (10.1) is vacuous.) Let X be such a module, generated (as a K-vector space) by $g_1, \ldots, g^m$.† Associate with X in the usual way a dynamical system $\Sigma_X = (F, G, H)$, where

$$F: X \to X: x \mapsto z \cdot x,$$

(10.7) $$G: K^m \to X: (\alpha_1, \ldots, \alpha_m) \mapsto \sum_{k=1}^{m} \alpha_k g_k, \qquad \alpha_k \in K,$$

$$H: X \to K^m: g_k \mapsto e_k.$$

Then Σ_X is completely reachable and completely observable. Hence Σ_X is canonical, and so (by Theorem (6.9)) Σ_X is essentially the same object as its input/output map $\omega \mapsto W \cdot \omega$, which (by Theorem (9.4)) is the same as the transfer-function matrix W_X of Σ_X. We emphasize: the uniqueness theorem of canonical realizations is needed to make sure that the invariant properties of W_X do not depend on the particular system Σ_X chosen to represent X as long as Σ_X is canonical.

From our point of view, the crucial conclusion may be phrased as follows:

(10.8) **Proposition.** *The invariant factors (see (8.2)) of a finite torsion $K[z]$-module X with annihilating polynomial ψ are identical with the invariant factors of the polynomial matrix ψW_X, where W_X is the transfer-function matrix associated with X via (10.7).*

In short, knowing ψ and W of the module X, we can give an algorithm for computing the invariant factors of X. This is one of the main reasons for our interest in transfer-function matrices. Definition (10.7) shows that we can always attach a W to X. For an alternative method of computing the invariant factors via multilinear functions, see [BOURBAKI, Algèbre, Chapter 7, §4, N° 4].

In view of (10.8), we are justified (in fact, forced) to introduce one more definition of invariant factors:

(10.9) **Definition.** *The **invariant factors** of a proper rational matrix W over $K[z]$ are given by the ψ_i constructed in step 4 of (10.5).*

(10.10) **Exercise.** Let $W = \text{diag}(\theta_1/\sigma_1, \ldots, \theta_m/\sigma_m)$. A careful calculation shows that (assuming $(\theta_i, \sigma_i) = 1$, $i = 1, \ldots, m$)

$$\dim W \triangleq \dim X_{f_W} = \sum_{i=1}^{q} \deg \psi_i = \sum_{i=1}^{m} \deg \sigma_i,$$

even though, of course, $\psi_i \neq \sigma_i$ in general.

---

† Recall that "X = finite R-module and $R = K[z]$" implies "dim X (as K-vector space) = finite". See Corollary (8.5).

Now we shall apply this machinery to a number of different but related problems.

a.   Invariant factors of a square matrix.   We want to compute the invariant factors of an $n \times n$ matrix (or K-endomorphism) $F: X \to X$ in accordance with Definition (8.7). First we associate the dynamical system $\Sigma_F = (F, I, I)$ with F. Σ is clearly canonical. Hence Σ is in one-to-one correspondence with its transfer-function matrix $W_F = (zI - F)^{-1}$. By the invariant-factor theorem and (10.2), we have

$$(10a.1) \qquad\qquad zI - F = A(z)\Lambda(z)B(z),$$

where $\Lambda = \text{diag } (\lambda_1, \ldots, \lambda_n)$.   None of the λ_i are zero because

$$\det (zI - F) = \chi(z)$$

is nonzero (as a polynomial), and so rank $(zI - F) = n$.

Taking inverses, we get

$$(10a.2) \qquad\qquad (zI - F)^{-1} = B^{-1}(z)\Lambda^{-1}(z)A^{-1}(z).$$

Since $\lambda_i | \lambda_n$, λ_n is a common denominator for $(zI - F)^{-1}$.   Since

$$\det A^{-1} = (\det A)^{-1} \in K,$$

actually λ_n is the *least* common denominator.   Hence we get from (10a.2)

$$(10a.3) \qquad\qquad \lambda_n(z)(zI - F)^{-1} = \hat{A}(z)\hat{\Lambda}(z)\hat{B}(z),$$

where $\hat{A} = B^{-1}$, $\hat{B} = A^{-1}$, and

$$\hat{\Lambda} = \text{diag } (\lambda_n/\lambda_1, \lambda_n/\lambda_2, \cdots, 1).$$

Except for the reversed order of terms in $\hat{\Lambda}$, this is a representation of the polynomial matrix $\lambda_n(z)(zI - F)^{-1}$ in the sense of the invariant-factor theorem.   Since λ_n is the least common denominator of $(zI - F)^{-1}$, we see by the preceding theory that $\lambda_n = \psi_F$.

Much more is true.   By means of Proposition (10.8) and the notations of step 4 of (10.5), the invariant factors of W_F are computed as the denominators of $\hat{\lambda}_{n-i+1}/\psi_W$, after cancellation of common factors.   So we get

$$1/\psi_i = \hat{\lambda}_{n-i+1}/\psi_W = \hat{\lambda}_{n-i+1}/\lambda_n = 1/\lambda_{n-i+1}.$$

This proves the following

(10a.4)   Proposition.   *The invariant factors of $(zI - F)^{-1}$ in the sense of Definition (10.9) are equal to the invariant factors of $zI - F$ in the sense of Theorem (A.16).   Hence the invariant factors of a square matrix F in the sense of Definition (8.7) may be computed as the invariant factors of $zI - F$.*

The conclusion of (10a.4) is one of the classical definitions of invariant factors (see [GANTMAKHER, 1959, chap. 7, sec. 6]).

This state of affairs was apparently first pointed out and clarified by [KALMAN, 1965a]. The treatment given there is based on elementary polynomial algebra and (consequently) requires tortuous and delicate arguments. As we have just seen, these technical difficulties disappear if the module viewpoint is used.

For future purposes, let us note also the following obvious

(10a.5) **Corollary.**   *A square matrix F is cyclic iff*

$$\lambda_1 = \cdots = \lambda_{n-1} = 1, \lambda_n = \chi_F,$$

where the λ_i are the invariant factors of $zI - F$.

Note that this implies $\psi_F = \lambda_n = \chi_F$, which is criterion (7.23).

b. Computation of canonical realizations.   Representation (10.3) provides the critical information about Σ_W from which we can write down a canonical realization almost by inspection. This was first done in [KALMAN, 1965b, 1966b]. The key tool is the following

(10b.1) **Lemma.**   *Let F be a cyclic $n \times n$ matrix over K with characteristic polynomial χ. Then*

$$\chi(z)(zI - F)^{-1} = v(z)w'(z) \mod \chi,$$

where:
  (a)   *The components $(v_1, \ldots, v_n)$ of v and the components $(w_1, \ldots, w_n)$ of w are linearly independent over K.*
  (b)   *If $\hat{v}\hat{w}' = vw' \mod \chi$, then $\hat{v} = \epsilon v$ and $\hat{w} = \epsilon^{-1}w$, where $\epsilon = $ unit in $K[z]/K[z]\chi$.*

Proof.   The existence of this representation is a special case of the invariant-factor theorem. In fact, by (10a.4) and (10a.5), "cyclic" means that if

$$\chi W = \text{numerator of } (zI - F)^{-1}$$

then $L_W = \text{diag } (1, 0, \ldots, 0)$ in (10.3). Hence, by (10.3), we may take $v = $ first column of P and $w' = $ first row of Q.

If $w'(z)a = 0$ (for all z) for some $a \in K^n$, then

$$\chi(z)(zI - F)^{-1}a = 0 \mod \chi.$$

Since $(zI - F)^{-1}$ is proper, this means that

$$\chi(z)(zI - F)^{-1}a = 0,$$

or, equivalently, that

$$(zI - F)^{-1}a = 0.$$

But $(zI - F)^{-1}$ is nonsingular as a rational matrix (even though it is of course of rank 1 in the "mod χ" sense), which implies that $a = 0$. Similarly, $b'v(z) = 0$ (for all z) implies $b = 0$. This proves (a).

Now take $a \in K^n$ such that both $w'a$ and $\hat{w}'a$ are units in $K[z]/K[z]\chi$. (The existence of such an a is obvious from the fact that we cannot have $(w'a, \chi) \neq 1$ for all $a \in K^n$, since the elements of w are a basis for $\mathcal{P}_{\deg \chi}$.) Then $\hat{v}\hat{w}' = vw'$ mod χ implies

$$\hat{v} = (\hat{w}'a)^{-1}(w'a) \cdot v,$$

which proves (b). ☐

In general, the explicit form of v and w will depend not only on the choice of the unit in (10b.1b), but also on the particular form of F.

(10b.2) Example. Consider the matrix

$$F = \begin{bmatrix} 0 & 0 \\ 0 & 1 \end{bmatrix},$$

which is cyclic because it has distinct eigenvalues. $\chi_F(z) = z^2 - z$. Then

$$\chi_F(z)(zI - F)^{-1} = \begin{bmatrix} z - 1 & 0 \\ 0 & z \end{bmatrix} = \begin{bmatrix} z - 1 \\ z \end{bmatrix} [-z + 1, z] \text{ mod } \chi.$$

Notice that in this case the elements of v and w all *happen* to be divisors of zero mod χ.

(10b.3) Example [KALMAN, 1966b]. If F is a companion matrix of type (4.9) in Chapter 2, a simple calculation shows that v and w may be chosen as

$$v(z) = \begin{bmatrix} 1 \\ z \\ \cdot \\ \cdot \\ \cdot \\ z^{n-1} \end{bmatrix}, \qquad w(z) = \begin{bmatrix} z^{n-1} + \alpha_1 z^{n-2} + \cdots + \alpha_{n-1} \\ \cdot \\ \cdot \\ \cdot \\ z + \alpha_1 \\ 1 \end{bmatrix},$$

where

$$\chi_F(z) = z^n + \alpha_1 z^{n-1} + \cdots + \alpha_n.$$

The claimed linear independence over K of the components of v and w is clear.

Let us now develop the explicit formulas for a canonical realization of a proper rational transfer-function matrix W. First we compute a representation (10.3) (or (10.2), whichever is easier). For each invariant factor ψ_i of W we choose a cyclic F_i such that $\chi_{F_i} = \psi_i$. (For instance, let F_i be a companion matrix of ψ_i.) Write $L = (l_1, \ldots)$, $p_i = i$th column of P, and $q'_i = i$th row of Q in (10.3). Let v_i and w_i be polynomial vectors as given in the lemma. Consider the systems of equations

(10*b*.4) $\qquad H_i v_i = (l_i/\mu_i) \, p_i \bmod \psi_i, \qquad i = 1, \ldots, m;$

(10*b*.5) $\qquad w_i' G_i = q_i' \bmod \psi_i, \qquad i = 1, \ldots, m.$

In view of "mod ψ_i" the polynomial vectors on the right-hand sides above can be taken to have components of degree less than deg ψ_i. This being understood, the lemma shows that the equations (10*b*.4) and (10*b*.5) have unique solutions H_i and G_i. But then, recalling that $\mu_i = (l_i, \psi)$, $\psi_i = \psi/\mu_i$, we get

$$\begin{aligned} l_i p_i q_i' &= \mu_i \, [(l_i/\mu_i) \, p_i q_i'], \\ &= \mu_i [H_i v_i w_i' G_i] \bmod \mu_i \psi_i; \end{aligned}$$

using the lemma again gives

$$\begin{aligned} l_i p_i q_i' &= \mu_i [\psi_i H_i (zI - F)^{-1} G_i] \bmod \mu_i \psi_i, \\ &= \psi [H_i (zI - F_i)^{-1} G_i] \bmod \psi. \end{aligned}$$

Hence, by (10.3),

$$\begin{aligned} \psi W &= \sum_{i=1}^{q} l_i p_i q_i' \bmod \psi, \\ &= \psi \sum_{i=1}^{q} H_i (zI - F_i)^{-1} G_i \bmod \psi. \end{aligned}$$

Examining the right-hand side and remembering that every element of W is a *proper* rational fraction, we see that we have equality not only mod ψ, but also in the ordinary sense. Hence

(10*b*.6) $\qquad \displaystyle W(z) = \sum_{i=1}^{q} H_i (zI - F_i)^{-1} G_i,$

and we have proved the desired

(10*b*.7) **Realization theorem.** *Every proper rational transfer-function matrix W may be realized as the direct sum of the systems*

$$\Sigma_i = (F_i, G_i, H_i),$$

where F_i is a cyclic matrix with characteristic polynomial ψ_i and G_i, H_i are computed via (10b.4,5), in which make use of the representation (10b.1) of $(zI - F_i)^{-1}$.

(10*b*.8) **Exercise.** Give a precise definition of the "direct sum" of linear systems as used in this theorem.

(10*b*.9) **Corollary.** *The realization given in (10b.7) is canonical.*

Proof. In view of (6.10), it suffices to show that our realization has the same dimension as the module X_{f_W} of its input/output map f_W. In other words, we must show dim Σ_W = dim X_W. But, by (10.6) and (8.5), dim X_W = deg $\psi_1 + \cdots +$ deg ψ_q. Obviously, dim Σ_i = deg ψ_i. So the direct sum of the Σ_i has the same dimension as X_W. $\square$

Numerical examples of the required computations may be found in [KALMAN, 1965a, 1966a]. Let us note also the following

(10b.10) **Example.** Let

$$W(z) = \begin{bmatrix} \dfrac{1}{z+1} & 0 & 0 \\ 0 & \dfrac{1}{z+2} & 0 \\ 0 & 0 & \dfrac{1}{z+3} \end{bmatrix}.$$

Clearly, $\psi(z) = (z+1)(z+2)(z+3)$. The invariant factors of ψW are

$$\Lambda = \text{diag } (1, \psi, \psi),$$

while the representation (10.3) is given by

$$L = \text{diag } (1, 0, 0),$$

$$P = \begin{bmatrix} (z+2)(z+3) & 1 & 0 \\ 2(z+1)(z+3) & 2 & 2 \\ (z+1)(z+2) & 1 & 2 \end{bmatrix},$$

$$Q = \begin{bmatrix} \tfrac{1}{2}(z+2)(z+3) & -\tfrac{1}{2}(z+1)(z+3) & \tfrac{1}{2}(z+1)(z+2) \\ -1 & 1 & -1 \\ 0 & -\tfrac{1}{2} & 1 \end{bmatrix}.$$

To verify that $PLQ = \psi W \bmod \psi$ requires computations of the following sort:

$$(z+2)^2(z+3)^2 = (z+2)(z+3)(z+1+1)(z+1+2),$$
$$= 2(z+2)(z+3) \bmod (z+1)(z+2)(z+3).$$

If we choose v and w as in (10b.3), then the solution of (10b.4) is given by

$$H = \begin{bmatrix} 6 & 5 & 1 \\ 6 & 8 & 2 \\ 2 & 3 & 1 \end{bmatrix}.$$

The computation of G is left to the reader.

(10b.11) **Remark.** A system diagram of the canonical realization is shown in Fig. 10.2. This picture differs drastically from the one usually given in the literature, which is shown in Fig. 10.3. Since the realization of scalar transfer functions is a well-known solved problem, Fig. 10.3 certainly *does* represent a realization of the transfer-function matrix W; in

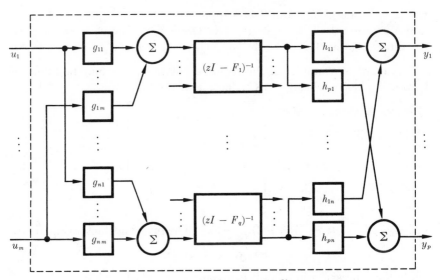

Figure 10.2   General structure of a linear system.

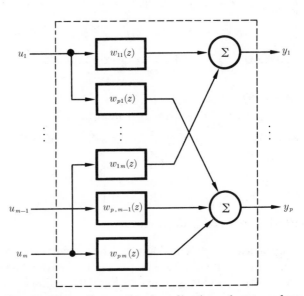

Figure 10.3   Conventional realization of a transfer function matrix.

fact, this is just a consequence of linearity. However, this realization is very seldom canonical. Hence the simple form of Fig. 10.3 results from an arbitrary choice, and its visual appeal may bear no relation whatever to the actual system Σ giving rise to W. Figure 10.2 is more appealing as a model for real systems because such systems (computers, animal brains, and so on) "generally" exhibit a very high degree of internal interconnectedness which is absent from Fig. 10.3.

(10*b*.12) **Historical remark.** These results were first obtained by [KALMAN, 1965*a*]. The invariants ψ_i were computed much earlier by [McMILLAN, 1952], using the representation (10.2) of a polynomial matrix and proceeding essentially as we did in (10.5). McMillan's work contains some lacunae, however; for instance, he did not notice that the intrinsically relevant representation is (10.3), not (10.2). (Unnecessary complications may arise when some λ's in (10.2) are zero mod ψ.) The consistent use of the "mod ψ_i" point of view in [KALMAN, 1965*a*] suggests the need for and possibility of a sharper algebraic treatment. This is precisely how the theory presented in this chapter was originally discovered.

c. Description of equivalent canonical realizations. Since canonical realizations differ only by the choice of basis in X, the equivalence class of all canonical realizations of a given input/output map f is in one-to-one correspondence with the class of all nonsingular matrices of size dim f. (This is one of the implications of Theorem (6.9)!) Using Lemma (10*b*.1), we can give a somewhat more explicit description of equivalent realizations.

Consider the scalar case $m = p = 1$. Then $W_f = \theta/\psi$, where it is assumed, of course, that $(\theta, \psi) = 1$. Pick a fixed cyclic F with $\chi_F = \psi$ and corresponding v and w, as in Lemma (10*b*.1). Write $\theta = \epsilon(\epsilon^{-1}\theta)$, where ϵ is any unit in $K[z]/K[z]\psi$. Then h and g of a canonical realization (F, g, h') are determined by

(10*c*.1) $\epsilon = h'v \bmod \psi,$
(10*c*.2) $\epsilon^{-1}\theta = w'g \bmod \psi$

(see (10*b*.4) and (10*b*.5)), since $m = p = 1$ implies "cyclic." Since $(\theta, \psi) = 1$, θ is a unit in $K[z]/K[z]\psi$. It is also easy to see that $w'g$ and $h'v$ are units whenever (F, g) is completely reachable and (F, h') is completely observable. So equations (10*c*.1) and (10*c*.2) are true for all canonical realizations. Remembering that the choice of F was arbitrary, we have the following "structure theorem":

(10*c*.3) **Proposition.** *Let $m = p = 1$. Then the class of all canonical realizations of a given input/output map f with minimal polynomial* *ψ is isomorphic to the set*

{all cyclic matrices F over K with $\chi_F = \psi$} $\times$ all {units in $K[z]/K[z]\psi$}.

In view of the uniqueness theorem (6.9) for canonical realizations, the second set above must be isomorphic to the set of all basis changes which leave F fixed, in other words, to the set of all nonsingular matrices which commute with F. This is, in fact, a classical result (see [JACOBSON, 1953, sec. 3.15]):

(10c.4)   **Proposition.**   *A square matrix A over K commutes with a cyclic matrix F over K if and only if $A = \alpha(F)$ and $\alpha \in K[z]$. Moreover, A is nonsingular if and only if α is a unit in $K[z]/K[z]\chi_F$.*

Proof.   Consider the $K[z]$-module X_F induced by F (Proposition (5.1)). Since F is cyclic, so is X_F (this was our definition of cyclic for F; see (8.7)). If A commutes with F, then $A(z \cdot x) = (AFx) = (FAx) = z \cdot (Ax)$; that is, A is a module homomorphism. Let g be a cyclic generator of X_F and write $A: g \mapsto \alpha \cdot g = \alpha(F)g$. Since X_F is cyclic, A is completely determined by α.

Moreover, if $(\alpha, \chi) \neq 1$ (that is, if α is not a unit in $K[z]/K[z]\chi$), then there is a δ such that $\delta\alpha = \beta\chi$; consequently,

$$\delta(F)\alpha(F) = (\beta\chi)(F) = 0,$$

and A is singular. So A is nonsingular iff α is a unit.   □

The generalization of Proposition (10c.3) to the case of q cyclic pieces is left to the reader. The corresponding analog of (10c.4) has been worked out in detail by [JACOBSON, 1953, secs. 3.16–3.18].

d. Degree of a rational matrix.   In a loose analogy with the degree of a polynomial, [MCMILLAN, 1952] defined the degree of a proper rational matrix as

(10d.1) $$\deg W = \sum_{i=1}^{q} \deg \psi_i.$$

$\deg W$ is, of course, the same as our definition of dim $W = \dim X_{f_W}$ (see (8.5)). [DUFFIN and HAZONY, 1963] investigated some properties of $\deg (\cdot)$ using ad hoc methods. A comparison between various definitions proposed by them in preference to (10d.1) is worked out in [KALMAN, 1965a]. As shown there, all results concerning the matrix degree are simple consequences of intuitively obvious (and easily proved) properties of realizations. The following are some sample results:

(10d.2)   **Proposition.**   *If V is a submatrix of W, then*

$$\deg V \leq \deg W.$$

Proof.   If Σ is a minimal realization of W, Σ is also a realization of V after deletion of some input or output terminals.   Since Σ *may* be non-minimal as a realization of V, the strict inequality sign can actually occur.   □

(10*d*.3)   **Proposition.**   deg $W\hat{W} \leq$ deg $W +$ deg $\hat{W}$.

Proof.   If Σ and $\hat{\Sigma}$ are minimal realizations of W and $\hat{W}$, then the cascade combination of Σ and $\hat{\Sigma}$ realizes $W\hat{W}$, but possibly nonminimally.   □

e. Information-lossless and reproducible systems.†   To conclude this section we consider a problem which requires representation (10.2) rather than (10.3) of ψW.

We need the following special definitions (not used later): a linear system is *information lossless* iff, given any initial state $x(\tau)$ and the output sequence $y(\tau), \ldots, y(\tau + l)$ (and, of course, also assuming that F_{Σ}, G_{Σ}, and H_{Σ} are known), the input sequence $u(\tau), \ldots, u(\tau + l - d)$ can be uniquely determined.   (Here $d \geq 1$ is suitable delay to allow for computations on the output sequence.)   Similarly, Σ is *reproducible* iff for all $x(\tau)$ and all desired output sequences $y(\tau), \ldots, y(\tau + l - d)$ there exists an input sequence $u(\tau), \ldots, u(\tau + l)$ generating this output sequence.   It will be clear later that "information lossless" and "reproducible" are complementary properties; we shall consider only the first.   Our problem is not merely to determine whether or not a given system Σ is information lossless, but also to find a system $\Sigma^{\#}$ that constructs the desired input sequence from a given output sequence.   In a rough sense, such a $\Sigma^{\#}$ is the inverse of Σ.

By linearity, we may (and shall) assume that $x(\tau) = 0$.

Scalar case m = p = 1.   If we are interested only in the input mod ψ, then the problem is essentially one of state determination and can be solved within the framework of Section 2.6.   If we are interested in determining the *whole* input sequence, then we must solve the equation

(10*e*.1) $$W\omega = (\theta/\psi)\, \omega = \gamma.$$

The solution is immediate from the fact that $(\theta, \psi) = 1$ iff Σ is canonical; $\psi\gamma$ is a polynomial divisible by θ, and the quotient is ω.

General case.   If W has rank $< m$ (over the field of rational functions $K(z)$ in z), then there is a nonzero vector $\omega \in K^m[z]$ such that $W\omega = 0$.   Hence rank $\psi W = m$ is a necessary condition, and then $p \geq m$.   By the invariant-factor theorem, $\lambda_i \neq 0$, $i = 1, \ldots, m$, and this is sufficient.   In fact, consider (10.2) and let $\hat{\gamma} = A^{-1}\gamma$, and $\hat{\omega} = B\omega$.   (Recall (10.5) but remember that the operations are now over $K[z]$, not $K[z]/$

† This subsection is due primarily to Dr. P. L. Faurre.

$K[z]\psi$.)   Then we must have

(10e.2) $\lambda_i \hat{\omega}_i = \psi \hat{\gamma}_i,$ $i = 1, \ldots, m;$

$0 = \hat{\gamma}_i,$ $i = m + 1, \ldots, p.$

The first equation is solved as in the scalar case, and then ω is known, since $\det B =$ unit in $K[z]$. The second equation, called "parity check" in coding theory, may be used to correct certain types of errors in the assumed values of γ.

Since rank $W = m$, there is a nonsingular $m \times n$ submatrix W_1 obtained by deleting suitable rows of W corresponding to the outputs which are not needed in the determination of ω. So W_1^{-1} is a rational matrix, but it is (usually) not proper rational and therefore does not represent a dynamical system. If we allow a suitable delay d in decoding the input from the output, then $z^{-d} W_1^{-1}$ is proper rational and represents the desired system $\Sigma^{\#}$. (To minimize d it is necessary to consider all $m \times p$ matrices C over K such that CW has rank m. Then $d_{\min}$ is a function of C, but the explicit dependence of d on C is not known at present. The difficulty arises from the fact that maximum degree of elements in A and B is quite arbitrary, and, of course, these matrices are not uniquely determined by ψW.)

Since $q \leq r =$ rank $\psi W = m$ and the strict inequality sign can occur (as in (10b.10), for example), we see that the questions discussed here have very little to do with the internal structure of Σ, but depend only on the transmission properties of the system.

10.11   The algorithm of B. L. Ho

The solution of the realization problem via the invariant factor theorem in Section 10.10b provides complete information concerning the structure of the canonical realizations. Unfortunately, this method can be applied only at the cost of the very tedious and complex computations, because that is the nature of the classical algorithm for the determination of invariant factors.

But suppose we *do not want to know* the invariant factors, and we require only the three matrices F, G, and H of the realization. Then it is reasonable to ask for a simpler algorithm. The first solution of this problem was given by B. L. Ho in the spring of 1965 (see [Ho, 1966; Ho and KALMAN, 1965–1966]) and independently also by [YOULA and TISSI, 1966]. Subsequent research has shown that this algorithm, completely unknown in system theory before 1965, is of central importance. The basic tool is the invariant-factor theorem over K (rather than $K[z]$, as in Section 10.10). Consequently, the results of this section may be expected to apply not only to an arbitrary field K, but also to quite general rings. There are still many open (and probably fruitful) research problems in this area.

Our exposition is based on [Ho and KALMAN, 1965–1966, 1969], as well as on [ZEIGER, 1967b].

Consider a linear system specified by its input/output map f. By linearity, knowledge of f is equivalent to knowledge of the infinite sequences

$$(f(e_k)(1),\ f(e_k)(2),\ \ldots),\qquad k = 1,\ \ldots,\ m,$$

where $f(e_k)(i) \in K^p$. We can also associate with f an infinite sequence $\{A_1, A_2, \ldots\}$ of $p \times m$ matrices over K defined by

(11.1) $A_i \colon K^m \to K^p,$

$$\colon (\alpha_1, \ldots, \alpha_m) \mapsto \sum_{k=1}^{m} \alpha_k f(e_k)(i), \qquad \alpha_k \in K,$$

for $i = 1, 2, \ldots$.   So f is also equivalent to the doubly infinite block matrix

(11.2)
$$\mathfrak{K}(f) = \begin{bmatrix} A_1 & A_2 & A_3 & \cdots \\ A_2 & A_3 & A_4 & \cdots \\ \cdot & \cdot & \cdot & \\ \cdot & \cdot & \cdot & \\ \cdot & \cdot & \cdot & \end{bmatrix},$$

which we shall call the *Hankel matrix* of f.   (A matrix whose (i, j)th element is given by ( )$_{i+j}$ is often called a Hankel matrix.)   Note that while $\mathfrak{K}(f)$ is "block symmetric," it is not necessarily symmetric in the ordinary sense.   We shall be interested mainly in $N' \times N$ block submatrices appearing in the upper left-hand corner of $\mathfrak{K}(f)$:

(11.3)
$$\mathfrak{K}_{N'N}(f) = \begin{bmatrix} A_1 & \cdots & & A_N \\ \cdot & & & \cdot \\ \cdot & & & \cdot \\ \cdot & & & \cdot \\ A_{N'} & \cdots & & A_{N'+N-1} \end{bmatrix}.$$

The shift operators on Ω and Γ induce a shift operator on $\mathfrak{K}(f)$ defined as

(11.4) $\sigma^k \mathfrak{K}(f) = \mathfrak{K}(\sigma^k f) = \begin{bmatrix} A_{1+k} & A_{2+k} & \cdots \\ A_{2+k} & A_{3+k} & \cdots \\ \cdot & \cdot & \\ \cdot & \cdot & \\ \cdot & \cdot & \end{bmatrix},$

where

$$\sigma^k f : \omega \mapsto f(z^k \cdot \omega) = z^k \cdot f(\omega)$$

(the equality follows by the commutative diagram (2.1c)).

In summary, *the Hankel matrix* $\mathfrak{IC}(f)$ *is a mathematical object equivalent to* f *which may be used instead of* f *as the "input" for various computations.* For instance, the shift operator σ is represented by the map $\mathfrak{IC}(f) \to \mathfrak{IC}(zf)$, which is explicitly known, since $\mathfrak{IC}(zf)$ is a submatrix of $\mathfrak{IC}(f)$. Various further applications of $\mathfrak{IC}(f)$ may be found in [Ho and KALMAN, 1969].

We shall now describe B. L. Ho's algorithm. It provides an explicit formula for the matrices (F, G, H) of a canonical realization. The computational requirement is essentially the invariant-factor algorithm applied to the K-matrix $\mathfrak{IC}_{N+1,N}(f)$, where N need not be taken larger than dim f. Two proofs will be given. The first uses conventional linear algebra, but in an intricate way. The other, due to Zeiger, requires only manipulation of commutative diagrams of the type introduced in Section 10.6. The second proof shows especially clearly how the algorithm *could have been* deduced from abstract realization theory in the style of Section 10.6. It should be added, however, that the original discovery of B. L. Ho was based on a reasoning quite similar to the first proof.

The algorithm is developed via three lemmas.

(11.5)   **Lemma.**   *Suppose* f *has a finite-dimensional realization. Then the induced sequence* $(A_1, A_2, \ldots)$ *of* f *(see* (11.1)) *satisfies the relation*

$$A_{r+j+1} = - \sum_{i=1}^{r} \beta_i A_{i+j}, \qquad j = 0, 1, \ldots,$$

for some $\beta_1, \ldots, \beta_r \in K$, *where* r *may be taken to be* $\geq \deg \psi_f$.

Proof.   The definition of the input/output map f and (1.5) show immediately that (F, G, H) realizes f iff

(11.6) $A_i = HF^{i-1}G, \qquad i = 1, 2, \ldots.$

Let θ be any annihilating polynomial of the $K[z]$-module X_f induced by f. (In particular, we could take $\theta = \psi_f$.)   If

$$\theta_f(z) = z^n + \alpha_1 z^{n-1} + \cdots + \alpha_n,$$

then writing $\beta_i = \alpha_{n-i+1}$ provides the desired relation via the Cayley-Hamilton theorem (or via module theory). (We do not have to know that $r < \deg \psi_f$ is impossible; the proof of this fact is left to the reader as an easy exercise related to reachability and observability.) $\square$

In the sequel we shall not have to calculate the β_i. We need only their *existence*. In fact, one realization can be written down immediately from (11.5).   We need the following special notations:

$$E_n^m = \begin{cases} m \times n \text{ matrix} & [I_m^m \quad 0_{n-m}^m] & \text{if } m < n, \\ m \times n \text{ matrix} & \begin{bmatrix} I_n^n \\ 0_n^{m-n} \end{bmatrix} & \text{if } m > n, \\ m \times m \text{ unit matrix } [I_m^m] & & \text{if } m = n; \end{cases}$$

I_n^m and 0_n^m are the $m \times n$ unit and zero matrix respectively.

(11.7)   **Lemma.**   *If f has a finite-dimensional realization then it is realized by*

(11.8) $$H = E_{pr}^p,$$
(11.9) $$G = \mathcal{H}_{rr}(f)E_m^{mr},$$

and

(11.10) $$F = C = \begin{bmatrix} 0_p & I_p & 0_p & \cdots & 0_p \\ 0_p & 0_p & I_p & \cdots & 0_p \\ \cdot & \cdot & \cdot & & \cdot \\ \cdot & \cdot & \cdot & & \cdot \\ \cdot & \cdot & \cdot & & \cdot \\ 0_p & 0_p & 0_p & \cdots & I_p \\ -\beta_1 I_p & -\beta_2 I_p & -\beta_3 I_p & \cdots & -\beta_r I_p \end{bmatrix},$$

where r and $\beta_1, \ldots, \beta_r$ are as given by (11.5) and I_p and 0_p are the $p \times p$ identity and zero matrix.

Proof.   First we observe that, by (11.5),

$$(\sigma^i \mathcal{H})_{rr}(f) = C^i \mathcal{H}_{rr}(f) \qquad \text{for all } i = 0, 1, 2, \ldots.$$

(This shows the importance of describing the system via its Hankel matrix.   The shift operator is automatically represented by the F-matrix of a realization; we see also the close connection between the shift operator and the "block companion matrix" C.)
    Clearly,

$$A_1 = E_{pr}^p \mathcal{H}_{rr}(f)E_m^{mr} = HG,$$

and so for any $i = 1, 2, \ldots$ we have

(11.11) $$\begin{aligned} A_i &= E_{pr}^p[\sigma^{i-1}\mathcal{H}_{rr}(f)]E_m^{mr}, \\ &= E_{pr}^p C^{i-1}\mathcal{H}_{rr}(f)E_m^{mr}, \\ &= HC^{i-1}G, \end{aligned}$$

which verifies (11.6).   Hence (11.8-10) realizes f. $\square$

    The next crucial observation is:

(11.12)   **Lemma.**   *Suppose f has a finite-dimensional realization Σ.*

Then

$$\text{rank } \mathfrak{IC}_{N'N}(f) \leq \dim f \leq \dim \Sigma$$

for all positive integers N and N'.

Proof. Let $\Sigma = (F, G, H)$ be any realization of f. Define

$$\mathfrak{R}_N = [G, FG, \ldots, F^{N-1}G]$$

and

$$\mathfrak{O}_{N'} = [H', F'H', \ldots, (F')^{N'-1}H'].$$

Then, by (11.6), we have

$$\mathfrak{IC}_{N'N} = \mathfrak{O}'_{N'}\mathfrak{R}_N$$

for *all* realizations and all positive N and N'. Now take Σ to be canonical. Then

$$\text{rank } \mathfrak{IC}_{N'N}(f) \leq \min \{\text{rank } \mathfrak{R}_N, \text{rank } \mathfrak{O}_{N'}\},$$
$$\leq \dim f.$$

(The second equality sign holds at least as long as $N, N' \geq \deg \psi_f$.) $\square$

(11.13) **Comment.** The proof becomes clearer and more interesting if we recall that the matrix $\mathfrak{R}_N$ plays a key role in providing conditions for complete reachability (Theorem (3.18) of Chapter 2). By duality, the matrix $\mathfrak{O}_N$ plays the same role with regard to complete observability.

We are now ready to describe the desired algorithm.

(11.14) **Realization algorithm of B. L. Ho.** *The following steps lead to a canonical realization of an arbitrary finite-dimensional input/output map f:*

1. *Choose an r such that (11.5) holds.*

2. *Using the invariant factor algorithm over a field K (or any other equivalent procedure†) find a nonsingular $pr \times pr$ matrix P and a nonsingular $mr \times mr$ matrix M over K such that*

$$(11.15) \qquad P[\mathfrak{IC}_{rr}(f)]M = \begin{bmatrix} I_n^n & 0_{mr-n}^n \\ 0_n^{pr-n} & 0_{mr-n}^{pr-n} \end{bmatrix} = E_n^{pr}E_{mr}^n.$$

3. *Now write down a canonical realization of f as follows:*

$$(11.16) \qquad F = E_{pr}^n P[(\sigma\mathfrak{IC})_{rr}(f)]ME_n^{mr},$$
$$(11.17) \qquad G = E_{pr}^n P[\mathfrak{IC}_{rr}(f)]E_m^{mr},$$
$$(11.18) \qquad H = E_{pr}^p[\mathfrak{IC}_{rr}(f)]ME_n^{mr}.$$

The square brackets in these formulas are intended to emphasize

---

† An elegant description of the "usual" algorithm may be found in [ANDRÉE, 1949].

that the basic data is always $\mathcal{H}_{rr}(f)$ or its shift, and the computed data consist of P and M. The E matrices correspond to "editing" operations, not to computations.

First proof (*B. L. Ho*). The dimension of $\Sigma = (F, G, H)$ given by (11.16) to (11.18) is clearly equal to rank $\mathcal{H}_{rr}(f)$. Hence, by Lemma (11.12), our triple (F, G, H) is a canonical realization of f, provided that it is a realization of f. So it suffices to show that (11.16) to (11.18) verify (11.6).

For simplicity, we write $\mathcal{H}$ for $\mathcal{H}_{rr}(f)$ and $\sigma^k\mathcal{H}$ for $(\sigma^k\mathcal{H})_{rr}(f)$ and drop some unessential indices in the E's.† The matrix

$$\mathcal{H}^{\#} = ME_nE^nP$$

is a *pseudo-inverse* of $\mathcal{H}$ in the sense that $\mathcal{H}\mathcal{H}^{\#}\mathcal{H} = \mathcal{H}$. Using $\mathcal{H}^{\#}$, we simply check that (11.16) to (11.18) defines a realization. This requires many steps:

$$
\begin{aligned}
A_i &= E^p[\sigma^{i-1}\mathcal{H}]E_m && \text{(definition of } \sigma\text{),} \\
&= E^pC^{i-1}\mathcal{H}E_m && \text{(by (11.14)),} \\
&= E^pC^{i-1}\mathcal{H}\mathcal{H}^{\#}\mathcal{H}E_m && \text{(definition of } \#\text{),} \\
&= E^pC^{i-1}\mathcal{H}ME_nE^nP\mathcal{H}E_m && \text{(definition of } \mathcal{H}^{\#}\text{),} \\
&= E^pC^{i-1}\mathcal{H}ME_nG && \text{(definition of } G\text{),} \\
&= E^p\mathcal{H}D^{i-1}ME_nG && \text{(}D \text{ is the "block transpose"}
\end{aligned}
$$

(11.19)
$$
D = \begin{bmatrix}
0_m & 0_m & 0_m & \cdots & -\beta_1 I_m \\
I_m & 0_m & 0_m & \cdots & -\beta_2 I_m \\
0_m & I_m & 0_m & \cdots & -\beta_3 I_m \\
\cdot & \cdot & \cdot & & \cdot \\
\cdot & \cdot & \cdot & & \cdot \\
\cdot & \cdot & \cdot & & \cdot \\
0_m & 0_m & 0_m & \cdots & -\beta_r I_m
\end{bmatrix}
$$
of C and $C\mathcal{H} = \mathcal{H}D$),

$$
\begin{aligned}
&= E^p\mathcal{H}\mathcal{H}^{\#}\mathcal{H}D^{i-1}ME_nG && \text{(definition of } \#\text{),} \\
&= E^p\mathcal{H}ME_nE^nP\mathcal{H}D^{i-1}ME_nG && \text{(definition of } \mathcal{H}^{\#}\text{),} \\
&= HE^nP\mathcal{H}D^{i-1}ME_nG && \text{(definition of } H\text{),} \\
&= HE^nPC^{i-1}\mathcal{H}ME_nG && \text{(}\mathcal{H}D = C\mathcal{H}\text{).}
\end{aligned}
$$

It remains now only to verify that

(11.20) $E^nPC^{i-1}\mathcal{H}ME_n = (E^nPC\mathcal{H}ME_n)^{i-1} = F^{i-1}.$

For the case $i = 3$:

---

† Of course, the reader should verify that all matrix products below are well defined.

$$
\begin{aligned}
F^2 &= E^n P C \mathcal{3C} M E_n E^n P C \mathcal{3C} M E_n && \text{(definition of } F\text{),} \\
&= E^n P C \mathcal{3C} \mathcal{3C}^{\#} C \mathcal{3C} M E_n && \text{(definition of } \mathcal{3C}^{\#}\text{),} \\
&= E^n P C \mathcal{3C} \mathcal{3C}^{\#} \mathcal{3C} D M E_n && (C\mathcal{3C} = \mathcal{3C}D), \\
&= E^n P C \mathcal{3C} D M E_n && \text{(definition of } \#\text{),} \\
&= E^n P C^2 \mathcal{3C} M E_n && (\mathcal{3C}D = C\mathcal{3C}).
\end{aligned}
$$

In the general case (11.20) is proved in a similar way, by induction on i.   □

(11.21)   **Comment.**   The chief limitation of B. L. Ho's algorithm is that everything hinges on the abstract statement

> f *has a finite-dimensional realization.*

By Lemma (11.12), we may use the more explicit equivalent statement

(11.22)   *There exists an integer* n *such that* rank $\mathcal{3C}_{N'N}(f) \leq n$ *for all positive integers* N *and* N'.

Of course, even this statement can never be decided in an empirical way, since we would have to examine infinitely many values of N and N'. We can bypass this difficulty to some extent as follows. By refining the proof of (11.14) we shall show (see also [Ho and KALMAN, 1965–1966, 1969]) that the condition

$$
(11.23) \qquad \text{rank } \mathcal{3C}_{N'N}(f) = \text{rank } \mathcal{3C}_{N'+1,N}(f) = \text{rank } \mathcal{3C}_{N',N+1}(f)
$$

implies that the system Σ given by (11.16) to (11.18) satisfies (11.6) for $i = 1, \ldots, N + N'$. It follows further from (11.23) that Σ is then a subsystem of the canonical realization of f (the latter could be even infinite dimensional). The second proof of (11.14) is arranged to clarify this crucial point.

Second proof (Zeiger-Kalman).†   The idea of the proof is to compute the canonical realization of f directly from the three basic facts:

   1. f is constant.
   2. f has finite rank (has a finite-dimensional factorization).
   3. A canonical factorization splits f into an onto and a one-to-one map.

Although the proof is long, it is certainly very easy.

   Step 1   *Finiteness.*   Assume that f is factored as

$$
\Omega \xrightarrow{g} X \xrightarrow{h} \Gamma,
$$

where $X = K^n$. Let Ω_k be the K-vector space consisting of the last k terms of sequences in Ω, that is, of all polynomials of degree $k - 1$ in

---

† For an earlier version of this proof, see [ZEIGER, 1967*b*] which, however, contains serious gaps.

$K^m[z]$. Similarly, let Γ_k be the K-vector space consisting of the first k terms of sequences in Γ. Then we have the trivial injection $\Omega_k \to \Omega$ (regard a polynomial in Ω_k as a polynomial in Ω) and trivial projection $\Gamma \to \Gamma_k$ (take the first k terms of a sequence in Γ). Let

$$X_k = \text{range } (g_k: \Omega_k \to \Omega \to X)$$

and

$$X'_k = \text{kernel } (h_k: X \to \Gamma \to \Gamma_k).$$

Then we have an ascending and a descending chain of subspaces of X:

(11.24) $$\{0\} \subseteq X_1 \subseteq X_2 \subseteq \cdots \subseteq X,$$

and

(11.25) $$X \supseteq X'_1 \supseteq X'_2 \supseteq \cdots \supseteq \{0\}.$$

Since X is finite dimensional as a K-vector space, it follows that there is a positive integer N such that $X_N = X_k$ for all $k \geq N$, and similarly there is a positive integer N' such that $X'_{N'} = X'_k$ for all $k \geq N'$. If the map $\Omega \to X$ is onto, then $X_N = X$, and if $X \to \Gamma$ is one-to-one, then $X_{N'} = \{0\}$.

Clearly f induces a map

$$f_{N'N}: \Omega_N \to \Gamma_{N'},$$

whose matrix representation is

$$\mathfrak{IC}_{N'N}(f) = \mathfrak{IC}(f_{N'N}).$$

We want to replace the problem of canonically factoring f with the problem of canonically factoring $\mathfrak{IC}_{N'N}(f)$. This is equivalent to step 2 in (11.14). So we must find K-linear maps corresponding to the dashed arrows in the following commutative diagram of K-linear maps:

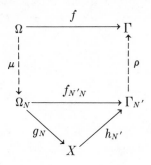

Consider the vector polynomials

$$e_k z^j, \qquad k = 1, \ldots, m \text{ and } j = 0, 1, \ldots$$

in Ω, where $e_k = (\delta_{ik}) \in K^m$ (see (3.5)). They are a basis for Ω regarded as a free K-module. Since the map $g_N: \Omega_N \to X$ is onto, every equivalence class $[e_k z^j]_f$ contains some element of $\bar{\omega}_{jk}$ of Ω_N. Let $\mu: e_k z^j \mapsto \bar{\omega}_{jk}$, and then extend μ to Ω by linearity. To construct ρ, observe that $h: X \to \Gamma$ and $h_N: X \to \Gamma_{N'}$ are one-to-one, and let

$$\rho: \gamma' \mapsto \gamma = (h \circ h_{N'}^{-1})(\gamma')$$

whenever $\gamma' \in \text{range } h_{N'} \subset \Gamma_{N'}$. Since $\Gamma_{N'}$ is a K-vector space, range $h_{N'}$ is a direct summand in $\Gamma_{N'}$, and it is clear that ρ can be extended linearly to $\Gamma_{N'}$. Note the strictly dual arguments used on the "input" and "output" sides of the diagram. (*Exercise* (Zeiger): Show that $X_r = X_{r+1}$ implies $X_r = X_k$ for all $k \geq r$; similarly for X'_k.)

Since N, $N' \leq r$, these numbers may be regarded as providing a more accurate estimate than r of the size of the submatrix of $\mathcal{K}(f)$ needed for numerical computation.

Step 2   Canonical factorization of $f_{N'N}$. Just as in the first proof, the invariant-factor theorem over K† shows that the matrix representation of $f_{N'N}$ is canonically factored as

$$\mathcal{K}_{N'N}(f) = A\Lambda B,$$

where

$$A = pN' \times pN' \quad \text{(nonsingular)},$$
$$B = mN \times mN \quad \text{(nonsingular)},$$
$$\Lambda = E_n^{pN'} E_{mN}^n.$$

The problem is now reduced to a trivial one: the factorization of Λ.

We can summarize these results by the following commutative diagram of K-linear maps:

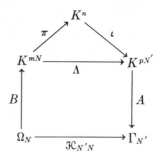

where for all $k_i \in K$

$$\pi: K^{mN} \to K^n: (k_1, \ldots, k_{mN}) \mapsto (k_1, \ldots, k_n)$$

---

† Over a field, the invariant factor theorem is very simple, since all the λ_i's are units. In fact, A, B, and Λ are computed via the usual algorithms for the determination of the rank [ANDRÉE, 1949].

is the natural projection (retain the first n coordinates and throw away the rest) and

$$\iota \colon K^n \to K^{pN'} \colon (k_1, \ldots, k_n) \mapsto (k_1, \ldots, k_n, 0, \ldots 0)$$

is the natural injection (append a suitable number of zeros).

Step 3   Determination of G and H.   Let $\iota \colon K^m \to \Omega_N$ be the natural injection $k \mapsto$ (polynomial $\theta = k$ of degree 0). Then G is obtained as the map

$$K^m \xrightarrow{\iota} \Omega_N \xrightarrow{B} K^{mN} \xrightarrow{\pi} K^n.$$

(Although we have not yet proved that G is part of a realization, this is clearly the natural definition, since G: one-step input $\mapsto$ corresponding state.)

Similarly, H is the map

$$K^n \xrightarrow{\iota} K^{pN'} \xrightarrow{A} \Gamma_{N'} \xrightarrow{\pi} K^p$$

where π: power series $\mapsto$ first term.

The preceding expressions for G and H do not agree with (11.17) and (11.18), because $P = A^{-1}$ and $M = B^{-1}$ in (11.15). To obtain B. L. Ho's formula for G, we observe that our basic commutative diagram implies the commutative diagram

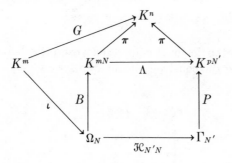

We can now read off the expression

$$G = \pi \circ P \circ \mathcal{K}_{N'N} \circ \iota,$$

which is

(11.17*) $$G = E^n_{pN'} P \mathcal{K}_{N'N} E^{mN}_m$$

modulo the notations. To compute H we consider a third commutative diagram

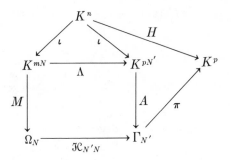

from which we get

$$H = \pi \circ \mathfrak{IC}_{N'N} \circ M \circ \iota;$$

this is equal to

(11.18*) $$\qquad\qquad H = E^p_{pN'}\mathfrak{IC}_{N'N}ME^{mN}_m.$$

Step 4   *Determination of F.*   Now we must incorporate the assumption that f is constant. As a special case of the commutative diagram (2.1c) we have the commutative diagram

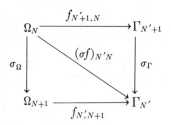

The existence of the diagonal map is clear from the definition of the shift operator. Using the constructions in step 1 we factor this diagram into the form

(11.27)

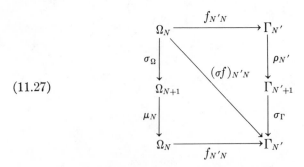

and then we "paste" a diagram of the preceding type on the upper and lower edge; this preserves commutativity. (Notice that this diagram is the abstract way of expressing stationarity which in the matrix $\mathfrak{IC}(f)$ is obtained by the special Hankel pattern of blocks.)   The result is

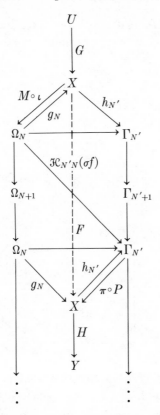

where we write, as before, $U = K^m$, $X = K^n$, and $Y = K^p$.   We are to find a map $F : X \to X$ as shown by the dashed arrow.   Since g_N (in the upper region) is onto and $h_{N'}$ (in the lower region) is one-to-one, we may apply Zeiger's fill-in lemma (6.2) and deduce the existence of a *unique* such map F.   From the diagram it follows further that

$$(11.28) \qquad F = \pi \circ P \circ \mathfrak{IC}_{N'N}(\sigma f) \circ M \circ i,$$

which is

$$(11.16^*) \qquad F = E_{pN'}^{n} P(\sigma \mathfrak{IC})_{N'N} M E_{m}^{mN}.$$

From the asymmetry of the diagram we would expect that in general F is not invertible.   In fact, F is invertible iff the maps σ, μ, and ρ are invertible, which is not true in general.   Notice also that F cannot be computed in general without inverting A and B.   This is because $f_{N'N}$ is in general not invertible.   In short, *the formulas of B. L. Ho are the most efficient way of obtaining the desired F, G, and H; no other formula can work in general.*

Step 5 *Verification that we have a realization.* The fact that $A_1 = HG$ follows at once from the commutative diagram

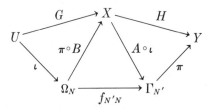

(upper edges $= HG$, lower edges $= A_1$). From the diagram used to compute F we get

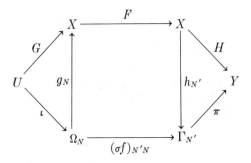

which proves $A_2 = HFG$. By iterating this construction, we prove $A_i = HF^{i-1}G$ for all $i > 2$. $\qquad\qquad\square$

(11.29)   Comment. B. L. Ho's algorithm is very remarkable in that it avoids determination of the invariant factors of the module X_f induced by f; yet it works for all patterns of invariant factors. By contrast, the primitive realization provided by Lemma (11.7) depends on the knowledge of an annihilating polynomial θ of X_f; precisely because we do not know whether or not $\theta = \psi_f$ and what the cyclic structure is, the dimension of the realization must be chosen so that the formulas are valid in the worst possible case, which is $\psi_1 = \cdots = \psi_q$. Hence we must set $n = [\min(m, p)][\deg\theta]$, and therefore the realization provided by (11.8) to (11.10) is almost always nonminimal. B. L. Ho's theorem shows further that the inefficient realization of (11.8) to (11.10) can be reduced further merely with knowledge of the rank of $\mathcal{3C}(f)$, since the dimension of the minimal realization is encoded as rank $\mathcal{3C}_{N'N}(f)$ for N and N' sufficiently large. In other words, formulas (11.16) to (11.18) of the minimal realization may be viewed as suitably "edited" versions of (11.8) to (11.10). Notice also that the simple and seemingly useful canonical form (11.10) of F gets lost in the editing process (recall here also the comments at the end of Section 10.10b).

(11.30) **Comment.** The second proof shows rather clearly the irrelevance of the invariant factors in B. L. Ho's algorithm. Here we replace $r \times r$ blocks by $N' \times N$ blocks and discard Lemma (11.5) (and even the use of an annihilating polynomial) in favor of the ascending and descending chain conditions, (11.24) and (11.25). By destroying the block structure of $\mathcal{K}(f)$ we could even refine these chains further, so that we would be dealing only with nonsingular submatrices of $\mathcal{K}(f)$. (As far as numerical analysis is concerned, this would differ only slightly from the process of determining the rank of $\mathcal{K}_{rr}(f)$.)

(11.31) **Comment.** It is important to notice that, although Lemma (11.5) is false for $r < \dim f$, this fact is basically irrelevant for the realization algorithm. For instance, let F be a cyclic $n \times n$ matrix, and consider the Hankel $\mathcal{K}$ matrix defined by the sequence $\{I, F, F^2, \ldots\}$. Since $\mathcal{K}$ is realized by (F, I, I), it is clear that we can take $N = N' = 1$, while $r = n$ (by definition of cyclic). So module theory is not at all needed for proving (11.14). On the other hand, the structure theory of modules as developed in Sections 10.8 and 10.10 is rather helpful in visualizing the power of a proposed algorithm.

From all these preliminary results, one basic question arises. Suppose we are given an input/output map f and the corresponding sequence $\{A_1, A_2, \ldots\}$, but we know or assume nothing of the finite dimensionality of f. (In other words, we have no way of performing step 1 of (11.14).) Suppose we pick r (or the pair (N, N')) arbitrarily and then compute a dynamical system Σ via steps 2 and 3 of (11.14). What are the properties of such a Σ? In particular, does Σ realize a *part* of the sequence $\{A_1, A_2, \ldots\}$? Is Σ canonical? The sharpest answer known today (1967) is given by the main result of this section:

(11.32) **Realizability criterion.** *Let* $\{A_1, A_2, \ldots\}$ *be an arbitrary infinite sequence of* $p \times m$ *matrices over* K *and let* $\mathcal{K}$ *be the corresponding Hankel matrix. Then* Σ *given by* (11.16*) *to* (11.18*) *realizes the sequence up to and including the term* A_{N_0}, *that is,* (11.6) *holds for* $i = 1, \ldots, N_0$ (i) *if and* (ii) *only if there exist positive integers* N *and* N' *such that*
(a) $N + N' = N_0$, *and*
(b) rank $\mathcal{K}_{N'N}$ = rank $\mathcal{K}_{N'+1,N}$ = rank $\mathcal{K}_{N',N+1}$.

Proof. (ii) *Necessity.* Suppose Σ given by (11.16*) to (11.18*) realizes the finite sequence $\{A_1, \ldots, A_{N_0}\}$. Then

$$\dim f_\Sigma \leq \dim \Sigma,$$

$$= \operatorname{rank} \mathfrak{IC}_{N'N} \leq \begin{Bmatrix} \operatorname{rank} \mathfrak{IC}_{N'+1,N} \\ \operatorname{rank} \mathfrak{IC}_{N',N+1} \end{Bmatrix} \leq \dim f_\Sigma,$$

where the last inequality is a consequence of Lemma (11.12), using the fact that all elements of $\mathfrak{IC}_{N'+1,N}$ as well as those of $\mathfrak{IC}_{N',N+1}$ belong to the sequence $\{A_1, \ldots, A_{N_0}\}$.

(i) *Sufficiency.* Observe that in the second proof of (11.14), the only point at which we used the assumption that "f is finite dimensional" was in the commutative diagram (11.27), where we had to show the existence of the maps $\mu_N \colon \Omega_{N+1} \to \Omega_N$ and $\rho_{N'} \colon \Gamma_{N'} \to \Gamma_{N'+1}$. The condition

$$\operatorname{rank} \mathfrak{IC}_{N'N} = \operatorname{rank} \mathfrak{IC}_{N'+1,N}$$

implies that the last block row of $\mathfrak{IC}_{N'+1,N}$ is linearly dependent on the preceding block rows, that is, on $\mathfrak{IC}_{N'N}$. Hence the output $y(N'+1)$ can be computed by taking linear combinations of the outputs $y(1), \ldots, y(N')$. So there is a map

$$\rho_{N'} \colon \gamma_{N'} \mapsto \gamma_{N'+1} = (\gamma_{N'}, y(N'+1)).$$

Similarly, the condition

$$\operatorname{rank} \mathfrak{IC}_{N'N} = \operatorname{rank} \mathfrak{IC}_{N',N+1}$$

ensures the existence of the map μ_N. That these maps are compatible with the commutative diagram (11.27) follows immediately from the special arrangement of the A_i in the Hankel matrix.

It remains to prove that under these conditions Σ given by (11.16*) to (11.18*) provides a realization of $\{A_1, \ldots, A_{N_0}\}$. This is done exactly as in step 4 of the second proof of (11.14), which depends only on the specification of $f_{N'N}$ and the commutative diagram (11.27).   □

(11.33) **Corollary.** *If conditions (a) and (b) of (11.32) are satisfied, then Σ given by (11.16*) to (11.18*) is canonical.*

Proof. Σ is an actual (rather than partial) realization of one input/output map: its own. So Σ is a realization of f_Σ computed according to the algorithm (11.14). Hence Σ is canonical.   □

The claim of this corollary is by no means trivial. See Example (11.40) below.

The contents of Theorem (11.32) may be expressed more directly with the aid of some new terminology. We say that Σ is a *partial realization* of an input/output map iff (11.6) is true for $i = 1, \ldots, N_0$.†

---

† Or we may say that Σ realizes a *partially defined input/output map*. The distinction between these two points of view is immaterial here.

The integer N_0 is the *order* of the partial realization.   If Σ is a partial realization of f, we call $f - f_\Sigma$ the *remainder* of f (*relative to* Σ).

It is clear that *every input/output map has a finite-dimensional canonical partial realization for every order N*.   To prove this, we take *arbitrary* constants $\beta_1, \ldots, \beta_{N_0} \in K$ and then define a continuation of the partial sequence $\{A_1, \ldots, A_{N_0}\}$ by means of the recursion formula of Lemma (11.5).   The upper bound on the dimension of this realization is, of course, min (mN_0, pN_0).

A *minimal partial realization* is a partial realization of minimal dimension.   Clearly "minimal" implies "canonical."   However, "canonical" does not imply "minimal," and, in fact, we shall see that "minimal" realizations are not necessarily unique.   With the new terminology, Theorem (11.32) is equivalent to the following

(11.34)   **Theorem.**   *Σ given by* (11.16)* *to* (11.18*) *is a canonical partial realization of order N_0 of an input/output map f if and only if $N_0 = N + N'$ and the rank condition* (11.32*b*) *holds.   Moreover, if the rank condition fails then every partial realization of f has dimension $>$ rank $\mathfrak{IC}_{N'N}$.*

There is one interesting case in which (11.32) is automatically satisfied:

(11.35)   **Proposition.**   *Consider the scalar case $m = p = 1$ and suppose that* rank $\mathfrak{IC}_{nn}(f) = n$.   *Then* (11.16) *to* (11.18) *provide a unique canonical partial realization of f with $N_0 = 2n$, and this partial realization of f is minimal.*

Proof.   In this case condition (11.32*b*) is trivially satisfied.   Hence (11.16) to (11.18) provide a partial realization of f, as described.   For any partial realization Σ of f, dim $\Sigma \geq$ rank $\mathfrak{IC}_{N'N}(f)$ for all N and N'.   So our partial realization is certainly minimal. $\square$

If the rank condition fails, then the minimal partial realization of an input/output map f may be nonunique, since we must then specify a new matrix A_{N_0+1} to be able to apply (11.16*) to (11.18*).   That is, A_{N_0+1} may or may not be uniquely determined by the requirement that the realization have minimal dimension.

If the rank condition does hold but rank $\mathfrak{IC}_{N'N}$ *is less than* dim f, then computing (11.16*) to (11.18*) will yield a partial realization, but it may not be very good.   However, we can at least be sure that a partial realization of f does not make it *very much more* difficult to realize the remainder of f.

(11.36)   **Proposition.**   *If f has a finite-dimensional realization, then the* *remainder after a partial realization of f also has a finite-dimensional realization.*

Proof. Let $\{A_1, A_2, \ldots\}$ be the infinite sequence corresponding to f. Let Σ^* be a partial realization of f and let $\{B_1, B_2, \ldots\}$ be the infinite sequence corresponding to f_{Σ^*}. Then the remainder $f - f_{\Sigma^*}$ has the infinite sequence $\{0, \ldots, 0, A_{N_0+1} - B_{N_0+1}, \ldots\}$. Since by hypothesis both f and f_{Σ^*} are finite dimensional, Lemma (11.12) shows that

$$\text{rank } \mathfrak{IC}_{N'N}(f) \leq \dim f$$

and

$$\text{rank } \mathfrak{IC}_{N'N}(f_{\Sigma^*}) \leq \dim f_{\Sigma^*}$$

for all positive N and N'. Since $\mathfrak{IC}$ is an additive functional on the sequences defined by f, we have

$$\begin{aligned}
\text{rank } \mathfrak{IC}_{N'N}(f - f_{\Sigma^*}) &= \text{rank } [\mathfrak{IC}_{N'N}(f) - \mathfrak{IC}_{N'N}(f_{\Sigma^*})], \\
&\leq \text{rank } \mathfrak{IC}_{N'N}(f) + \text{rank } \mathfrak{IC}_{N'N}(f_{\Sigma^*}), \\
&\leq \dim f + \dim f_{\Sigma^*} < \infty. \qquad \square
\end{aligned}$$

An alternate proof, which uses only system-theoretic concepts, is sketched in Fig. 10.4. The reader should supply the verbal details.

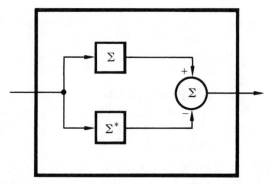

Figure 10.4   Diagrammatic proof of Proposition (11.36). Σ = realization of f, Σ^* = partial realization of f. System inside thick box: realization of $f - f\Sigma^*$.

Unfortunately, there is no simple way to say whether a partial realization of f is useful or not. On one hand, it could conceivably be viewed as a useful approximation of the minimal realization of f (see Example (11.40) below). On the other hand, consider the rational transfer-function matrix

$$W(z) = W_1(z) + z^{-h}W_2(z)$$

(see Proposition (11.37) below).   Choose h suitably large—say, much larger than the degree of the common denominator of W_1.   Then the effect of the second term will be absent from $\mathfrak{IC}_{rr}$, $\deg \psi_{W_i} \leq r \ll h$. Hence we get a minimal realization Σ_1 of W_1 from (11.16) to (11.18). Since

$$W(z) - W_{\Sigma}(z) = W(z) - W_1(z) = z^{-h} W_2(z),$$

it is clear that the minimal realization of the remainder has dimension equal to $\dim z^{-h} W_2(z)$.   (*Exercise:* Prove this fact via the invariant-factor theory of Section 10.10.)

For a deeper treatment of the theory of minimal realizations of partial input/output functions see [KALMAN, 1968] and Appendix B.

Now let us take a glimpse at the applications of B. L. Ho's algorithm.   First we note the trivial but interesting

(11.37)   **Proposition.**   *Let $\{A_1, A_2, \ldots\}$ be the infinite sequence associated with the input/output map f via (11.1).   Then the transfer function of f is given by*

$$W_f(z) = \sum_{i=1}^{\infty} A_i z^{-i}.$$

Proof.   This is merely a restatement of the definitions.   □

This result is important because it shows that the fundamental parameters $\{A_1, A_2, \ldots\}$ required by the realization algorithm may also be obtained from the transfer function of f.   To do this, we must represent W_f by a *convergent* power series in a neighborhood of ∞ ; the A_i are the coefficients of this series.   *So the algebraic computations required by B. L. Ho's algorithm are the same regardless of whether the data is given in the time-domain (f) or in the frequency-domain (W).*

It turns out that the algorithm remains exactly the same even if we pass from discrete-time to continuous-time systems.   We shall sketch the arguments involved.   First, let $K = \mathbf{R}$.   The impulse-response map $L(\cdot)$ of a continuous-time, constant, real linear system is given by

$$L(t) = \begin{cases} He^{Ft}G & \text{if } t \geq 0, \\ 0 & \text{if } t < 0 \end{cases}$$

(see Section 10.13 for a detailed discussion).   Clearly $L(\cdot)$ is a real entire function in $[0, \infty]$ which has the power-series expansion

$$He^{Ft}G = \sum_{i=0}^{\infty} \frac{A_{i+1} t^i}{i!},$$

where the A_i are given by (11.6).   So the input data to B. L. Ho's algorithm can be obtained by computing the derivatives of $L(\cdot)$ at $t = 0$.

(This is not a very practical method, but see Remark (11.38) below.)
The transfer-function matrix of a continuous-time, constant, real linear
system is given by

$$W(s) = H(sI - F)^{-1}G,$$

which happens to be identical with the corresponding expression for a
discrete-time system.   Hence the input data to the algorithm can also
be obtained in the frequency domain.

(11.38)   **Remark.**   Since the algorithm is concerned with matching the
first N_0 terms of the sequence $\{A_1, A_2, \ldots\}$, it is clear from the preceding discussion that, heuristically speaking, the algorithm is mainly
concerned with the *high-frequency input/output behavior.*   This is subject
to many practical difficulties, such as noisy observations and bias errors
of very low frequency.   Such problems are well known and of long standing (see [KALMAN, 1958] for closely related comments), but they have
never been thoroughly explored.

(11.39)   **Remark.**   The algorithm may also be interpreted as providing
a rational approximation of (convergent or formal) power series in a
single complex variable, by matching the first N_0 terms exactly.   Some
of the algebraic features of our method are, in fact, related to the classical theory of *Padé approximations* [Ho and KALMAN, 1969] (see also
Appendix B).

We conclude this section with a numerical example:

(11.40)   **Example.**   Consider the input/output function generating the
scalar partial sequence

(11.41)   1, 1, 1, 2, 1, 3, ?, ??, ...

Since $\mathfrak{K}_{11}$ is of full rank, a realization of the first two terms is given by

$$\Sigma_1 = (1, 1, 1),$$

and the corresponding infinite sequence is

$$1, 1, 1, 1, \ldots .$$

The remainder sequence is

(11.42)   0, 0, 0, 1, 0, 2, ? $- 1$, ?? $- 1$, ...

Its 4×4 Hankel matrix is

$$\mathfrak{K}_{44} = \begin{bmatrix} 0 & 0 & 0 & 1 \\ 0 & 0 & 1 & 0 \\ 0 & 1 & 0 & 2 \\ 1 & 0 & 2 & ? - 1 \end{bmatrix},$$

and it has full rank, independently of ?.   Hence, by Proposition (11.35), the remainder sequence (11.41) has a 4-dimensional realization. Since each realization of (11.42) has $A_7 = ? - 1$ and $A_8 = ?? - 1$, there are exactly ∞^2 nonisomorphic minimal realizations of (11.42). (At the same time we have an example of a partial sequence with $N_0 = 2n$ and rank $\mathfrak{IC}_{nn} < n$ which has nonunique minimal realizations.) Now we know that (11.41) may be realized using at most 5 dimensions; compare this with Proposition (11.36) and its proof. Actually, (11.41) has a *minimal* realization of 3 dimensions. In fact, for (11.41)

$$\mathfrak{IC}_{33} = \begin{bmatrix} 1 & 1 & 1 \\ 1 & 1 & 2 \\ 1 & 2 & 1 \end{bmatrix},$$

which has full rank. A suitable pair (P, M) satisfying (11.15) is given by

$$P = \begin{bmatrix} 1 & 0 & 0 \\ -1 & 1 & 0 \\ -1 & 0 & 1 \end{bmatrix} \quad \text{and} \quad M = \begin{bmatrix} 1 & -1 & -1 \\ 0 & 0 & 1 \\ 0 & 1 & 0 \end{bmatrix}.$$

Then (11.16) through (11.18) give

$$F = \begin{bmatrix} 1 & 1 & 0 \\ 0 & -1 & 1 \\ 1 & 0 & -1 \end{bmatrix},$$

$$G = \begin{bmatrix} 1 \\ 0 \\ 0 \end{bmatrix},$$

$$H = [1 \quad 0 \quad 0].\dagger$$

For this realization,

$$? = 2, \ ?? = 3, \ ??? = 5, \ ???? = 2, \ ?^5 = 9.$$

(It is not easy to guess that this is the "natural" continuation of (11.41)!) Finally, notice that if we apply formulas (11.16) through (11.18) to (11.41) with $r = 2$, we obtain, for instance

$$F = I, \quad G = \begin{bmatrix} 1 \\ 0 \end{bmatrix}, \quad H = [1 \quad 0],$$

and all the A_i are 1, as for the case $r = 1$. So, if the rank condition is not satisfied, the system defined by (11.16) to (11.18) may be neither canonical nor a partial realization.

(11.43) **Exercise.** Prove that the minimal realization of the infinite

† The corresponding transfer function is $\dfrac{z^2 + 2z + 1}{z^3 + z^2 - z - 2}$.

sequence
$$1, 2, 3, 4, \ldots$$
is two-dimensional.

(11.44)   **Exercise** (*easy*).   Realize the *Fibonacci sequence*
$$0, 1, 1, 2, 3, 5, 8, 13, \ldots$$

(11.45)   **Exercise** (*difficult*).   Prove that the sequence of prime numbers
$$2, 3, 5, 7, 11, 13, \ldots$$
has no finite-dimensional realization.

10.12   The semigroup and primes of a finite-state linear system

This section is purely expository; we shall study an example of the computation of primes in the sense of Krohn-Rhodes.   Since this theory applies only to finite-state systems, we must consider only those fields K for which $X = K^n$ is a finite set.   So *we assume that* $K = $ *finite field.* (See [ALBERT, 1956] for the theory of finite fields.)

The first step is to define the semigroup S_Σ of a linear system Σ. This requires that we reconcile certain automaton-theoretic concepts with the corresponding module-theoretic ones.

We begin by reviewing some definitions from Section 7.2.   As usual, we denote the input and output alphabets by U and Y.   An *automaton-theoretic input/output map* is $f\colon U^* \to Y$, where U^* is the *free semigroup* over U (the set of all finite sequences of symbols from U with concatenation as multiplication).   The *Myhill equivalence relation* $\equiv_f$ induced by f on U^* is defined by

(12.1) $\omega \equiv_f \omega'$ iff $f(\alpha \circ \omega \circ \beta) = f(\alpha \circ \omega' \circ \beta)$ for all $\alpha, \beta \in U^*$.

The set $S_f = \{ \{\omega\}_f \colon \omega \in U^* \}$ of Myhill equivalence classes induced by f is a semigroup with multiplication defined as

(12.2) $\{\omega\}_f \cdot \{\omega'\}_f = \{\omega \circ \omega'\}_f$.

We call S_f the *semigroup* of f.   Analogously, we define semigroup structure on $\Omega = K^m[z]$, using the multiplication (12.2) on the set $\Omega/\!\equiv_f$. When Σ is a canonical realization of f we *must* set $S_\Sigma = S_{f_\Sigma}$; any other definition would be inconsistent with the one-one correspondence between input/output maps and minimal realizations.   In the general case it is natural to express S_Σ via the state transitions in Σ (this will be discussed below).

For our present purposes, instead of considering $K[z]$, the ordinary polynomial ring, it is desirable to consider the ring of *polynomials of formal degree*, or simply *formal polynomials*, which we shall also denote by $K[z]$. This allows us to consider polynomials with leading coefficient zero. The precise setup is the following: $K[z]$ is now the collection of pairs (π, d), where π is an ordinary polynomial and $d = d_\pi$, the *formal degree* of π, is any integer $d \geq \deg \pi$, the ordinary degree of π. These pairs form a ring, with addition and multiplication defined by keeping track of the leading term even though it may have coefficient zero:

$$(\pi, d) + (\pi', d') = (\pi + \pi', \max (d, d'))$$
$$(\pi, d) \cdot (\pi', d') = (\pi\pi', d + d').$$

The reader should formally verify the ring axioms (see Appendix, (A.1) and (A.2)).†

The set $K[z]$ of formal polynomials is isomorphic to the set K^* of all finite strings (products) consisting of symbols from K: to each string κ of length l_κ in K^* there corresponds a formal polynomial $(\pi_\kappa, l_\kappa - 1)$ whose coefficients are the elements of κ. (In particular, the string of l zeros is the formal polynomial $(0, l - 1)$.) This set isomorphism extends to the semigroup structure: multiplication in K^* is expressed via multiplication in $K[z]$ as

$$(12.3) \qquad (\pi, d) \circ (\pi', d') = (z^{d'+1}\pi + \pi', d + d' + 1).$$

(The identity with respect to concatenation is $(0, -1)$, but we shall need it only in the proof of $(12.9) \Leftrightarrow (12.7)$.) Finally, m-tuples of formal polynomials are, of course, defined as

$$\begin{bmatrix} (\pi_1, d) \\ \cdot \\ \cdot \\ \cdot \\ (\pi_m, d) \end{bmatrix}, \text{ where } d \geq \deg \pi_i, \, i = 1, \, \ldots, \, m.$$

(12.4) **Definition.** *The* **semigroup structure** *on*

$$\Omega = K^m[z] \approx (K^m)^* = \text{the free semigroup on } K^m$$

is defined as follows:

(a) *We regard $K^m[z]$ as m-tuples of formal polynomials.*
(b) *We endow $K^m[z]$ with the concatenation multiplication (12.3).*

Given a fixed dynamical system Σ, we consider various objects $(\Omega, X, \varphi, F, G, \ldots)$ associated with it. (To simplify the notations, we

---

† Alternate procedure: we define a vector-space structure on U^* and then view U^* as the input space of a linear system.

shall omit the subscript Σ whenever there is no danger of confusion.)
We define first the family of input/output maps

$$\mathfrak{F} = \{f_x \colon x \in X\}$$

where

$$f_x \colon \Omega \to X \colon (\omega, d) \mapsto \varphi(1; -d, x, (\omega, d))$$

(remember that φ is the transition map (1.1) of Σ). The family $\mathfrak{F}$ displays all possible ways in which Ω can act on Σ. We define $\equiv_{\mathfrak{F}}$ to mean $\equiv_{f_x}$ with respect to every $f_x \in \mathfrak{F}$. We then have

(12.5)   **Definition.** *The **semigroup** S_Σ of a dynamical system Σ is the concatenation semigroup of equivalence classes $\Omega_\Sigma / \equiv_{\mathfrak{F}}$.*

(12.6)   **Exercise.**   Prove that $S_\Sigma = S_f$ when Σ is a minimal realization of f.

Definition (12.5) is meaningful for arbitrary dynamical systems. We want to see what simplifications result when (1) Σ is linear (and finite dimensional, discrete time) and (2) Σ is finite state.

Probably the most convenient method is to compute S_Σ via the state-transition equations (1.5). The statement

$$\omega \equiv_{\mathfrak{F}} \nu,$$

which is abstractly defined as

$$f_x(\alpha \circ \omega \circ \beta) = f_x(\alpha \circ \nu \circ \beta) \qquad \text{for all } \alpha, \beta \in \Omega, \text{ all } x \in X,$$

is transliterated through (1.5) into

(12.7) $$F^{d_\beta+1}[F^{d_\omega+1}(F^{d_\alpha+1}x + x_\alpha) + x_\omega] + x_\beta$$
$$= F^{d_\beta+1}[F^{d_\nu+1}(F^{d_\alpha+1}x + x_\alpha) + x_\nu] + x_\beta,$$

where

(12.8) $$x_\omega = \sum_{-d_\omega \leq t \leq 0} F^{-t}G\omega(t).$$

Simplifying (12.7) gives

(12.9) $$\omega \equiv_{\mathfrak{F}} \nu \qquad \text{iff} \qquad \begin{cases} x_\omega = x_\nu & \text{and} \\ F^{d_\omega+1}x = F^{d_\nu+1}x \text{ for all } x \in X. \end{cases}$$

The notation defined by (12.8) means: x_ω is the state set up at $t = 1$ by the input (ω, d). Clearly, x_ω depends on ω but is independent of d_ω if $d_\omega > \deg \omega$. (System-theoretic reason: as long as the input is zero, the zero state continues to remain zero.) In view of (12.9), the map $\omega \mapsto x_\omega$ is constant on the equivalence classes $\{\omega\}_{\mathfrak{F}}$. Hence we have a well-defined map

(12.10) $$\rho \colon S \to X \colon \{\omega\}_{\mathfrak{F}} \mapsto x_\omega.$$

To see the relationship with the automaton-theoretic terminology, the reader should verify that ρ *is simply the projection of a Myhill equivalence class into the Nerode equivalence class containing it.*

Now we use the assumption that Σ is finite state. We note first that the minimal polynomial ψ of Σ has the unique representation

$$\psi(z) = z^r \theta(z), \qquad \text{where} \qquad \theta(0) \neq 0, r \geq 0.$$

Now let us utilize a fact from the theory of finite fields: for every polynomial $\theta \in K[z]$, K = finite, there is a $q \in \mathbf{Z}$ such that $\theta | (z^q - 1)$. The least such q is the *period* $p(\theta) = P$ of θ. So, by the Cayley-Hamilton theorem, we have

(12.11) $$F^{k+P} = F^k \qquad \text{for all } k \geq r.$$

The relation (12.11) induces a finite cyclic abelian semigroup C of index r and period P, written additively. (See Section 8.2, where the same result is obtained *without* reference to finite fields.) In other words, C is defined as follows:

(12.12) $\qquad C: \begin{cases} \text{elements: } 0, 1, \ldots , (P + r - 1); \\ \text{addition: as in the integers, subject to } P + r = r; \\ \text{largest subgroup: } r, (r + 1), \ldots , (P + r - 1). \end{cases}$

If a, $b \in \mathbf{Z}$ and $a = b$ subject to the relation $P + r = r$, we write $a = b \pmod{C}$. In view of (12.9) and (12.11), we then have a well-defined map

(12.13) $$\sigma: S \to C: \{\omega\}_{\mathfrak{F}} \mapsto 1 + d_\omega \pmod{C}.$$

To make the notations more symmetrical, we write

$$\sigma: \{\omega\}_{\mathfrak{F}} \mapsto \sigma(\{\omega\}_{\mathfrak{F}}) = l_\omega \in C.$$

Because (12.9) expresses an equivalence, it is now clear that *the set S is isomorphic with the set $X \times C$, according to the rule*

$$\rho \times \sigma: \{\omega\}_{\mathfrak{F}} \mapsto (x_\omega, l_\omega).$$

It is interesting to ask if we can use this isomorphism also to express the semigroup structure of S. That is, can we build the semigroup structure of S from the abelian group structure on X (the state space) and the cyclic semigroup structure of C?

First, the definition of concatenation (12.3) shows that

$$x_{\omega \circ \nu} = F^{d_\nu + 1} x_\omega + x_\nu;$$

second,

$$l_{\omega \circ \nu} = d_{\omega \circ \nu} + 1 = d_\omega + d_\nu + 2 = l_\omega + l_\nu \pmod{C}.$$

So the multiplication in S_Σ is written explicitly as

(12.14) $\qquad (x_\nu, l_\nu) \circ (x_\omega, l_\omega) = (F^{l_\omega} x_\nu + x_\omega, l_\nu + l_\omega).$

This shows that S is the semidirect product of X by C via the connecting homomorphism represented by the map F. (See, for instance, Section 7.3 or [HALL, 1959, sec. 6.5].) We have now proved† the desired

(12.15) **Theorem.** *The semigroup S_Σ of a linear, finite-dimensional, discrete-time dynamical system Σ over a finite field is the semidirect product*

$$S_\Sigma = X_\Sigma \times_{F_\Sigma} C_\Sigma,$$

defined by (12.14), *where C_Σ is a finite cyclic semigroup of period $p(\theta_\Sigma)$ and index r_Σ, $z^{r_\Sigma} \theta_\Sigma = \chi_\Sigma$ is the minimal polynomial of Σ, and $p(\theta_\Sigma)$ is the period of θ_Σ.*

(12.16) **Corollary.** *If* $\det F_\Sigma \neq 0$, *then C_Σ is a cyclic group of order $p(\psi_\Sigma)$, and S_Σ is also a group.*

Proof. Since $\det F_\Sigma \neq 0$ implies $r_\Sigma = 0$, C_Σ is clearly a cyclic group. Each element of S_Σ has an inverse

$$(x_\omega, l_\omega)^{-1} = (-F_\Sigma^{-l_\omega} x_\omega, -l_\omega).$$

Hence S_Σ is a group. $\qquad\qquad\qquad\qquad\qquad\qquad\qquad\square$

Now we can make contact with the decomposition theory of Chapters 7 and 8.

(12.17) **Exercise.** Realize a finite-state linear system Σ as a finite automaton consisting of the cascade combination of the factor-semigroup machine corresponding to S_Σ/X_Σ followed by the normal-subgroup machine corresponding to X_Σ. (See Section 9.1.)

(12.18) **Remark.** Exercise (12.17) leads to a rather "unnatural" realization of a linear system. This is not surprising, since the aim of the Krohn-Rhodes theory is to give existence theorems for series-parallel realizations, which are almost always nonminimal. (Are they nonminimal in Exercise (12.17)?) Here module theory is clearly superior, since it leads to minimal realizations. On the other hand, the Krohn-Rhodes theory is able to treat nonlinear problems—which is much more difficult.

(12.19) **Exercise.** Prove that *the primes of a finite abelian group are the cyclic groups* $\mathbf{Z}/p\mathbf{Z}$, $p = prime$.

(12.20) **Exercise.** Prove that *the primes of a finite cyclic semigroup*

---

† This result was probably first proved by Paul Zeiger (via a direct application of group theory) in an unpublished memorandum dated November 1964.

are the same as the primes of a finite cyclic group plus units. (See Lemma (2.3) of Chapter 9.)

(12.21) **Exercise.** Prove that *the primes of a semidirect product are contained in the union of the primes of the components.*

These facts prove the main result of this section:

(12.22) **Theorem.** *The primes of a linear, finite-dimensional, discrete-time dynamical system over a finite field are cyclic groups of prime order.*

(12.23) **Corollary.** *Every cyclic group of prime order occurs as a prime of some such system.*

Proof. It suffices to take $K \approx \mathbf{Z}/p\mathbf{Z}$, $p = $ prime, $n = 1$ for then X is an abelian group of order p and hence a prime in the sense of Krohn and Rhodes. $\qquad\square$

(12.24) **Remark.** All the results of this section depend only on two simple facts: (1) The state-transition equation $x(t + 1) = Fx(t)$ holds, (2) X is a finite set. Since the special properties of a finite field are not used, it is clear that *all our results remain valid if U is an arbitrary (not necessarily commutative) finite unitary ring.*

10.13   Realization of nonconstant, continuous-time input/output maps

In this section we shall be concerned with *continuous-time, nonconstant* systems and their realization theory. Initially, this will require a setup quite different from that used in the discrete-time case, but in the end our investigation will resemble more and more the theory developed in Section 10.6.

Let us recall from Section 2.2 the definition of a *finite-dimensional, continuous-time, linear, smooth dynamical system* Σ *over the reals* $\mathbf{R}$. In differential form Σ is given by the vector equations

(13.1)
$$\frac{dx}{dt} = F(t)x + G(t)u(t),$$

(13.2)
$$y(t) = H(t)x(t),$$

where $t \in T = \mathbf{R}$, $x \in \mathbf{R}^n$, $u \in \mathbf{R}^m$, $y \in \mathbf{R}^p$, and $F(\cdot)$, $G(\cdot)$, $H(\cdot)$, and $u(\cdot)$ are real-valued continuous functions;† the adjective "smooth" will always mean of class C^q, $q \geq 0$. In state-transition form Σ is given by (13.2) and

---

† Many of the results of this section have interesting variants in the case when Σ is defined over a subinterval of T. The details are left to the reader.

(13.3) $x(t) = \varphi(t; \tau, x, u(\cdot))$,

$$= \Phi(t, \tau)x + \int_{\tau}^{t} \Phi(t, s)G(s)u(s)\ ds,$$

where $\Phi(\cdot, \cdot)$ is the transition matrix (map) associated with $F(\cdot)$. We also continue to use the notation $\Sigma = (F(\cdot), G(\cdot), H(\cdot))$. The qualifications "continuous-time" and "real" are understood to hold throughout this section and will not be expressly mentioned from here on. The infinite-dimensional case ($n = \infty$) will not be of special interest to us.

In this context the appropriate definitions of Ω and Γ are:

$\Omega_{\tau} = \{$vector space of all continuous functions
$\qquad u_{\tau}(\cdot)\colon (-\infty, \tau] \to \mathbf{R}^m\colon t \mapsto u_{\tau}(t)$ of compact support$\}$,
$\Gamma_{\tau} = \{$vector space of all continuous functions
$\qquad y_{\tau}(\cdot)\colon [\tau, \infty) \to \mathbf{R}^p\colon t \mapsto y_{\tau}(t)\}$.

The number $\tau \in \mathbf{R}$ now plays the role of the "present time" or "initial time." (We may no longer use the normalization $\tau = 0$, since our systems will not be constant.) As a rule we shall use the symbol t for the future and s for the past.

We have then:

(13.4) **Definition.** *A causal, linear, smooth, zero-state input/ output map f_{τ}^c (of a linear system Σ) with respect to the present time τ is a linear map*

$$f_{\tau}^c\colon \Omega_{\tau} \to \Gamma_{\tau}\colon u_{\tau}(\cdot) \mapsto y_{\tau}(\cdot)$$

defined for each $\tau \in T$ by

(13.5) $$y_{\tau}(t) = \int_{-\infty}^{\tau} L^c(t, s)u_{\tau}(s)\ ds, \qquad t \geq \tau,$$

where the function

$$L^c\colon (T \times T)^c \to \{p \times m \text{ matrices}\}$$

is continuous in both arguments and $(T \times T)^c = \{(t, s)\colon t \geq s\}$. We call L^c the **causal impulse-response map** *(of the underlying dynamical system).*

We emphasize that this definition, like that of Section 10.2, automatically incorporates the requirement of causality (the past affects the future, but not conversely). Our terminology differs essentially from the usual one in the engineering literature in that we do *not* require the so-called "causality condition"

(13.6) $L^c(t, s) = 0 \qquad$ for all $t < s$,

but take the point of view that L^c is *not necessarily defined* on $T \times T - (T \times T)^c$ and that the values of L^c outside $(T \times T)^c$ (if defined) do not affect the input/output map f_τ^c (see (13.5)). We shall see later that imposing the causality requirement via (13.5) is preferable to using (13.6). We observe also that the lower limit $-\infty$ in the integral (13.5) is merely a notation convention; because of the definition of Ω_τ, the actual evaluation of the integral is always over a compact interval.

We repeat: our definition of the input/output maps f_τ^c intentionally stresses the special role played by the present time τ, which was previously normalized to 0. We leave it to the reader to work out detailed formulas for system behavior generalizing those used in Sections 10.1 to 10.12.

In a strictly analogous way, by reversing the ordering on the time scale, we can define also *anticausal* input/output maps. The reader is referred to [WEISS and KALMAN, 1965] for the origins of this notion and to [KALMAN, 1969] for a more complete treatment.

If $\{f_\tau^c \colon \tau \in T\}$ is the family of zero-state input/output maps of the system Σ defined by (13.2) and (13.3), a simple calculation shows that

(13.7) $L_\Sigma^c(t, s) = H_\Sigma(t)\Phi_\Sigma(t, s)G_\Sigma(s)$ on $(T \times T)^c$;

we call L_Σ^c the *causal impulse-response map* of Σ. We say that Σ *realizes* L^c (or the family $\{f_\tau^c \colon \tau \in T\}$) iff $L_\Sigma^c = L^c$ on $(T \times T)^c$; in particular, Σ *realizes* a single input/output map f_τ^c iff $L_\Sigma^c = L^c$ for all $t \geq \tau \geq s$. It is now easy to formulate the following well-known

(13.8)   **Existence criterion** [KALMAN, 1963c, theorem 1]. *A causal impulse-response map may be realized by a finite-dimensional, linear, smooth dynamical system (i) if and (ii) only if there exist continuous maps*

$$P(\cdot) \colon T \to \{p \times n \text{ matrices}\},$$
$$Q(\cdot) \colon T \to \{n \times m \text{ matrices}\}$$

such that

(13.9) $L^c(t, s) = P(t)Q(s)$ *for all* $t \geq s$.

Proof.   (i)   If (13.9) holds, then $\Sigma = (0, Q(\cdot), P(\cdot))$ is such a realization. In fact, since the transition matrix of $F(\cdot) = 0$ is the identity, in this case (13.9) is equivalent to (13.7).

(ii)   If a finite-dimensional Σ realizes L^c, then (13.7) holds. If we let $P(t) = H_\Sigma(t)\Phi_\Sigma(t, r)$ and $Q(s) = \Phi_\Sigma(r, s)G_\Sigma(s)$ (where $r \in T$ is arbitrary but fixed), then (13.7) implies (13.9), by the composition property of the transition map.   □

(13.10)   **Definition.** *A **canonical** (or **natural**) **realization** of a (single) causal, linear, smooth, zero-state input/output map f_τ^c is a finite-dimensional, linear, smooth dynamical system Σ which realizes*

f_τ^c *and which is, in addition, completely reachable as well as completely observable at* $t = \tau$.

Similarly, L^c *is* **canonically realized by** Σ *iff these properties are true for every* $\tau \in T$.

The reader should now review Sections 2.2 and 2.6 for the definition and discussion of the concepts "reachable" and "observable."

Just as before, a finite-dimensional canonical realization Σ of f_τ^c corresponds to a canonical factorization $f_\tau^c = h_\tau^c \circ g_\tau^c$, where

(13.11)
$$g_\tau^c \colon \Omega_\tau \to X_\Sigma,$$
$$\colon u_\tau(\cdot) \mapsto \int_{-\infty}^{\tau} \Phi_\Sigma(\tau, s) G_\Sigma(s) u_\tau(s) \, ds;$$

(13.12)
$$h_\tau^c \colon X_\Sigma \to \Gamma_\tau,$$
$$\colon x \mapsto y_\tau(\cdot) \colon [\tau, \infty) \to \mathbf{R}^p,$$
$$\colon t \mapsto H_\Sigma(t) \Phi_\Sigma(t, \tau) x.$$

(13.13) **Exercise.** Check that "g_τ^c = onto" is the same as "Σ is completely reachable at τ" and "h_τ^c = one-to-one" is the same as "Σ is completely observable at τ."

Since we are not assuming constancy, we need a more general definition of system equivalence. The following properties are to be conserved under equivalence: the time scale, the causal impulse-response map, linearity, and smoothness. These requirements lead to the following

(13.14) **Definition.** *Two n-dimensional, linear, smooth dynamical systems* Σ *and* $\hat{\Sigma}$ *are* **algebraically equivalent with respect to an open interval** $J \subset T$ *iff there is a differentiable linear map* $A(\cdot) \colon J \to \{n \times n \text{ matrices}\}$, *with* $\det A(t) \neq 0$ *on* J, *such that the biunique assignment*
$$(\tau, x) \leftrightarrow (\tau, A(\tau)x) = (\tau, \hat{x})$$
is compatible with the structure of Σ *and* $\hat{\Sigma}$; *in other words, the relations*

(13.15)
$$
\begin{aligned}
&(a) & \Phi_{\hat{\Sigma}}(t, s) &= A(t)\Phi_\Sigma(t, s)A^{-1}(s), \\
&(b) & F_{\hat{\Sigma}}(t) &= A(t)F_\Sigma(t)A^{-1}(t) + \dot{A}(t)A^{-1}(t), \\
&(c) & G_{\hat{\Sigma}}(t) &= A(t)G_\Sigma(t), \\
&(d) & H_{\hat{\Sigma}}(t) &= H_\Sigma(t)A^{-1}(t),
\end{aligned}
$$

must hold for all $(t, s) \in (J \times J)^c \subset (T \times T)^c$.

The equivalence is **constant** *iff* $A(\cdot) = \text{const.}$

Note that (13.15) is the natural generalization of (1.6) and (1.7).

(13.16) **Remark.** The terminology "algebraic equivalence" is intended to emphasize that this is a relatively weak kind of isomorphism. Algebraic equivalence does *not* preserve the normed structure of the state space, because $A(\cdot)$ may not be bounded (unless J is compact). *Normed algebraic equivalence* would be algebraic equivalence plus

$$\|A(t)\| < c_1 \quad \text{and} \quad \|A^{-1}(t)\| < c_2 \quad \text{on } J,$$

with c_1 and c_2 independent of J. Without this stronger equivalence, internal-stability properties are not preserved [KALMAN, 1962*b*; SILVERMAN and ANDERSON, 1968]. We shall not explore this topic further here.

Relations (13.15) immediately imply that

$$(13.17) \qquad H_{\hat{\Sigma}}(t)\Phi(\hat{\Sigma}t, s)G_{\hat{\Sigma}}(s) = H_{\Sigma}(t)\Phi_{\Sigma}(t, s)G_{\Sigma}(s) \qquad \text{for all } t, s \in J.$$

So we have, as a special case of (13.17), the

(13.18) **Proposition.** *Causal input/output maps and the causal impulse-response maps are invariant under algebraic equivalence over T.*

As mentioned already in Section 10.6, the following theorem was the first major result in realization theory:

(13.19) **First uniqueness theorem** [KALMAN, 1963*c*, theorem 7(ii)].
Let Σ and $\hat{\Sigma}$ be any two (finite-dimensional) canonical realizations of the family $\{f_{\tau}^c \colon \tau \in J\}$ of causal input/output maps, with $J \subset T$ an open interval.
    Then Σ and $\hat{\Sigma}$ are algebraically equivalent over J. In short: a canonical realization is unique modulo algebraic equivalence.

Appendix B contains the first proof of this theorem [KALMAN, spring 1962, unpublished]; it is based directly on properties of reachability and observability. For convenience we also present here a short modern proof, in the spirit of Section 10.6.

Proof of Theorem (13.19).   We begin by observing that *it is sufficient to prove the theorem for the special case when $F_{\Sigma}(\cdot) = F_{\hat{\Sigma}}(\cdot) = 0$.* This is a consequence of the transitivity of algebraic equivalence plus the following well-known

(13.20) **Lemma.** *Every finite-dimensional, linear, smooth dynamical system Σ is algebraically equivalent over T to a system $\hat{\Sigma}$ with $F_{\hat{\Sigma}}(\cdot) = 0$. The equivalence transformation is given by $A(\cdot) = \Phi_{\Sigma}(r, \cdot)$, where $r \in T$ is arbitrary but fixed.*

If $F_{\Sigma}(\cdot) = 0$, we say that Σ is *normalized*.

Proof of lemma.   By the well-known formula for the adjoint of a linear differential system, $A(\cdot)$ satisfies the differential equation

$$\frac{d}{dt}\Phi_{\Sigma}(r, t) = \Phi_{\Sigma}(r, t)F_{\Sigma}(t).$$

Hence $A(\cdot)$ is obviously of class C^1 and $F_\Sigma(\cdot) = 0$ by (13.15b).   □

So we assume now that both Σ and $\hat{\Sigma}$ are normalized. (The reader may find it instructive to dispense with Lemma (13.20) and push through the ensuing arguments by keeping track of the factor $\Phi(r, t)$.)

If Σ is normalized and canonical, then f_τ^c has a canonical factorization through X_Σ for each $\tau \in J$. By the hypothesis of complete reachability at τ, there is a $\tau' < \tau$ dependent on τ such that any event (τ, x) is reachable from $(\tau', 0)$. (See Theorem (2.24), Chapter 2.) Hence the map

$$g_\tau^c: \Omega_\tau \to X_\Sigma,$$
$$: u_\tau(\cdot) \mapsto x = \int_{\tau'}^\tau G_\Sigma(s)u_\tau(s)\, ds$$

is onto. By the hypothesis of complete observability at τ, the condition

$$H_\Sigma(t)x = 0 \qquad \text{for all } t \in [\tau,\ \infty)$$

implies $x = 0$. (See the definition of complete observability in Section 2.6.) So the map

$$h_\tau^c: X_\Sigma \to \Gamma_\tau,$$
$$: x \mapsto h_\tau^c(\cdot): [\tau,\ \infty) \to \mathbf{R}^p,$$
$$: t \mapsto H_\Sigma(t)x$$

is one-to-one. Finally, we see at once that $h_\tau^c \circ g_\tau^c = f_\tau^c$ (compare the proof of (13.8)), and our assertion is proved.

So Σ and $\hat{\Sigma}$ provide two canonical factorizations $h_\tau^c \circ g_\tau^c = \hat{h}_\tau^c \circ \hat{g}_\tau^c = f_\tau^c$ through the vector spaces X_Σ and $X_{\hat{\Sigma}}$. By Lemma (6.8), *there is an isomorphism α_τ such that*

(13.21) $$\hat{g}_\tau = \alpha_\tau \circ g_\tau$$

and

(13.22) $$\hat{h}_\tau \circ \alpha_\tau = h_\tau.$$

For each τ, α_τ is represented by a constant matrix A_τ. Hence (13.21) and (13.22) are equivalent to

(13.23) $$G_{\hat{\Sigma}}(s) = A_\tau G_\Sigma(s) \qquad \text{for all } s \in (-\infty, \tau], \text{ all } \tau \in J$$

and

(13.24) $$H_{\hat{\Sigma}}(t)A_\tau = H_\Sigma(t) \qquad \text{for all } t \in [\tau,\ \infty), \text{ all } \tau \in J.$$

Actually, the map A_τ is constant for all $\tau \in J$. To prove this consider, for instance, (13.24). Take any $\sigma \in J$ with $\sigma < \tau$. Then

$$H_{\hat{\Sigma}}(t)(A_\tau - A_\sigma) = 0 \qquad \text{for all } t \in [\tau,\ \infty).$$

If $A_\tau \neq A_\sigma$, then there is some $x \neq 0$ such that $x' = (A_\tau - A_\sigma)x \neq 0$.

Then $H\hat{\Sigma}(t)x' = 0$ on $[\tau, \infty)$. This contradicts complete observability of $\hat{\Sigma}$ at τ. So A_τ is constant for any $\tau > \sigma$ and $\sigma \in J$; that is, A_τ is constant over all of J.

Since A is constant, it is clearly of class C^1. It follows that (13.15*a*) and (13.15*b*) are trivially satisfied, while (13.15*c*) and (13.15*d*) are just (13.23) and (13.24). This completes the proof of (13.19). ☐

Unfortunately, Theorem (13.19) does not constitute a full solution of the realization problem. (Such a claim may perhaps be inferred from [KALMAN, 1963*c*, theorem 8].) The reason is that *a canonical realization of the whole family* $\{f_\tau^c : \tau \in T\}$ *may fail to exist in the class of smooth dynamical systems*. This is clear even from the simplest examples; for instance, from

(13.25) **Example.** Consider the causal impulse-response function

$$L^c(t, s) = \begin{cases} ts & \text{for } t \geq s \geq 0, \\ 0 & \text{for } t \geq s \leq 0. \end{cases}$$

The family $\{f_\tau^c : \tau \in J\}$ has a canonical realization $\Sigma = (0, t, t)$ over $J = (0, \infty)$. On the other hand, no realization can be canonical on an interval $J' = (-\epsilon, 0)$ ($\epsilon > 0$ but otherwise arbitrary) without being zero: since L^c is identically zero on $(J' \times J')^c$, in any positive-dimensional realization the reachable states must be unobservable and the observable states must be unreachable. There is no zero-dimensional realization on the interval $(0, \infty)$, since here $L^c(t, \cdot) \neq 0$.

(13.26) **Comment.** The only possibility of getting a reasonably well-rounded realization theory without special assumptions on L^c is to generalize the notion of a dynamical system in such a way that the *dimension* of the state space is allowed to vary with time. It can be shown that, in general, such systems cannot be represented by differential equations valid on *all* of $T = \mathbf{R}$, even if we were to allow generalized functions (delta functions) as coefficients. For a deeper discussion of this topic, which is by no means fully researched at present, see [KALMAN, 1969].

To see why a nice differential equation may not provide a canonical realization of a *causal* input/output map we note that causality introduces an asymmetry with respect to time which is wholly absent from the concept of a differential equation. (Solutions of an *ordinary* differential equation always exist (if they exist at all) in an open interval containing the initial time τ; see also the comments following (2.11) of Chapter 2.)

This complicated state of affairs has caused much confusion in the literature during the period 1963–1967. In order to guarantee the existence of a canonical realization in the class of smooth dynamical systems, we have to abandon causality. In accordance with an initial attempt in this direction by [WEISS and KALMAN, 1965], we shall adopt the following

(13.27) **Definition.** *A* **weighting pattern** *L is any continuous map*

$$L: T \times T \to \{p \times m \text{ matrices}\}.$$

The **weighting pattern** L_Σ **of a finite-dimensional, linear, smooth dynamical system** Σ *is the map*

$$L_\Sigma: (t, s) \mapsto H_\Sigma(t)\Phi_\Sigma(t, s)G_\Sigma(s).$$

A **realization** *of L is any linear, smooth dynamical system* Σ *such that* $L_\Sigma = L$. *A (globally)* **reduced realization** *of L is a realization of minimal dimension. (If L has no finite-dimensional realization, we regard all realizations as reduced.)*

Note that (13.17) is equivalent to the

(13.28) **Proposition.** *The weighting pattern of any finite-dimensional, linear, smooth dynamical system is preserved under algebraic equivalence over T.*

We want to obtain a theorem that the finite-dimensional globally reduced realizations of a fixed L belong to a single equivalence class under algebraic equivalence. This result was announced in [WEISS and KALMAN, 1965, p. 155, bottom] and a proof was published subsequently and independently by [YOULA, 1966]; it is actually a minor variant of Theorem (13.19). The key fact is that "globally reduced" is equivalent to "canonical." This, in turn, follows as an exercise in linear independence.

To formulate the problem in the style of Section 10.6, we introduce some new terminology. Let $J \subset T$ be an arbitrary *finite* open interval, and define

$\Omega_J = \{$vector space of all continuous functions $u_J(\cdot): J \to \mathbf{R}^m\}$,
$\Gamma_J = \{$vector space of all continuous functions $y_J(\cdot): J \to \mathbf{R}^p\}$.

The analog of (13.4) restricted to the time set J is given by

(13.29) **Definition.** *An* **abstract, linear, smooth, zero-state input/output map** f_J^* *over J is a linear map*

$$\begin{aligned}
f_J^*: \Omega_J &\to \Gamma_J, \\
: u_J(\cdot) &\mapsto y_J(\cdot): J \to \mathbf{R}^p, \\
: t &\mapsto \int_J L(t, s)u_J(s)\, ds,
\end{aligned}$$

where L is a given weighting pattern.

A **realization** *of* f_J^* *is any linear, smooth dynamical system* Σ

such that $L_\Sigma = L$ on $J \times J$. A **reduced realization** *of f_J^* is a realization of minimal dimension. A* **canonical realization** *of f_J^* is a realization such that for any $r \in J$ (arbitrary but fixed) the map*

(13.30) $$g_J^*: \Omega_J \to X_\Sigma : u_J(\cdot) \mapsto \int_J \Phi_\Sigma(r, s) G_\Sigma(s) u_J(s) \, ds$$

is onto and the map

(13.31) $$h_J^*: X_\Sigma \to \Gamma_J,$$
$$: x \mapsto y_J(\cdot): J \to \mathbf{R}^p,$$
$$: t \mapsto H_\Sigma(t) \Phi_\Sigma(t, r) x$$

is one-to-one.

Notice that $f_J^* = h_J^* \circ g_J^*$; in view of Lemma (13.20) the choice of r is immaterial and is therefore suppressed in the notation.

To extend Definition (13.29) to the case $J = T$, we let

$\Omega = \{$vector space of all continuous functions of compact support $u(\cdot): T \to \mathbf{R}^m\}$,
$\Gamma = \{$vector space of all continuous functions $y(\cdot): T \to \mathbf{R}^p\}$.

Then the natural generalization of (13.4) is the following

(13.32) **Definition.** *An* **abstract, linear, smooth, zero-state input/output map** *f_T^* over T is a linear map*

$$f_T^*: \Omega \to \Gamma : u(\cdot) \mapsto \int_T L(\cdot, s) u(s) \, ds,$$

where L is a given weighting pattern.

The definitions of a **realization, reduced realization,** *and* **canonical realization** *are analogous to those for f_J^*.*

The existence of reduced realizations of f_J^* [or f_T^*] is an immediate consequence of the existence criterion (13.8): we simply take two maps $P(\cdot)$ and $Q(\cdot)$ such that (13.9) holds over $J \times J$ [or over $T \times T$] and the integer n is a minimum. The only remaining problem is to relate the minimality of n in the factorization of L to a canonical factorization of f_J^* [or f_T^*].

(13.33) **Lemma.** *Let $\Sigma^0 = (0, G(\cdot), H(\cdot))$ be any finite-dimensional reduced and normalized realization of f_J^* [or f_T^*]. Then there is a finite closed interval $I \subset J \subset T$ such that the columns of $H(\cdot)$ and the rows of $G(\cdot)$ are linearly independent functions over I.*

Proof. *Step* 1. Suppose that the columns of $H(\cdot)$ are linearly dependent over an open interval J (which is not necessarily finite). Then there is some $x \neq 0$ such that $H(t)x \equiv 0$ on J. Let C be any constant nonsingular matrix such that

$$(Cx)_i = \begin{cases} 0 & \text{for } 1 \le i < n, \\ 1 & \text{for } i = n. \end{cases}$$

(For instance, C may be a suitably chosen rotation matrix.)   Then

$$\Sigma° = (0, \tilde{G}(\cdot), \tilde{H}(\cdot)) = (0, CG(\cdot), H(\cdot)C^{-1})$$

is a realization.   Since $\tilde{H}(\cdot)Cx = H(\cdot)x = 0$, the last column of $\tilde{H}(\cdot)$ is zero on J.   Hence we may drop the last row of $\tilde{G}(\cdot)$ without affecting the realization condition $L(\cdot, \cdot) = H(\cdot)G(\cdot) = \tilde{H}(\cdot)\tilde{G}(\cdot)$.   So $\Sigma°$ is not a minimal realization, which is a contradiction, and the *columns of $H(\cdot)$ are linearly independent over J*.   By the same reasoning, *the rows of $G(\cdot)$ are linearly independent over J*.   We did not use finiteness; hence these claims are also true for $J = T$.

   Step 2.   By the proof in the previous case, we may assume that the columns of $H(\cdot)$ are linearly independent over J.   Let $I_0 =$ arbitrary finite *closed* interval in J.   Consider the subspace

$$X_0 = \{x: H(t)x = 0, t \in I_0\}.$$

If dim $X_0 = 0$, the lemma is proved.   If not, there is some $t_1 \in J$ such that $H(t_1)x_0 \neq 0$ for some $x_0 \neq 0$ in X_0, because of the linear independence of the columns of $H(\cdot)$ over J.   So we let

$$X_1 = \{x: H(t)x = 0, t \in I_0 \cup I_1\},$$

where I_1 is any finite closed interval in J containing t_1.   Then dim $X_0 >$ dim X_1, because X_1 omits the ray generated by x_0.   But dim $X_0 <$ dim $\Sigma° < \infty$, by the definition of a realization.   Hence dim $X_k = 0$ for some $k = q \le 1 + $ dim $\Sigma°$.   So the linear independence of columns of $H(\cdot)$ is assured not only over J but also over some finite union I' of finite closed intervals in J.   The same argument is valid also with respect to the rows of $G(\cdot)$, yielding a finite union I'' of finite closed intervals.   Finally, we let I be a suitable finite interval in J containing $I' \cup I''$.†   □
   Let us note now the obvious

(13.34)   **Proposition.**   *For any abstract linear, smooth, zero-state input/ output map f_J^* [or f_T^*] the notions of "canonical" and "reduced" are equivalent.*

   Intuitively, this result means that, in the *linear* case, we may use minimality of dimension ($=$ reduced) as a substitute for "canonical." In other words, we have a generalization of (6.10).

† Notice that this lemma plays a role analogous to the finiteness argument in step 1 of the second proof of Theorem (11.14).

Proof. In view of Lemma (13.20), there is no loss in generality if we assume that the realization of f_J^* [or f_T^*] is normalized.   Let it be

$$\Sigma^\circ = (0, G(\cdot), H(\cdot)).$$

"Reduced" $\Rightarrow$ "canonical."   The linear independence of the columns of $H(\cdot)$ over $I \subset J$ (Lemma (13.33)) shows immediately that the map h_J^* defined by (13.31) is one-to-one.   On the other hand, if the map g_J^* defined by (13.30) is not onto, there is an $x_0 \in \mathbf{R}^n$, $x_0 \neq 0$ which is orthogonal to range $g_J^* \subset \mathbf{R}^n$.   Letting $u_J(\cdot) \colon t \mapsto G'(t)x_0$ gives

$$0 = \Big\langle x_0, \int_J G(s)G'(s)x_0\, ds \Big\rangle = \int_J \|G'(s)x_0\|^2\, ds.$$

This implies $G'(s)x_0 \equiv 0$ on J, which shows that the rows of $G(\cdot)$ are linearly dependent over $I \subset J$, contradicting Lemma (13.33).   For f_T^*, we use the same argument but define $u(\cdot) \colon t \mapsto \theta(t)G'(t)x_0$, where $\theta \colon T \to \mathbf{R}$ is C°, $\theta(t) \neq 0$ on int I and supp $\theta = I$.   (Then $u(\cdot) \in \Omega$, because Ω contains all maps $T \to \mathbf{R}^n$ of compact support.)

"Canonical" $\Rightarrow$ "reduced."   This is exactly the converse of the preceding argument. $\square$

In view of the isomorphism (6.8) of canonical factorizations of a linear map, we have now immediately the

(13.35) **Second uniqueness theorem.** *Every abstract, linear, smooth, zero-state input/output map f_J^* [or f_T^*] which has a finite-dimensional realization has a reduced realization.   Every reduced realization is canonical.   Reduced realizations of a fixed f_J^* [f_T^*] belong to a single equivalence class under algebraic equivalence over J [over T].*

In short: a reduced realization of an abstract input/output map is unique modulo algebraic equivalence.

We conclude this section with some observations concerning the construction of regulators, as a sequel to our discussion in Section 2.6.

With a suitable definition of a "canonical realization" in the general case as a *piecewise smooth* dynamical system [KALMAN, 1969] we obtain the expected

(13.36) **Causal uniqueness theorem.** *A canonical realization of a family $\{f_\tau^c \colon \tau \in J =$ arbitrary open interval$\}$ of causal input/output maps is unique modulo algebraic equivalence over J, and such a system is at the same time completely reachable and completely observable.*

It does not follow at all, however, that such systems are either completely controllable or completely constructible for each $\tau \in J$.

(13.37) **Example.** Let $g(\cdot)$ be a C^∞ function $\mathbf{R} \to \mathbf{R}$ which is nonzero on $(-\infty, 0)$ and identically zero on $[0, \infty)$. Let $h(t) = g(-t)$. Then the causal impulse-response function

$$L^c(t, s) = h(t)g(s), \qquad t \geq s,$$

is canonically realized by

$$\Sigma = (0, g(\cdot), h(\cdot)).$$

However, in this system no state is controllable for $\tau \geq 0$ and no state is constructible for $\tau \leq 0$. To put it differently, the *anticausal impulse-response function* (see [WEISS and KALMAN, 1965])

$$L^a(t, s) = H_\Sigma(t)\Phi_\Sigma(t, s)G_\Sigma(s),$$
$$= h(t)g(s) \qquad \text{for } t \leq s$$

is identically zero.

By reversing the ordering in time, that is, considering the anticausal case instead of the causal case, we obtain the following mirror image of (13.36):

(13.38) **Anticausal uniqueness theorem.** *A canonical realization of a family $\{f_\tau^a: \tau \in J\}$ of anticausal input/output maps is unique modulo algebraic equivalence over J, and such a system is at the same time completely controllable and completely constructible.*

Any realization Σ of a causal impulse-response map L^c defines of course an anticausal impulse-response map L_Σ^a, equal to the right-hand side of (13.7) over the half-plane

$$(T \times T)^a = \{(t, s): t \leq s, \, t, s \in T\}.$$

We call L_Σ^a an *anticausal extension* of L^c. If Σ happens to be a realization of L^c which is canonical over T, then, by the first uniqueness theorem, Σ and its L_Σ^a are unique, as was the case in Example (13.37). If Σ is *not* a canonical realization of its own L_Σ^a, then in general, Σ cannot be controlled and contains unconstructible states. Since Σ is nevertheless unique (algebraic equivalence does not affect controllability or constructibility properties), we could still solve the regulator problem relative to some subspace X_1 of X_Σ. (Note that X_1 is not unique, because $X_1 = \{\text{controllable states}\} \cap \{\text{constructible states}\}$, and the latter subspace is *not* coordinate-free.)

In general, however, the situation is much more complicated. L^c may have more than one anticausal extension, and therefore it may be impossible to identify in a reliable way the controllable and constructible states of a system that is specified solely in terms of its causal impulse-response map. This state of affairs was never understood in the classical development of system theory (1940 to 1960), in which a "system" is often regarded as synonymous with "input–output properties." (For instance, [ZADEH, 1952] has taken an extreme position, which must be viewed in its historical context.) A comprehensive discussion may be found in [KALMAN, 1969].

To summarize, the following situation confronts system theorists wanting to solve the regulator problem when only the causal impulse-response map of the system is given:

1. At each instant, the state is known in some given internal coordinate system. This is such a strong assumption that all theoretical difficulties are eliminated, but the probability of this happening in practical situations is almost zero.

2. The solution must be based on incomplete information concerning the controllability and constructibility properties of the system. This is an exceedingly complex problem, and it has received little theoretical attention so far.

3. By special assumptions on the nature of the system, the anti-causal properties may be uniquely inferred from the causal properties, so that the system is described effectively by its weighting pattern and the second uniqueness theorem applies.

The assumption that the system is constant falls into the last category. So 3 seems to be the most fruitful research direction at the present time. For instance, it has recently been shown that most of the results of the algebraic theory of the preceding sections also apply to analytic nonconstant systems [KALMAN, 1969; Ho and KALMAN, 1969].

APPENDICES

10.A Review of module theory

This appendix contains some basic facts of modern algebra which have been frequently used throughout the chapter. It is assumed that the reader is familiar with elementary concepts on the level of, say, [HERSTEIN, 1964] or [HU, 1965]. He will find here all the definitions and facts not explicitly discussed in the body of Chapter 10.

We begin by recalling the notion of a *ring R*. We mean by this a set R in which two associative laws of composition are given, the first written additively and the second multiplicatively, such that R is an abelian group under addition and a (not necessarily abelian) monoid under multiplication†; these two laws of composition are required to be compatible with each other in the sense that the *distributivity law* holds:

(A.1) $$\alpha(\beta + \gamma) = \alpha\beta + \alpha\gamma,$$
(A.2) $$(\alpha + \beta)\gamma = \alpha\gamma + \beta\gamma,$$

† We are not interested here in rings without identity.

for all α, β, $\gamma \in R$. Excluding the trivial case $R = \{0\}$ (where $0 =$ identity with respect to addition, or simply *zero* of R), $1 \neq 0$ (where $1 =$ identity with respect to multiplication, or simply *identity* of R), and so 0 cannot have a multiplicative inverse. If an element $r \in R$ has a multiplicative inverse, we call r a *unit* of R. If r and s are nonzero elements of R and $rs = 0$, we call r and s (proper) *divisors of zero*. A unit in R cannot be a zero divisor. A ring is *commutative* iff its multiplication is commutative. A commutative ring without zero divisors is an *integral domain*, or simply a *domain*. If $R - \{0\}$ is an abelian group under multiplication, we call R a *field*. In other words: a field is a commutative ring in which every element is a unit except 0. A field is always an integral domain.

(A.3) **Example** [ZARISKI and SAMUEL, 1958, chap. I, secs. 17–18]. Consider the *set of all polynomials* $\{\pi(z)\}$ *in the indeterminate z with coefficients in a field K*. Explicitly,

$$\pi(z) = \alpha_0 + \alpha_1 z + \cdots + \alpha_n z^n, \; \alpha_i \in K, \; n < \infty.$$

If $\alpha_n \neq 0$, n is the *degree* of π. We define addition and multiplication of polynomials in the usual way:

$$\pi + \pi' = \pi'', \qquad \text{where} \qquad \pi''(z) = \sum_k (\alpha_k + \alpha_k') z^k;$$

$$\pi\pi' = \pi'', \qquad \text{where} \qquad \pi''(z) = \sum_k \Big(\sum_{i+j=k} \alpha_i \alpha_j' \Big) z^k.$$

Then the set $\{\pi(z)\}$ is a ring which is an integral domain. We denote this ring by $K[z]$. The units of $K[z]$ are polynomials of degree 0 (which are viewed as isomorphic with K).

(A.4) **Example** [LANG, 1965, chap. VI, sec. 3]. A *formal power series with coefficients in K and indeterminate z* is an infinite sequence

$$(\alpha_0, \alpha_1, \ldots),$$

which we denote formally as

$$\sum_0^\infty \alpha_i z^i.$$

Formal power series are a ring (and are then denoted by $K[[z]]$) if we define addition and multiplication as in Example (A.3). The units of $K[[z]]$ are power series such that $\alpha_0 \neq 0$.

Let us recall next the notion of an abstract (left) *vector space X over a field K*, or simply a *K-vector space*. We mean by this a set X which is an abelian group and a map (called *scalar multiplication*)

$$K \times X \to X : (\alpha, x) \mapsto \alpha \cdot x$$

such that for all α, $\beta \in K$ and all x, $y \in X$ we have

(A.5) $\alpha \cdot (x + y) = \alpha \cdot x + \alpha \cdot y,$

(A.6) $(\alpha + \beta) \cdot x = \alpha \cdot x + \beta \cdot x,$

(A.7) $(\alpha\beta) \cdot x = \alpha \cdot (\beta \cdot x),$

(A.8) $1 \cdot x = x.$

The last axiom is merely a normalization convention. The set of all n-tuples $K = (k_1, \ldots, k_m)$, $k_i \in K$, with componentwise addition and scalar multiplication is a vector space over K which is always denoted by K^n. A vector space belongs to the species of *operator structures:* one object (the field K) *operates* or *acts* on another (the abelian group X). The elements of K are called *scalars* or *operators*. Note that K may be regarded as a vector space over itself, because the ring axioms imply the vector-space axioms. Linear maps $f: X \to Y$ between two K-vector spaces X and Y will be often called *K-linear maps* or *K-homomorphisms*, to emphasize the underlying field. The dot notation for the scalar product is sometimes omitted if the context is clear.

The notion of an *R-module*† is a generalization of that of a K-vector space: the field K is replaced by a ring R, but otherwise the axioms (A.5) to (A.8) remain in force. So a module is a more natural concept than a vector space: the ring operations are needed in the vector-space axioms, but the existence of a multiplicative inverse is not. Interesting special case: a ring R is an R-module over itself, with the scalar product defined as the ring product. Specialization: a K-module (where $K =$ a field) is a vector space. Since we are interested only in commutative rings, we need not distinguish between left and right R-modules. We shall always write the scalar product on the left.

A map $f: X \to Y$ between R-modules X and Y is an R-(module)-*homomorphism* (or is *R-linear*) iff

(A.9) $f(x + y) = f(x) + f(y), \qquad x, y \in X;$

(A.10) $f(\alpha \cdot x) = \alpha \cdot f(x), \qquad \alpha \in R, x \in X.$

This definition is strictly analogous to the usual definition of a K-linear map.

An R-module X is said to be *finite* or, more explicitly, *finitely generated* with *generators* $g_1 \in X, \ldots, g_m \in X$, iff every element of X can be represented as

$$x = \sum_i r_i(x) \cdot g_i, \qquad r_i \in R.$$

The $r_i(x)$ are not necessarily unique. If they are unique for all x, then we call X *free* and the set $\{g_1, \ldots, g_m\}$ a *basis* for X. In other words,

---

† The word "module" (without mention of a ring R) is obsolete terminology for "abelian group " with the group operation denoted by $+$.

in a free module

$$\sum_i r_i g_i = 0 \text{ implies } r_i \equiv 0.$$

So a free module is almost the same thing as a vector space.

Given any $K[z]$-module X, we may automatically regard X as a K-vector space by restricting the scalars to the subfield $K^{(\circ)}[z]$ of $K[z]$ consisting of polynomials of degree 0; clearly, $K^{(\circ)}[z] \approx K$.

(A.11)   **Example.**   The set $K^m[z]$ of polynomial m-tuples (that is, m-vectors with elements in the ring $K[z]$) is an abelian group under componentwise addition. $K^m[z]$ can be given the structure of a K-vector space by defining a scalar multiplication as

$$\alpha \cdot \begin{bmatrix} \pi_1 \\ \cdot \\ \cdot \\ \cdot \\ \pi_m \end{bmatrix} = \begin{bmatrix} \alpha\pi_1 \\ \cdot \\ \cdot \\ \cdot \\ \alpha\pi_m \end{bmatrix}, \qquad \alpha \in K.$$

Moreover, $K^m[z]$ becomes a $K[z]$-module if we replace $\alpha \in K$ by $\pi \in K[z]$ in the preceding definition.   Then $K^m[z]$ is a free $K[z]$-module with precisely m generators: the vectors $e_1, \ldots, e_m$, with $e_k = (\delta_{ik})$.

We shall usually employ the same letter, say, X, to denote (1) a set, (2) an abelian group, (3) a vector space, and (4) a module.   This is convenient and serves to emphasize the fact that the same set may carry different kinds of structure in different contexts.

Some further facts about rings and modules:

A subset J of a ring R is a (*left*) *ideal* iff

1. J is stable under addition ($x, y \in J$ implies $x - y \in J$, which can also be expressed as $J - J \subset J$), and

2. J is stable under left multiplication by elements of R ($r \in R$, $s \in J$ implies $rs \in J$, which can also be expressed as $RJ \subset J$).

Right ideals and *two-sided ideals* are defined similarly.   Since we are interested here only in commutative rings, these distinctions are immaterial and all ideals will be written as left ideals.

Intersections of ideals (as sets) are ideals.   The *ideal generated by a set* $A = \{r_1, \ldots, r_q\}$ is the intersection of all ideals containing A.   An ideal J generated by a single element r of R is called a *principal ideal*.   Then we write $J = Rr$ (sometimes $J = (r)$, when the ring R is clear).   If every ideal in R is principal, then R is a *principal-ideal ring*.   A principal ideal ring which is also an integral domain is called a *principal-ideal domain*.

A *submodule* Y of an R-module X is a subgroup of X which is stable under scalar multiplication ($r \in R$, $y \in Y$ implies $ry \in Y$ or $RY \subset Y$).   Our notation shows immediately that J *is an ideal of a ring* R *iff* J *is a submodule of* R *viewed as a module over itself.*

Let R be a commutative ring and J one of its ideals. The family of sets $\{r + J : r \in R\}$ are the *residue classes of R with respect to J*. The notation is ambiguous; two residue classes $r + J$ and $r' + J$ are the same iff $r - r' \in J$. The residue classes form a ring R/J or $R - J$, called the *residue-class ring*, or *quotient ring*, or *difference ring*, in which addition and multiplication are defined in the "obvious" way:

$$(r + J) + (s + J) = (r + s + J),$$
$$(r + J)(s + J) = (rs + J).$$

Because of the ambiguous notation $(r + J)$, it is *not* obvious that these operations are well defined; the fact that they *are* follows from the properties of (two-sided) ideals. The reader should actually perform the appropriate verifications.

In the ring R/J we sometimes write equality as

$$a = b(J) \qquad \text{or} \qquad a = b \bmod J,$$

meaning that $a - b \in J$. If J is a principal ideal $J = Rq$, then we may also write the preceding relation as

$$a = b \bmod q \ (a \text{ equals } b \text{ modulo } q).$$

(A.12)   Example. *The ring $K[z]$ is a principal-ideal domain.* This fact is an easy consequence of the euclidean algorithm with respect to polynomials and the fact that the coefficients belong to a field.

(A.13)   Example. An arbitrary polynomial $\chi \in K[z]$ induces the corresponding quotient ring $K[z]/K[z]\chi$. This ring (as a set) is isomorphic with the set of all polynomials of degree less than $\deg \chi$ because in the residue class $\theta + K[z]\chi$ of any polynomial $\theta \in K[z]$ there is always a unique polynomial $\tilde{\theta}$ of such (least) degree. In other words, each $\theta \in K[z]$ has the unique representation

$$\theta = \tilde{\theta} + \rho_\theta \chi, \qquad \deg \tilde{\theta} < \deg \chi, \qquad \rho_\theta \in K[z].$$

So the polynomials $\{\tilde{\theta}\}$ are isomorphic (as a set) with the set of residue classes $\{\theta + K[z]\chi\}$; the $\tilde{\theta}$ qualify as "canonical" representatives of the residue classes $\theta + K[z]\chi$. We can make $\{\tilde{\theta}\}$ into a ring: addition is the same as for polynomials, and multiplication is defined as $\tilde{\theta} \cdot \tilde{\psi} = \widetilde{\theta\psi}$. Read: "multiply $\tilde{\theta}$ and $\tilde{\psi}$ as ordinary polynomials and then find the element of least degree in the residue class $\tilde{\theta}\tilde{\psi} + K[z]\chi$." The ring $K[z]/K[z]\chi$ is clearly isomorphic with the ring $\{\tilde{\theta}\}$. It is often more convenient to think of $K[z]/K[z]\chi$ as $\{\tilde{\theta}\}$, since the ring multiplication is then simpler. This identification is commonly used, often without special mention. $\{\tilde{\theta}\}$ is usually called the *residue-class ring (with respect to χ)*.

The ring $\{\tilde{\theta}\} \approx K[z]/K[z]\chi$ is not necessarily a domain; it may have divisors of zero, which are precisely those polynomials that have a factor in common with the polynomial χ. For if α, ψ, χ, $\chi/\alpha \in K[z]$, then writing $\theta = (\chi/\alpha)\psi$, it follows that $\alpha\theta = \psi\chi = 0 \bmod \chi$.

In the ring $K[z]$ the only units are the nonzero polynomials of degree 0. In $K[z]/K[z]\chi$ the units are precisely those polynomials that are prime relative to χ. For if θ and χ are relatively prime, then, by the euclidean algorithm, $1 = \alpha\theta + \beta\chi$ with α, $\beta \in K[z]$; so $\alpha\theta \equiv 1 \bmod \chi$, which means that $\theta^{-1} = \alpha \bmod \chi$.

We have now proved the basic

(A.14) **Proposition.** *Let χ be an arbitrary element of $K[z]$. Then every element of the ring $K[z]/K[z]\chi$ is either a unit or a divisor of zero.*

A polynomial $\pi \in K[z]$ is *irreducible* (or *prime*) iff it cannot be factored as $\pi = \rho\sigma$ with ρ, $\sigma \in K[z]$, but ρ, $\sigma \neq$ unit in $K[z]$ (that is, ρ, $\sigma \notin K$). This and (A.14) prove at once the following

(A.15) **Proposition.** *If $\chi \in K[z]$ is irreducible, then the ring $K[z]/K[z]\chi$ is a field. In other words, if $K[z]/K[z]\chi$ is a domain, it is even a field.*

(A.16) **Exercise.** Show that $\mathbf{Z}[z]$ is not a principal-ideal domain. (*Hint:* check the ideal generated by (q, z), $q > 1$.)

The most important classical result about principal-ideal domains (in particular, about $K[z]$) is the

(A.17) **Invariant-factor theorem.** *Let R be a principal-ideal domain and Π an arbitrary $p \times m$ matrix with elements in R. Then Π has the representation*

$$\Pi = A\Lambda B,$$

where:

(*a*) $A = p \times p$ *matrix with elements in R and* $\det A = $ *unit in R. (This means that A is a unit in the ring of $p \times p$ matrices over R.)*

(*b*) $B = m \times m$ *matrix with elements in R and* $\det B = $ *unit in R.*

(*c*) $\Lambda = p \times m$ *matrix all of whose elements are zero except those on the main diagonal, which are $\lambda_1, \ldots, \lambda_r, 0, \ldots, 0$. The following divisibility relations hold: $\lambda_i | \lambda_{i+1}$, $i = 1, \ldots, r - 1$. The λ_i are uniquely determined by Π up to units in R. They are the* **invariant factors** *of Π. The integer r is the* **rank** *of Π.*

(*d*)   *The λ_i may be computed directly as $\lambda_i = \Delta_i/\Delta_{i-1}$, where $\Delta_0 = 1$ and $\Delta_i = $ greatest common divisor of all $i \times i$ minors of Π.*

(*e*)   *The matrices A and B are, in general, not unique.*

The proof of the theorem rests on a classical algorithm for reducing Π to Λ by elementary matrix operations. We shall call this process the *invariant-factor algorithm*. Complete proofs may be found in [GANTMAKHER, 1959, chap. VI; ALBERT, 1956, chap. III]. Shorter proofs are given in the references of Section 10.8, following Theorem (8.1), which is the module-theoretic version of (A.17). Early engineering applications of the theorem to vibration problems in aeronautics appeared in [FRAZER, DUNCAN, and COLLAR, 1946], who give many interesting examples. See also [KALMAN, 1965a] for discussion of related questions.

If $R = K = $ a field, then all the λ_i may be taken as 1 (since in a field every element is a unit except 0), and Theorem (A.17) amounts to giving an algorithm for determining the rank r of a K-matrix Π. The rank is an invariant of an arbitrary K-homomorphism $X \to Y$ under changes of basis in the K-vector spaces X and Y. In general, we say that Π is *equivalent* to Λ and write $\Pi \sim \Lambda$.

10.B Partial realization of an input/output map (scalar case)

As a sequel to our discussion of partial realizations (see Theorem (11.32)), we present here the detailed solution of the problem of existence, uniqueness, and computation of minimal partial realizations in the special case when $m = p = 1$. The general case is treated in [KALMAN, 1968].

(B.1) **Theorem.** *Let* $\{A_1, A_2, \ldots\}$ *be an arbitrary infinite sequence with elements in a fixed field* K *and let* $\mathcal{H}$ *be its Hankel matrix. For each fixed* N_0 *one of the following three cases will arise:*

(a) *The minimal partial realizations are unique (that is, a single equivalence class under system isomorphism).*

(b) *The problem is overspecified: the minimal partial realization of order* N_0 *is unique and is at the same time the unique minimal partial realization for some order* $M_0 < N_0$.

(c) *The problem is underspecified: there is an integer* $P_0 > N_0$ *such that every minimal partial realization of order* N_0 *is at the same time the unique minimal partial realization of order* P_0 *for some arbitrary extension* $B_{N_0+1}, \ldots, B_{P_0}$ *of the given sequence. In short, in this case there is a* $(P_0 - N_0)$-*parameter family of minimal realizations.*

Case (a) *arises if and only if* $N_0 = 2n$ *and*

$$\text{rank } \mathcal{H}_{nn} = \text{rank } \mathcal{H}_{n+1,n} = n.$$

Case (b) *arises if and only if*

$$\text{rank } \mathcal{H}_{nn'} = \text{rank } \mathcal{H}_{n+1,n'} = n' - q,$$

where $q > 0$ *and* $n' = n$ $[n' = n + 1]$ *when* $N_0 = 2n$ $[N_0 = 2n + 1]$. *Then* $M_0 = 2(n - q)$.

Case (c) *arises if and only if*

$$\text{rank } \mathcal{H}_{nn'} = n' - q,$$
$$\text{rank } \mathcal{H}_{n+1,n'} = n' - q + 1,$$

*where $q > 0$ and $n' = n$ $[n' = n + 1]$ when $N_0 = 2n$ $[N_0 = 2n + 1]$.
Then $P_0 = 2(n + q)$.*

*In all cases formulas (11.16) to (11.18), with $r = n$, M_0, P_0,
respectively, provide minimal partial realizations of order N_0, with
the understanding that in case (c) $P_0 - N_0$ arbitrary parameters
$B_{N_0+1}, \ldots, B_{P_0}$ must be added to the sequence $A_1, \ldots, A_{N_0}$. The
dimension of the minimal partial realization is n, $n - q$, and $n + q$
respectively.*

Proof. In view of the claims of the theorem, let us first consider the
case $N_0 = 2n$. (The primary reason for this is that we have to determine
$2n$ coefficients of the transfer function of the desired n-dimensional
realization Σ; in other words, the number of the unknowns is always even.)

Case (a). Here $\mathcal{H}_{nn}$ has full rank and so the rank condition in (11.32)
is trivially met. (Notice that in the special case $m = p = 1$, $\mathcal{H}_{n+1,n}$ is
the transpose of $\mathcal{H}_{n,n+1}$.) To prove uniqueness, suppose we have two
minimal partial realizations, Σ and $\hat{\Sigma}$, with corresponding sequences
$\{B_i\}$ and $\{\hat{B}_i\}$, such that

$$A_i = B_i = \hat{B}_i, \qquad i = 1, \ldots, N_0;$$
$$B_i = \hat{B}_i, \qquad i = N_0 + 1, \ldots, N;$$
$$B_{N+1} \neq \hat{B}_{N+1}.$$

By Lemma (11.12), both matrices

(B.2)
$$
\begin{bmatrix}
A_1 & \cdots & A_n & A_{n+1} \\
\vdots & & \vdots & \vdots \\
A_n & \cdots & A_{2n-1} & A_{2n} \\
\vdots & & \vdots & \vdots \\
B_{N-n+1} & \cdots & B_N & B_{N+1}
\end{bmatrix}
$$

and

(B.3)
$$
\begin{bmatrix}
A_1 & \cdots & A_n & A_{n+1} \\
\vdots & & \vdots & \vdots \\
A_n & \cdots & A_{2n-1} & A_{2n} \\
\vdots & & \vdots & \vdots \\
B_{N-n+1} & \cdots & B_N & \hat{B}_{N+1}
\end{bmatrix}
$$

have rank n. By hypothesis, the $n \times n$ submatrix $\mathcal{H}_{nn}$ of (B.2) in the upper left-hand corner has rank n. Hence the n-tuple $(B_{N-n+1}, \ldots,$ $B_N)$ is linearly dependent on the rows of $\mathcal{H}_{nn}$. Since (B.2) has rank n, B_{N+1} is uniquely determined by linear dependence. So $\hat{B}_{N+1}$ is equal to B_{N+1}, for otherwise (B.3) has rank $n + 1$.

Case (b). Suppose that rank $\mathcal{H}_{nn} = $ rank $\mathcal{H}_{n+1,n} = n - q$, $q > 0$. By Theorem (11.32), there is a realization Σ of dimension $n - q$, given by the formulas (11.16) to (11.18), with $r = n$. By Lemma (11.12), this realization is certainly minimal. Since Σ is $(n - q)$-dimensional, its minimal polynomial is of degree $\leq (n - q)$. Hence Σ may be computed also by setting $r = n - q$ in (11.16) to (11.18). This shows that Σ is completely determined by the partial sequence of order $M_0 = 2(n - q)$. On the other hand, if $r < n - q$ and $s \leq n - q$, then rank $\mathcal{H}_{rs} < n - q$, so that it is impossible to get a realization of order N_0 using a partial sequence of order less than M_0. The uniqueness of Σ is proved the same way as in Case (a).

Case (c). Suppose that

$$\text{rank } \mathcal{H}_{nn} = n - q < \text{rank } \mathcal{H}_{n+1,n} = n - q + 1, \qquad q > 0.$$

Then the last row of $\mathcal{H}_{n+1,n}$ is linearly independent of the preceding n rows. Hence the last row of $\mathcal{H}_{n+1,n+1}$ is linearly independent of the preceding n rows of length $n + 1$, no matter what the value of A_{2n+1} is. This means that rank $\mathcal{H}_{n+1,n+1} = n - q + 2$ *for all extensions of the partial sequence* $\{A_1, \ldots, A_{2n}\}$ *to* $\{A_1, \ldots, A_{2n+1}\}$. So we have a new lower bound for the dimension of the minimal partial realization of order $2n$.

Suppose now that there exist numbers A_{2n+1} and A_{2n+2} such that

$$\text{rank } \mathcal{H}_{n+1,n+1} = \text{rank } \mathcal{H}_{n+2,n+1} = n - q + 2.$$

This means that *some partial sequence of order* $2(n + 1)$ *which agrees with the original sequence up to* A_{2n} *has a realization* Σ *of dimension* $n - q + 2$. If $q > 1$, it follows that Σ has a minimal polynomial of degree $\leq n$; that is, Σ is determined by means of (11.16) to (11.18), with $r = n$, and dim $\Sigma = $ rank $\mathcal{H}_{nn}$. But rank $\mathcal{H}_{nn} = n - q < n - q + 2$. This proves: *for all extensions of the partial sequence* $\{A_1, \ldots, A_{2n}\}$ *to* $\{A_1, \ldots,$ $A_{2n+2}\}$ *we have*

$$\text{rank } \mathcal{H}_{n+1,n+1} = n - q + 2 < \text{rank } \mathcal{H}_{n+2,n+1} = n - q + 3.$$

Now Case (c) is reduced to $N_0' = 2n + 2$ and $q' = q - 1$, $q > 0$. The proof is completed via finite induction.

The proof in the case $N_0 = 2n + 1$ is reduced to the case $N_0 = 2n$ by straightforward arguments, similar to those employed above. $\square$

In view of Proposition (11.35), our system-theoretic problem of *minimal partial realization of order N_0 of a sequence $\{A_1, A_2, \ldots\}$ with elements in K*

is equivalent to the following function-theoretic

(B.4) **Problem** (*Padé approximation*). *Find two polynomials σ, $\chi \in K[z]$, $\deg \sigma < \deg \chi$, such that the coefficients of the formal power series*

$$\frac{\sigma(z)}{\chi(z)} = B_1 z^{-1} + B_2 z^{-2} + \cdots$$

agree with a given sequence $\{A_1, A_2, \ldots\}$ up to and including the N_0th term and such that $\deg \chi = minimum$.

Our results may be transcribed to this problem in the following form:

(B.5) **Theorem.** *Write $N_0 = 2n$ or $2n + 1$. The solution of Problem (B.4) is given by*

$$\frac{\sigma(z)}{\chi(z)} = H(zI - F)^{-1}G$$

where F, G, and H are computed from (11.16) to (11.18), with $\deg \chi = n + q$, where the "deficiency index"

$$q = \begin{cases} N_0 - n - \operatorname{rank} \mathfrak{IC}_{n,N_0-n} & \text{if } \operatorname{rank} \mathfrak{IC}_{n+1,N_0-n} > \operatorname{rank} \mathfrak{IC}_{n,N_0-n}, \\ \operatorname{rank} \mathfrak{IC}_{n,N_0-n} - n & \text{if } \operatorname{rank} \mathfrak{IC}_{n+1,N_0-n} = \operatorname{rank} \mathfrak{IC}_{n,N_0-n}. \end{cases}$$

(B.6) **Historical remark.** The essential results of Problem (B.4) were obtained by Cauchy and Jacobi. (See [FROBENIUS, 1881] for a summary of the classical formulas, in particular, closed-form determinental expressions for σ and χ.) For many purposes, B. L. Ho's formulas (11.16) to (11.18) would be preferable to the determinants of Jacobi and Frobenius. More important perhaps is the fact that the classical results apparently have never been extended to the case of matrix sequences or numbers not from a field. By contrast, our methods do, in fact, encompass both generalizations, which have important practical applications in computer science and system theory [KALMAN, 1968].

10.C Original proof of the uniqueness theorem of canonical realizations

With some editorial improvements, we reproduce here the original proof [KALMAN, Spring 1962, unpublished] of Theorem (13.19). The basic ideas are quite similar to those used later (independently) by [YOULA, 1966].

A number of small lemmas will be needed; we shall develop them informally in the course of the proof.

Assume that $J = (a, b)$, where a is finite but b may be $+\infty$.

Remember that Σ, $\hat{\Sigma}$ are defined over all of $T = \mathbf{R} \supset J$.

Take any $\tau \in J$. By definition of a realization,

$$H_{\hat{\Sigma}}(t)\Phi_{\hat{\Sigma}}(t, s)G_{\hat{\Sigma}}(s) = H_{\Sigma}(t)\Phi_{\Sigma}(t, s)G_{\Sigma}(s)$$

for all t, s with $s \leq \tau \leq t$. By the composition property of transition maps (see (2.21) in Chapter 2), this is equivalent to

(C.1) $\qquad H_{\hat{\Sigma}}(t)\Phi_{\hat{\Sigma}}(t, \tau)\Phi_{\hat{\Sigma}}(\tau, s)G_{\hat{\Sigma}}(s) = H_{\Sigma}(t)\Phi_{\Sigma}(t, \tau)\Phi_{\Sigma}(\tau, s)G_{\Sigma}(s)$.

Note that we do not make the assumption (and in fact do not yet know) that $n = \dim \Sigma = \hat{n} = \dim \hat{\Sigma}$.

By complete reachability of Σ over J at τ, *there is a $\tau' < \tau$ such that the matrix*

(C.2) $\qquad \int_{\tau'}^{\tau} \Phi_{\Sigma}(\tau, s)G_{\Sigma}(s)G_{\Sigma}'(s)\Phi_{\Sigma}'(\tau, s) \, ds$

is positive definite (Theorem (2.24) of Chapter 2). Hence (C.1) implies

(C.3) $\qquad H_{\hat{\Sigma}}(t)\Phi_{\hat{\Sigma}}(t, \tau)A(\tau ; \tau') = H_{\Sigma}(t)\Phi_{\Sigma}(t, \tau) \qquad$ for all $t \geq \tau$,

where

$$A(\tau ; \tau') = \int_{\tau'}^{\tau} \Phi_{\hat{\Sigma}}(\tau, s)G_{\hat{\Sigma}}(s)G_{\Sigma}'(s)\Phi_{\Sigma}'(\tau, s) \, ds$$
$$\times \left[\int_{\tau'}^{\tau} \Phi_{\Sigma}(\tau, s)G_{\Sigma}(s)G_{\Sigma}'(s)\Phi_{\Sigma}'(\tau, s) \, ds \right]^{-1}.$$

Substituting (C.3) into (C.1), we have

(C.4) $\qquad H_{\hat{\Sigma}}(t)\Phi_{\hat{\Sigma}}(t, \tau)[A(\tau ; \tau')\Phi_{\Sigma}(\tau, s)G_{\Sigma}(s) - \Phi_{\hat{\Sigma}}(\tau, s)G_{\hat{\Sigma}}(s)] = 0$.

The bracketed expression is identically zero for all $s \leq \tau$. For if not, there is some $x \neq 0$ such that

$$H_{\hat{\Sigma}}(t)\Phi_{\hat{\Sigma}}(t, \tau)x = 0 \qquad \text{for all } t \geq \tau.$$

This is impossible, because $\hat{\Sigma}$ is completely observable at τ. Hence

(C.5) $\qquad \Phi_{\hat{\Sigma}}(\tau, s)G_{\hat{\Sigma}}(s) = A(\tau ; \tau')\Phi_{\Sigma}(\tau, s)G_{\Sigma}(s) \qquad$ for all $s \leq \tau$.

Interchanging the role of Σ and $\hat{\Sigma}$ (and using the fact that $\hat{\Sigma}$ is completely reachable at τ and Σ is completely observable at τ) gives two more relations analogous to (C.3) and (C.5):

(C.6) $\qquad H_{\Sigma}(t)\Phi_{\Sigma}(t, \tau)B(\tau ; \tau'') = H_{\hat{\Sigma}}(t)\Phi_{\hat{\Sigma}}(t, \tau) \qquad$ for all $t \geq \tau$,

and

(C.7) $\qquad \Phi_{\Sigma}(\tau, s)G_{\Sigma}(s) = B(\tau ; \tau'')\Phi_{\hat{\Sigma}}(\tau, s)G_{\hat{\Sigma}}(s) \qquad$ for all $s \leq \tau$,

where

$$B(\tau ; \tau'') = \int_{\tau''}^{\tau} \Phi_{\Sigma}(\tau, s)G_{\Sigma}(s)G_{\hat{\Sigma}}'(s)\Phi_{\hat{\Sigma}}'(\tau, s) \, ds$$
$$\times \left[\int_{\tau''}^{\tau} \Phi_{\hat{\Sigma}}(\tau, s)G_{\hat{\Sigma}}'(s)G_{\hat{\Sigma}}'(s)\Phi_{\hat{\Sigma}}'(\tau, s) \, ds \right]^{-1}.$$

Now it is easy to prove that A and B are *nonsingular and, in fact, inverses of one another.* For instance, substitute (C.3) into (C.6); then

$$H_{\hat{\Sigma}}(t)\Phi_{\hat{\Sigma}}(t,\,\tau)[A(\tau;\,\tau')B(\tau;\,\tau'') \,-\, I] = 0.$$

Complete observability of $\hat{\Sigma}$ at τ implies immediately, in the same way as before, that the bracketed expression is identically zero. This shows that B is the right inverse of A (and that A is the left inverse of B), and in particular that $\hat{n} \leq n$. By symmetry $n \leq \hat{n}$, hence $n = \hat{n}$, and our assertion is verified.

All previous arguments are valid whenever τ' or τ'' are replaced by a smaller number τ'''. Moreover, if $\tau_1 < \tau_2$ then we can always take $\tau_1' < \tau_2'$. (Proof: examine (C.2) and note that its rank is nondecreasing as the lower limit of integration decreases.) Hence *we may replace both τ' and τ'' by $\tau''' = \min\{\tau'(a),\ \tau''(a)\} = c$, and this choice works for all $\tau \in J$.* For simplicity, we write $A(\tau)$ instead of $A(\tau;\,c)$ and $A^{-1}(\tau)$ instead of $B(\tau;\,c)$.

Now let us take any $\sigma \in J$ such that $\sigma < \tau$. We write (C.5) as

$$\Phi_{\hat{\Sigma}}(\tau,\,\sigma)\Phi_{\hat{\Sigma}}(\sigma,\,s)G_{\hat{\Sigma}}(s) = A(\tau)\Phi_{\Sigma}(\tau,\,\sigma)\Phi_{\Sigma}(\sigma,\,s)G_{\Sigma}(s).$$

Complete reachability of $\hat{\Sigma}$ at σ implies

$$\Phi_{\hat{\Sigma}}(\tau,\,\sigma) = A(\tau)\Phi_{\Sigma}(\tau,\,\sigma)A^{-1}(\sigma), \qquad \tau > \sigma, \qquad \tau,\,\sigma \in J$$

(as before, we multiply on the right by $G_{\hat{\Sigma}}(s)\Phi_{\hat{\Sigma}}(\sigma,\,s)$ and integrate with respect to s). But if $t,\,s \in J$ and $r \in J$ is such that $t,\,s \geq r$, then

$$(C.8) \qquad \begin{aligned} \Phi_{\hat{\Sigma}}(t,\,s) &= \Phi_{\hat{\Sigma}}(t,\,r)\Phi_{\hat{\Sigma}}^{-1}(s,\,r), \\ &= A(t)\Phi_{\Sigma}(t,\,r)A^{-1}(r)[A(s)\Phi_{\Sigma}(s,\,r)A^{-1}(r)]^{-1}, \\ &= A(t)\Phi_{\Sigma}(t,\,r)\Phi_{\Sigma}^{-1}(s,\,r)A^{-1}(s), \\ &= A(t)\Phi_{\Sigma}(t,\,s)A^{-1}(s), \end{aligned}$$

which proves (13.15a). Combining this with (C.3) and (C.5) and using the nonsingularity of Φ gives

$$H_{\hat{\Sigma}}(t)A(t) = H_{\Sigma}(t), \qquad t \in J$$

as well as

$$G_{\hat{\Sigma}}(t) = A(t)G_{\Sigma}(t), \qquad t \in J,$$

which proves (13.15c) and (13.15d).

Three of the four relations (13.15) have now been verified. It remains to prove that $A(\cdot): J \to \{n \times n \text{ matrices}\}$ is of class C^1. This may be done by direct differentiation of $A(\tau;\,a)$, but the following argument is more revealing.

Consider first the special case when $F_{\Sigma}(\cdot) = F_{\hat{\Sigma}}(\cdot) = 0$, so that $\Phi_{\Sigma}(t,\,s) = \Phi_{\hat{\Sigma}}(t,\,s) = I$ for all t and s. Then (C.8) shows that $A(\cdot)$ is constant on J. Hence it is surely of class C^1. Since algebraic equivalence is transitive, the proof of the general case reduces to the special case, by Lemma (13.20).

Now the theorem is established in the case when the left endpoint a of J is not $-\infty$. If it is, we consider the nested sequence of sets $J_{k+1} \supset J_k$, where

$$J_k = (a_k, b), \text{ with } a_k \to -\infty \text{ as } k \to \infty.$$

Since the previous results hold for each k, they also hold when $J = (-\infty, b)$ or $(-\infty, \infty)$.

With this modification, the proof of the uniqueness theorem is complete. $\qquad\qquad\square$

10.D Index of notations

GENERAL NOTATIONS

Symbol	Definition	Page
J	interval in **R** (see precise specification in each case)	
K	arbitrary field	
R	arbitrary ring	
C	complex numbers	
R	real numbers	
Z	(rational) integers	
$\to$	map (sets $\to$ sets)	
$\mapsto$	map (element $\mapsto$ element)	
$\overset{\sim}{\to}$	isomorphism	
$\oplus$	direct sum	
(a, b)	greatest common divisor ($a, b \in R$ = ring)	
$a\|b$	a divides b	
$a \nmid b$	a does not divide b	

SPECIAL NOTATIONS

Symbol	Definition	Page
e_k	standard basis vectors in K^m	244
f, f_τ, f_J	input/output maps	243
g	generator of a module	265
g_k	accessible generators of X_Σ	252
$A, A(t), \alpha$	isomorphism of state space	242
A_Y	annihilator of a subset Y of a module	264
$\{A_1, A_2, \ldots\}$	matrix infinite sequence induced by f	289
E_n^m	"editing" matrix	291
F, F_Σ	one-step state-transition map or matrix (of Σ)	241
G, G_Σ, G_f	input map or matrix (of Σ, of f)	241

References

A. A. ALBERT
[1956] *Fundamental Concepts of Higher Algebra*, University of Chicago Press (modern paperback reprint in Phoenix Science Series).

R. V. ANDRÉE
[1949] Computation of the inverse of a matrix, *Am. Math. Monthly*, 58:87–92.

M. A. ARBIB
[1964] *Brains, Machines, and Mathematics*, McGraw-Hill.
[1965] A common framework for automata theory and control theory, *SIAM J. Contr.*, 3:206–222.
[1966] Automata theory and control theory: a rapprochement, *Automatica*, 3:161–189.
[1967] Tolerance automata, *Kybernetik*, 3:223–233.
[1968a] (ed.) *The Algebraic Theory of Machines, Languages and Semigroups*, Academic Press.
[1968b] Automaton decomposition and semigroup extensions, chap. 3 of [ARBIB, 1968a].
[1968c] *Theories of Abstract Automata*, Prentice-Hall.

M. A. ARBIB AND H. P. ZEIGER
[1968] An automaton-theoretic approach to linear systems, preprints of IFAC Symposium, Sydney, Australia (Institution of Engineers, Australia), pp. 91–97.

M. ATHANS
[1966] On the uniqueness of the extremal controls for a class of minimum fuel systems, *IEEE Trans. Automatic Control*, AC-11:660–669.

M. ATHANS AND P. L. FALB
[1966] *Optimal Control: An Introduction to the Theory and Its Applications*, McGraw-Hill.

A. V. BALAKRISHNAN AND L. W. NEUSTADT (eds.)
[1964] *Computing Methods in Optimization Problems,* Academic Press.

E. K. BLUM
[1964] Minimization of functionals with equality constraints, *SIAM J. Contr.,* 3:299–316.

V. BOLTYANSKII
[1966] Sufficient conditions for optimality and the justification of the dynamic programming method, *SIAM J. Contr.,* 4:326–361.

N. BOURBAKI
[1962] Éléments de Mathématique; *Livre II; Algèbre, chap. 2: Algèbre linéaire* (3d ed.), Actualités Scientifiques et Industrielles No. 1236, Hermann.
[1964] *Ibid., chap. 7: Modules sur les anneaux principaux* (2d ed.) Actualités Scientifiques et Industrielles No. 1179, Hermann.

J. V. BREAKWELL, J. L. SPEYER, AND A. E. BRYSON
[1963] Optimization and control of nonlinear systems using the second variation, *SIAM J. Contr.,* 1:193–223.

A. E. BRYSON AND W. F. DENHAM
[1962] A steepest-ascent method for solving optimum programming problems, *J. Appl. Mech. (Trans. ASME, Ser. E),* 29:247–257.

D. BUSHAW
[1963] Dynamical polysystems and optimization, *Contr. to Diff. Equations,* 2:361–365.

M. CANON, C. CULLUM, AND E. POLAK
[1967] Constrained minimization problems in finite-dimensional spaces, *SIAM J. Contr.,* 4:528–547.

A. H. CLIFFORD AND G. B. PRESTON
[1961] *The Algebraic Theory of Semigroups,* vol. I, Mathematical Surveys, No. 7, Amer. Math. Soc.

E. A. CODDINGTON AND N. LEVINSON
[1955] *Theory of Ordinary Differential Equations,* McGraw-Hill.

L. COLLATZ
[1964] *Funktionalanalysis und numerische Mathematik,* Springer.

R. COURANT (revised by J. Moser)
[1962] *Calculus of Variations* (lecture notes), New York University, Courant Institute of Mathematical Sciences.

C. W. CURTIS AND I. REINER
[1962] *Representation Theory of Finite Groups and Associative Algebras,* Interscience-Wiley.

N. DACUNHA AND E. POLAK
[1967] Constrained minimization under vector-valued criteria in finite-dimensional spaces, *J. Math. Anal. and Appl.,* 19:103–124.

E. M. DAY AND A. D. WALLACE
[1967] Multiplication induced in the state space of an act, *Math. System Theory,* 1:305–314.

W. F. DENHAM
[1963] Steepest-ascent solution of optimal programming problems, Raytheon Report, BR-2393.

J. DIEUDONNÉ
[1960] *Foundations of Modern Analysis*, Academic Press.

J. L. DOOB
[1953] *Stochastic Processes*, Wiley.

R. J. DUFFIN AND D. HAZONY
[1963] The degree of a rational matrix function, *SIAM J.* **11**:645–658.

N. DUNFORD AND J. T. SCHWARTZ
[1958] *Linear Operators, Part I: General Theory*, Interscience.

J. H. EATON
[1962] An iterative solution to time-optimal control, *J. Math. Anal. Appl.*, **5**:329–344.

P. L. FALB
[1964] Infinite-dimensional control problems I: On the closure of the set of attainable states for linear systems, *J. Math. Anal. Appl.*, **9**:12–22.

[1967] Infinite-dimensional filtering: the Kalman-Bucy filter in Hilbert space, *Information and Control*, **11**:102–137.

[1968a] Stochastic differential equations in Hilbert space, to appear.

[1968b] *Direct Methods in Optimal Control*, McGraw-Hill.

P. L. FALB AND J. L. DEJONG
[1968] *On Successive Approximation Methods in Control and Oscillation Theory*, Academic Press.

P. L. FALB AND D. L. KLEINMAN
[1966] Remarks on the infinite-dimensional Riccati equation, *IEEE Trans. Automatic Control*, **AC-11**:534–537.

P. L. FALB AND E. POLAK
[1968] Conditions for optimality, in *Systems Theory*, L. Zadeh and E. Polak (eds.), McGraw-Hill.

P. FAURRE
[1965] EM 270a term paper, Stanford University, December, 1965.

R. A. FRAZER, W. J. DUNCAN, AND A. R. COLLAR
[1946] *Elementary Matrices and Some Applications to Dynamics and Differential Equations*, 3d ed., Cambridge University Press.

H. FREEMAN
[1965] *Discrete-Time Systems*, Wiley.

G. FROBENIUS
[1881] Über Relationen zwischen Näherungsbrüchen von Potenzreihen, *J. reine u. angew. Math.*, **90**:1–17.

R. V. GAMKRELIDZE
[1965] On some extremal problems in the theory of differential equations with applications to the theory of optimal control, *SIAM J. Contr.*, **3**:106–128.

F. R. GANTMAKHER
[1959] *The Theory of Matrices*, 2 vols., Chelsea.

A. GINSBURG
[1966] Six lectures on algebraic theory of automata, Center for the Study of Information Processing, Carnegie Institute of Technology.

S. GINSBURG
[1962] *An Introduction to Mathematical Machine Theory*, Addison-Wesley.

W. H. Greub
 [1967]   *Linear Algebra*, 3d ed., Springer.
H. Halkin
 [1963a]   On the necessary condition for optimal control of nonlinear systems, *J. Anal. Math. (Jerusalem)*, **12**:1–82.
 [1963b]   The principle of optimal evolution in *Nonlinear Differential Equations and Nonlinear Mechanics*, J. P. LaSalle and S. Lefschetz (eds.), Academic Press.
 [1964]   Topological aspects of optimal control of dynamical polysystems, *Contr. Diff. Equations*, **3**:377–385.
 [1967]   An abstract framework for the theory of process optimization, *Bull. Am. Math. Soc.*
M. Hall, Jr.
 [1959]   *The Theory of Groups*, McMillan.
P. R. Halmos
 [1958]   *Finite-Dimensional Vector Spaces*, 2d ed., Van Nostrand.
J. Hartmanis and R. E. Stearns
 [1966]   *The Algebraic Structure Theory of Sequential Machines*, Prentice-Hall.
I. N. Herstein
 [1964]   *Topics in Algebra*, Ginn-Blaisdell.
B. L. Ho
 [1966]   An effective construction of realizations from input/output descriptions, doctoral dissertation, Stanford University.
B. L. Ho and R. E. Kalman
 [1965–1966]   Effective construction of linear state-variable models from input/output functions, *Proc. Third Allerton Conf.*, pp. 449–459; *Regelungstechnik*, **14**:545–548.
 [1969]   The realization of linear, constant input/output maps. I. Complete realizations, *SIAM J. Contr.*, to appear.
S. T. Hu
 [1965]   *Elements of Modern Algebra*, Holden-Day.
N. Jacobson
 [1953]   *Lectures in Abstract Algebra*, vol. II: *Linear Algebra*, Van Nostrand.
C. D. Johnson
 [1965]   Singular solutions in problems of optimal control, in *Advances in Control Systems: Theory and Applications*, vol. II, C. T. Leondes (ed.), Academic Press.
C. D. Johnson and J. E. Gibson
 [1963]   Singular solutions in problems of optimal control, *IEEE Trans. Automatic Control*, **AC-8**:4–14.
R. Kalaba
 [1959]   On nonlinear differential equations, the maximum operation and monotone convergence, *J. Math. Mech.*, **8**:519–574.
R. E. Kalman
 [1958]   Design of a self-optimizing control system, *Trans. ASME*, **80**:468–478.

[1960a]   Contributions to the theory of optimal control, *Bol. Soc. Mat. Mexicana*, **5**:102–119.

[1960b]   On the general theory of control systems, *Proc. 1st IFAC Congress, Moscow;* Butterworths, London.

[1962a]   Canonical structure of linear dynamical systems, *Proc. Nat. Acad. of Sci. (USA)*, **48**:596–600.

[1962b]   On the stability of time-varying linear systems, *Trans. IRE PGCT*, **9**:420–422.

[1963a]   The theory of optimal control and the calculus of variations, chap. 16 in Proc. of Conference on *Mathematical Optimization Techniques* (Santa Monica, 1960), R. Bellman (ed.), University of California Press.

[1963b]   New methods in Wiener filtering theory, *Proc. 1st Symp. on Engineering Applications of Random Function Theory and Probability*, Purdue University, November 1960, pp. 270–388, Wiley.   (Abridged from RIAS Technical Report 61–1.)

[1963c]   Mathematical description of linear dynamical systems, *SIAM J. Contr.*, **1**:152–192.

[1963d]   Lyapunov functions for the problem of Lur'e in automatic control, *Proc. Nat. Acad. Sci. (USA)*, **49**:201–205.

[1964]   When is a linear control system optimal?, *J. Basic Engr. (Trans. ASME, Ser. D)*, **86D**:51–60.

[1965a]   Irreducible realizations and the degree of a rational matrix, *SIAM J. Contr.*, **13**:520–544.

[1965b]   Algebraic structure of linear dynamical systems.   I. The module of Σ, *Proc. Nat. Acad. Sci. (USA)*, **54**:1503–1508.

[1966a]   Toward a theory of difficulty of computation in optimal control, *Proc. 4th IBM Scientific Computing Symposium*, pp. 25–43.

[1966b]   On structural properties of linear, constant, multivariable systems, *Proc. 3rd IFAC Congress*, London, to appear.

[1967]   Algebraic aspects of the theory of dynamical systems, in *Differential Equations and Dynamical Systems*, J. K. Hale and J. P. LaSalle (eds.), pp. 133–146, Academic Press.

[1968]   On partial realizations of a linear input/output map, Guillemin Anniversary Volume, Holt, Winston and Rinehart.

[1969]   *Lectures on Algebraic System Theory*, Springer Lecture Notes.

R. E. KALMAN AND R. S. BUCY

[1961]   New results in linear prediction and filtering theory, *J. Basic Engr. (Trans. ASME, Ser. D)*, **83D**:95–100.

R. E. KALMAN, Y. C. HO, AND K. NARENDRA

[1963]   Controllability of linear dynamical systems, *Contr. to Diff. Equations*, **1**:189–213.

L. V. KANTOROVICH AND G. P. AKILOV

[1964]   *Functional Analysis in Normed Spaces*, Pergamon Press.

H. J. KELLEY

[1960]   Gradient theory of optimal flight paths, *ARS J.* **30**:947–953.

[1962]   Method of gradients, in *Optimization Techniques*, G. Leitman (ed.), Academic Press.

G. M. KRANC AND P. E. SARACHIK

[1963]   An application of functional analysis to the optimum control problem, *J. Basic Eng. (Trans. ASME, Ser. D)*, **85D**:143–150.

K. B. KROHN AND J. L. RHODES

[1965]   Algebraic theory of machines. I. The main decomposition theorem, *Trans. Am. Math. Soc.*, **116**:450–464.

K. B. KROHN, J. L. RHODES, AND B. TILSON

[1968]   The prime decomposition theorem of the algebraic theory of machines, chap. 5 of [ARBIB, 1968a].

R. KULIKOWSKI

[1959]   On optimum control with constraints, *Bull. Polish Acad. Sci., Ser. Tech. Sci.*, **7**:385–394.

H. C. KUROSH

[1956]   *The Theory of Groups*, 2 vols., Chelsea.

S. LANG

[1965]   *Algebra*, Addison-Wesley.

E. B. LEE AND L. MARKUS

[1961]   Optimal control for nonlinear processes, *Arch. Ration. Mech. Anal.*, **8**:36–58.

D. G. LUENBERGER

[1964]   Observing the state of a linear system, *IEEE Trans. Military Electronics*, **MIL-8**:74–80.

S. MACLANE AND G. BIRKHOFF

[1967]   *Algebra*, Macmillan.

W. C. McCULLOCH AND W. H. PITTS

[1943]   A logical calculus of the ideas immanent in nervous activity, *Bull. Math. Biophys.*, **5**:115–133.

R. McGILL AND P. KENNETH

[1963]   A convergence theorem on the iterative solution of nonlinear two-point boundary-value systems, XIV IAF Congress, Paris.

B. McMILLAN

[1952]   Introduction to formal realizability theory, *Bell System Tech. J.*, **31**:217–279, 541–600.

E. F. MOORE

[1956]   Gedanken-experiments on sequential machines, in *Automata Studies*, C. E. Shannon and J. McCarthy (eds.), pp. 129–153, Princeton University Press.

E. F. MOORE (ed.)

[1963]   *Sequential Machines—Selected Papers*, Addison-Wesley.

V. V. NEMYTSKII AND V. V. STEPANOV

[1960]   *Qualitative Theory of Differential Equations*, Princeton University Press.

A. NERODE

[1958]   Linear automaton transformations, *Proc. Amer. Math. Soc.*, **9**:541–544.

L. W. NEUSTADT
[1960] Synthesizing time-optimal control systems, *J. Math. Anal. Appl.*, 1:484–493.
[1965] Optimal control problems as extremal problems in a Banach space, *Proc. of the Symposium on System Theory*, Polytechnic Institute of Brooklyn, pp. 215–224.
[1966] An abstract variational theory with applications to a broad class of optimization problems. I. General Theory, *SIAM J. Contr.*, 4:505–528.
[1967] An abstract variational theory with applications to a broad class of optimization problems. II. Applications, *SIAM J. Contr.*, 5:90–137.

J. PESCHON (ed.)
[1963] *Disciplines and Techniques of Modern Systems Control*, Ginn-Blaisdell.

L. S. PONTRYAGIN, V. BOLTYANSKII, R. GAMKRELIDZE, AND E. MISHCHENKO
[1961] *The Mathematical Theory of Optimal Processes*, Interscience.

M. O. RABIN AND D. SCOTT
[1959] Finite automata and their decision problems, *IBM J. Res. Develop.*, 3:114–125.

E. H. ROTHE
[1948] Gradient mappings and extrema in Banach spaces, *Duke Math. J.*, 15:421–431.

E. ROXIN
[1962] The existence of optimal controls, *Mich. Math. J.*, 9:109–119.
[1965] On generalized dynamical systems defined by contingent equations, *J. Diff. Equations*, 1:188–205.

L. M. SILVERMAN AND B. D. O. ANDERSON
[1968] Controllability, observability, and stability of linear systems, *SIAM J. Contr.*, 6:121–130.

J. TODD (ed.)
[1962] *Survey of Numerical Analysis*, McGraw-Hill.

C. P. VAN DINE
[1965] An application of Newton's method to the finite-difference solution of nonlinear boundary-value systems, United Aircraft Res. Lab. Rept. UAR-D37.

B. L. VAN DER WAERDEN
[1931] *Moderne Algebra*, 2 vols., Springer (English translation, Ungar, 1953). See also latest German edition with title *Algebra;* vol. 1, 7th ed., 1966, Heidelberger Taschenbücher no. 12; vol. 2, 5th ed., 1967, Heidelberger Taschenbücher, no. 23.

L. WEISS AND R. E. KALMAN
[1965] Contributions to linear system theory, *Intern. J. Engr. Sci.*, 3:141–171.

B. H. WILLIS AND R. W. BROCKETT
[1965] The frequency domain solution of regulator problems, 1965 JACC, Troy, New York, 228–235.

T. G. WINDEKNECHT

[1967] Unpublished class notes on *General Systems Theory*, Case-Western Reserve University.

H. WITSENHAUSEN

[1965] Some iterative methods using partial order for solution of nonlinear boundary-value problems, MIT Lincoln Laboratory Technical Note 1965-18.

A. W. WYMORE

[1967] *A Mathematical Theory of Systems Engineering: The Elements*, Wiley.

D. C. YOULA

[1966] The synthesis of linear dynamical systems from prescribed weighting patterns, *SIAM J. Appl. Math.*, **14**:527–549.

D. C. YOULA AND P. TISSI

[1966] n-port synthesis via reactance extraction. Part I, *IEEE Intern. Convention Record.*

L. A. ZADEH

[1952] A general theory of linear signal transmission systems, *J. Franklin Inst.*, **253**:293–312.

L. A. ZADEH AND C. A. DESOER

[1963] *Linear System Theory*, McGraw-Hill.

O. ZARISKI AND P. SAMUEL

[1958] *Commutative Algebra*, vol. 1, Van Nostrand.

E. C. ZEEMAN

[1962] The topology of the brain and visual perception, in *The Topology of 3-Manifolds*, M. K. Fort (ed.), 240–256.

H. P. ZEIGER

[1965] Cascade synthesis of finite automata, *Proc. Sixth Ann. Conf. Switching Theory and Automata, IEEE.*

[1967a] Yet another proof of the cascade decomposition theorem for finite automata, *Math. System Theory*, **1**:225–228.

[1967b] Ho's algorithm, commutative diagrams, and the uniqueness of minimal linear systems, *Inf. and Control*, **11**:71–79.

[1968] Cascade decomposition of automata using covers, chap. 4 in [ARBIB, 1968a].

Author index

Subject index